SEDIMENT TRANSPORT – FLOW PROCESSES AND MORPHOLOGY

Edited by **Faruk Bhuiyan**

Sediment Transport – Flow Processes and Morphology
Edited by Faruk Bhuiyan

Published by InTech
Janeza Trdine 9, 51000 Rijeka, Croatia

Publishing Process Manager Masa Vidovic

Technical Editor Teodora Smiljanic

Cover Designer InTech Design Team

Additional hard copies can be obtained from orders@intechopen.com

Sediment Transport – Flow Processes and Morphology, Edited by Faruk Bhuiyan
p. cm.
ISBN 978-953-307-374-3

For EU product safety concerns:
IN TECH d.o.o., Prolaz Marije Krucifikse Kozulić 3, 51000 Rijeka, Croatia,
info@intechopen.com or visit our website at intechopen.com.

Meet the editor

Dr. Faruk Bhuiyan is a water resources specialist with a civil engineering background. He has more than 15 years of experience in teaching, research and consultancy related to sediment transport and river processes. As a commonwealth scholar he obtained Ph.D from the School of Environmental Sciences, University of East Anglia, UK. Dr. Bhuiyan's interests encompass field studies, laboratory experimentation and numerical modeling in several areas including large alluvial river processes (e.g., the Ganges and Brahmaputra-Jamuna Rivers), variably saturated groundwater flow modeling and river basin management issues. He visited and worked in a number of institutions including Hydraulic Research Wallingford Ltd. (UK), Delft University of Technology (The Netherlands), Loughborough University (UK) and University of Windsor (Canada). As a Visiting Professor in the Department of Civil and Environmental Engineering, University of Alberta, he conducted researches on fish micro-habitat, entrainment during hydropower operation, CFD modeling and sediment mound formation.

Contents

Preface

It is my pleasure to write the preface of the book "Sediment Transport - Flow and Morphological Processes" published by Intech Open Access Publisher. The transport of sediment in the turbulent flow comprises of complex phenomena. Although sediment transport due to water flow is directly related with long-term and short-term alteration of the earth's surface, which has significance in science, engineering and environmental applications, up until now the scientific progress in quantifying the relevant processes has been rather slow. This explains the reason for abundance of empiricism and independent field observations in this discipline. It has been several decades since the advent of our understanding on micro-level turbulence properties of fluid flows that we have started to apply our knowledge on turbulence for transport in geophysical boundary layers.

Intech Open Access Publisher has taken a good step to publish a series of books on the issues of sediment transport. The participation to the current book is by special invitation to authors selected based on their previous contributions in recognized scientific journals. Consequently, contents of the chapters are the reflections of the authors' research thoughts.

This book provides indications on current knowledge, research and applications of sediment transport processes. The first three chapters of the book present basic and advanced knowledge on flow mechanisms and transport. These are followed by examples of modeling efforts and individual case studies on erosion-deposition and their environmental consequences. I believe that the materials of this book would help a wide range of readers to update their insight on fluvial transport processes.

Finally, I would like to thank Intech Open Access Publisher for inviting me to contribute as a book editor. Special thanks are also due to the Publishing Process Manager for her cooperation and help during preparation of the book.

Dr Faruk Bhuiyan
Department of Water Resources Engineering
Bangladesh University of Engineering & Technology (BUET), Dhaka,
Bangladesh

1

A Sediment Graph Model Based on SCS-CN Method

P. K. Bhunya[1], Ronny Berndtsson[2],
Raj Deva Singh[1] and S.N. Panda[3]
[1]*National Institute of Hydrology, Roorkee, Uttarakhand*
[2]*Dept. of Water Resources Engineering, Lund University, Lund,*
[3]*Indian Institute of Technology, Kharagpur WB*
[1,3]*India*
[2]*Sweden*

1. Introduction

Sediment is fragmented material primarily formed by the physical and chemical disintegration of rocks from the earth's crust. For example, physical disintegration means where the material is broken down by human interference or can be due to the construction or engineering works. Similarly chemical disintegration is by chemicals in fluids, wind, water or ice and/or by the force of gravity acting on the particle itself. The estimation of sediment yield is needed for studies of reservoir sedimentation, river morphology and soil and water conservation planning. However, sediment yield estimate of a watershed is difficult as it results due to a complex interaction between topographical, geological and soil characteristics. In spite of extensive studies on the erosion process and sediment transport modelling, there exists a lack of universally accepted sediment yield formulae (Bhunya et al. 2010). The conditions that will transport sediment are needed for engineering problems, for example, during canal construction, channel maintenance etc. Interpreting ancient sediments; most sediments are laid down under processes associated with flowing water like rivers, ocean currents and tides.

Usually, the transport of particles by rolling, sliding and saltating is called bed-load transport, while the suspended particles are transported as suspended load transport. The suspended load may also include the fine silt particles brought into suspension from the catchment area rather than from, the streambed material (bed material load) and is called the wash load. An important characteristic of wash load is that its concentration is approximately uniform for all points of the cross-section of a river. This implies that only a single point measurement is sufficient to determine the cross-section integrated wash-load transport by multiplying with discharge. In estuaries clay and silt concentrations are generally not uniformly distributed.

Bed load refers to the sediment which is in almost continuous contact with the bed, carried forward by rolling, sliding or hopping. *Suspended load* refers to that part of the total sediment transport which is maintained in suspension by turbulence in the flowing water for considerable periods of time without contact with the stream bed. It moves with practically the same velocity as that of the flowing water. That part of the suspended load which is composed of particle sizes smaller than those found in appreciable quantities in the bed material. It is in near-permanent suspension and therefore, is transported through the stream without deposition. The discharge of the *wash load* through a reach depends only on the rate with which these particles become available in the catchment area and not on the transport capacity of the flow. Fluid flow and sediment transport are obviously linked to the formation of primary sedimentary structures. Here in this chapter, we tackle the question of how sediment moves in response to flowing water that flows in one direction.

2. Fluid flow and sediment transport

The action of sediment transport which is maintained in the flowing water is typically due to a combination of the force of gravity acting on the sediment and/or the movement of the fluid. A schematic diagram of these forces in a flowing water is shown in Figure 1. The bottom plate is fixed and the top plate is accelerated by applying some force that acts from left to right. The upper plate will be accelerated to some terminal velocity and the fluid between the plate will be set into motion. Terminal velocity is achieved when the applied force is balanced by a resisting force (shown as an equal but opposite force applied by the stationary bottom plate).

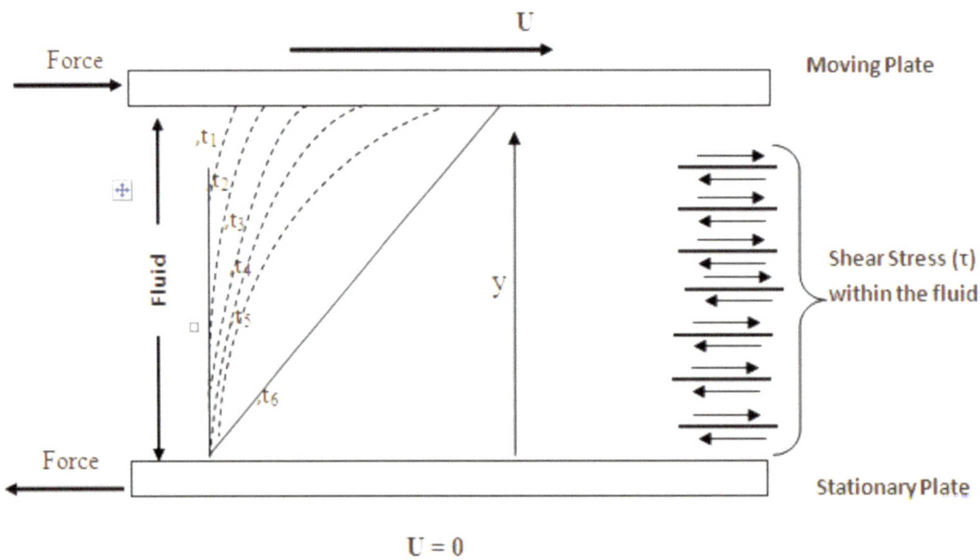

Fig. 1. Varying forces acting on flowing water along the flow depth

The shear stress transfers momentum (mass times velocity) through the fluid to maintain the linear velocity profile. The magnitude of the shear stress is equal to the force that is applied to the top plate. The relationship between the shear stress, the fluid viscosity and the velocity gradient is given by:

$$\tau = \mu \frac{du}{dy} \tag{1a}$$

Where u is the velocity, y is the fluid depth at this point as given in figure, μ is the fluid viscosity, and τ is the shear stress.
From this relationship we can determine the velocity at any point within the column of fluid. Rearranging the terms:

$$\tau / \mu = \frac{du}{dy} \quad \text{or} \quad \int (\tau / \mu) dy = \int du \quad \text{or} \quad u_y = (\tau / \mu) y + c \tag{1b}$$

where c (the constant of integration) is the velocity at y=0 (where u=0) such that:

$$u_y = \frac{\tau}{\mu} y$$

From this relationship we can see the following:

a. That the velocity varies in a linear fashion from 0 at the bottom plate (y=0) to some maximum at the highest position (i.e., at the top plate).
b. That as the applied force (equal to τ) increases so does the velocity at every point above the lower plate.
c. That as the viscosity increases the velocity at any point above the lower plate decreases.

Driving force is only the force applied to the upper, moving plate, and the shear stress (force per unit area) within the fluid is equal to the force that is applied to the upper plate. Fluid momentum is transferred through the fluid due to viscosity.

3. Fluid gravity flows

Water flowing down a slope in response to gravity e.g. in rivers, the driving force is the down slope component of gravity acting on the mass of fluid; more complicated because the deeper into the flow the greater the weight of overlying fluid. In reference to Figure 2 that shows the variation in velocity along the flowing water, D is the flow depth and y is some height above the boundary, FG is the force of gravity acting on a block of fluid with dimensions, (D-y) x 1 x 1; here y is the height above the lower boundary, θ is the slope of the water surface, it may be noted here that the depth is uniform so that this is also the slope of the lower boundary, and τ_y is the shear stress that is acting across the bottom of the block of fluid and it is the down slope component of the weight of fluid in the block at some height y above the boundary.

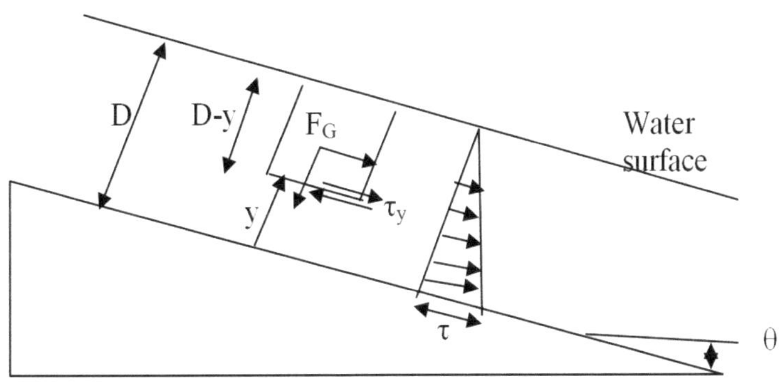

Fig. 2. Variation in velocity for depth

For this general situation, τ_y, the shear stress acting on the bottom of such a block of fluid that is some distance y above the bed can be expressed as follows:

$$\tau_y = \rho g(D-y)\times 1\times 1\times \sin(\theta) \tag{2}$$

The first term in the above equation i.e. $\rho g(D-y)\times 1\times 1$ is the weight of water in the block and Sin (θ) is the proportion of that weight that is acting down the slope. Clearly, the deeper within the water i.e. with decreasing y the greater the shear stress acting across any plane within the flow. At the boundary $y = 0$, the shear stress is greatest and is referred to as the *boundary shear stress* (τ_o); this is the force per unit area acting on the bed which is available to move sediment.

Setting y=0: $$\tau_0 = \rho g(D-y)\sin(\theta) \text{ and } \tau_y = \mu\frac{du}{dy} \tag{3a}$$

From the above equations, we get the following velocity distribution for such flows by substituting

$$du / dy = \rho g(D-y)\sin(\theta) / \mu \tag{3b}$$

Integrating with respect to y:

$$u_y = \int \frac{du}{dy}dy = \frac{\rho g Sin\theta}{\mu}\int (D-y)dy + c = \int \frac{\rho g Sin\theta}{\mu}(yD - y^2/2) + c \tag{4}$$

Where c is the constant of integration and equal to the velocity at the boundary (Uy=0) such that:

$$u_y = \frac{\rho g \sin\theta}{\mu}\left(yD - \frac{y^2}{2}\right) \tag{5}$$

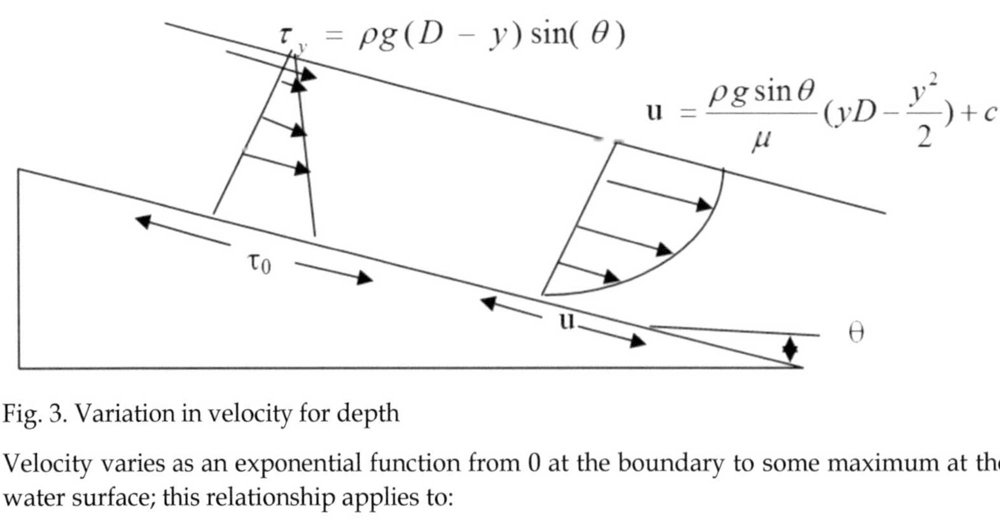

Fig. 3. Variation in velocity for depth

Velocity varies as an exponential function from 0 at the boundary to some maximum at the water surface; this relationship applies to:

a. **Steady flows:** not varying in velocity or depth over time.
b. **Uniform flows:** not varying in velocity or depth along the channel.
c. **Laminar flows:** see next section.

3.1 The classification of fluid gravity flows

3.1.1 Flow Reynolds' Number (R)

Reynolds's experiments involved injecting a dye streak into fluid moving at constant velocity through a transparent tube. Fluid type, tube diameter and the velocity of the flow through the tube were varied, and the three types of flows that were classified are as follows: (a) *Laminar Flow*: every fluid molecule followed a straight path that was parallel to the boundaries of the tube, (b) *Transitional Flow*: every fluid molecule followed wavy but parallel path that was not parallel to the boundaries of the tube, and (c) *Turbulent Flow*: every fluid molecule followed very complex path that led to a mixing of the dye. Reynolds's combined these variables into a dimensionless combination now known as the Flow Reynolds' Number (**R**) where:

$$R = \frac{\rho U D}{\mu} \tag{6a}$$

Where U is the velocity of the flow, ρ is the density of the fluid , D is the diameter of the tube, and μ is the fluid's dynamic viscosity. Flow Reynolds' number is often expressed in terms of the fluid's *kinematic viscosity* (ν) equally expressed as μ/ρ (units are m^2/s) and

$$R = \frac{UD}{\nu} \tag{6b}$$

The value of R determine the type of flows in the following manner:

a. *Laminar flows*: $R < 1000$
b. *Transitional flows*: $1000 < R < 2000$
c. *Turbulent flows*: $R > 2000$

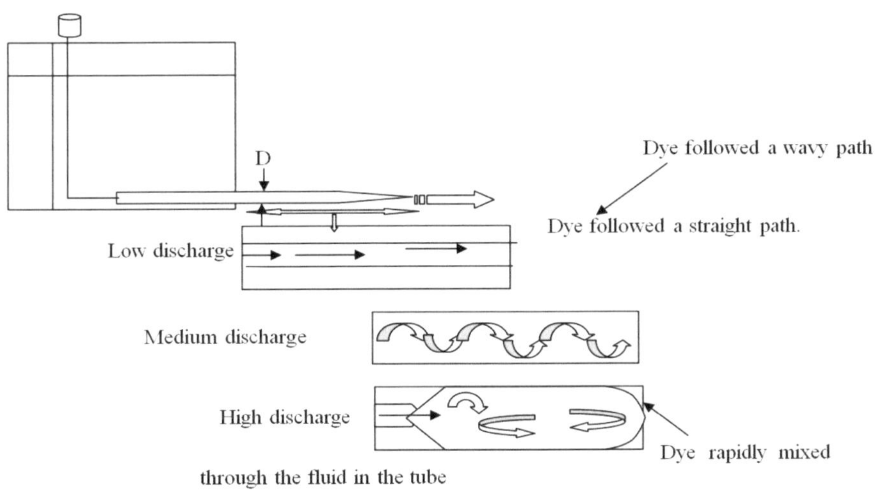

Fig. 4. Reynolds's experiments for different types of flows

In laminar flows, the fluid momentum is transferred only by viscous shear; a moving layer of fluid drags the underlying fluid along due to viscosity (see the left diagram, below). The velocity distribution in turbulent flows has a strong velocity gradient near the boundary and more uniform velocity (an average) well above the boundary. The more uniform distribution well above the boundary reflects the fact that fluid momentum is being transferred not only by viscous shear. The chaotic mixing that takes place also transfers momentum through the flow. The movement of fluid up and down in the flow, due to turbulence, more evenly distributes the velocity, low speed fluid moves upward from the boundary and high speed fluid in the outer layer moves upward and downward. This leads to a redistribution of fluid momentum.

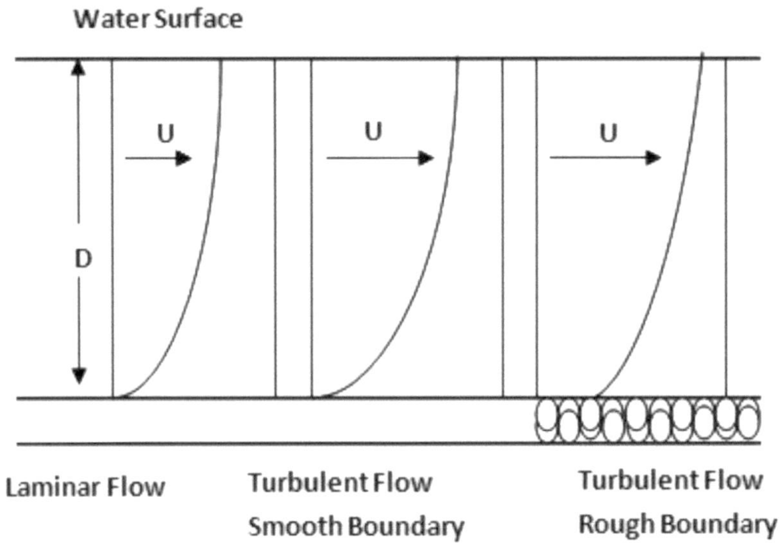

Fig. 5. Variation in velocity for depth at three different types of flows

Turbulent flows are made up of two regions. And there is an inner region near the boundary that is dominated by viscous shear i.e.,

$$\tau_y = \mu \frac{du}{dy} \tag{7}$$

And, an outer region that is dominated by turbulent shear which focus on transfer of fluid momentum by the movement of the fluid up and down in the flow.

$$\tau_y = \eta \frac{du}{dy} + \mu \frac{du}{dy} \tag{8}$$

Where η is the *eddy viscosity* which reflects the efficiency by which turbulence transfers momentum through the flow.

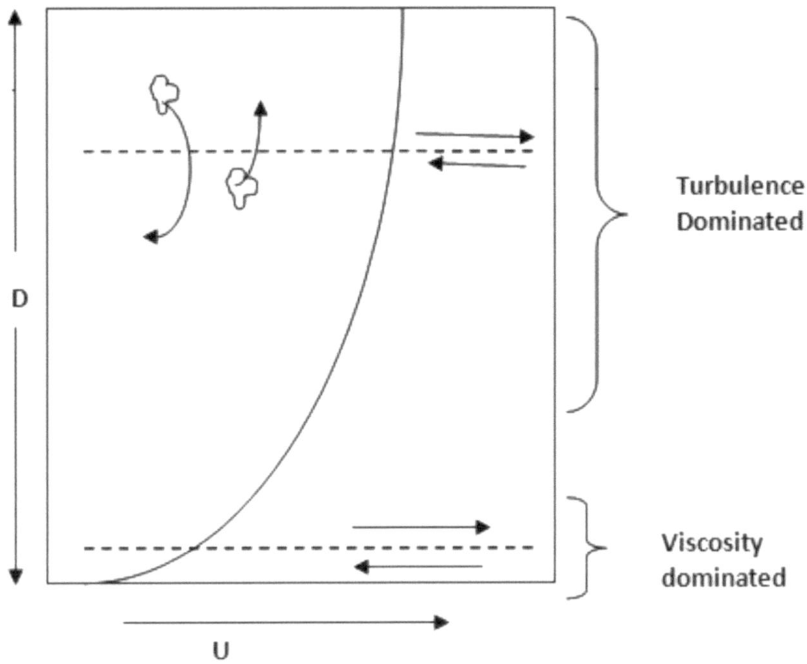

Fig. 6. Two regions of turbulent shear

As a result, the formula for determining the velocity distribution of a laminar flow cannot be used to determine the distribution for a turbulent flow as it neglects the transfer of momentum by turbulence. Experimentally, determined formulae are used to determine the velocity distribution in turbulent flows e.g. the Law of the Wall for rough boundaries under turbulent flows:

$$\frac{u_y}{U_*} = 8.5 + \frac{2.3}{\kappa}\log\frac{y}{y_o};\quad y_0\ (= d/30),\ U_* = \sqrt{\tau_{0/\rho}}\ \text{and}\ \tau_0 = \rho g D Sin(\theta) \tag{9}$$

Where κ is Von Karman's constant which is generally taken 0.41 for clear water flows lacking sediment, y is the height above the boundary, y_0 (= d/30) and d is grain size, and U* is the shear velocity of the flow. If the flow depth and shear velocity are known, as well as the bed roughness, this formula can be used to determine the velocity at any height y above the boundary.

$$u_y = U_*\left(8.5 + \frac{2.3}{\kappa}\log\frac{y}{y_0}\right) \tag{10a}$$

$$u_y = U_*\left(8.5 + \frac{2.3}{\kappa}\log\right) gDSin(\theta) \tag{10b}$$

The above formula may be used to estimate the average velocity of a turbulent flow by setting y to 0.4 times the depth of the flow i.e. y = 0.4D. Experiments have shown that the average velocity is at 40% of the depth of the flow above the boundary.

3.1.2 Flow Froude Number (F)

Classification of flows according to their water surface behaviour, is an important part of the basis for classification of flow regime

a. $F < 1$ has a sub critical flow (tranquil flow)
b. **$F = 1$** has a critical flow
c. **$F > 1$** has a supercritical flow (shooting flow)

Flow Froude Number (F) is defined as follow:

$$F = \frac{U}{\sqrt{gD}} \tag{11}$$

$\sqrt{gD}$ = the celerity (speed of propagation) of gravity waves on a water surface.

$F < 1$, $U < \sqrt{gD}$: water surface waves will propagate upstream because they move faster than the current. Bed forms are not in phase with the water surface.

$F > 1$, $U > \sqrt{gD}$: water surface waves will be swept downstream because the current is moving faster than they can propagate upstream. Bed forms are in phase with the water surface.

In sedimentology the Froude number, is important to predict the type of bed form that will develop on a bed of mobile sediment.

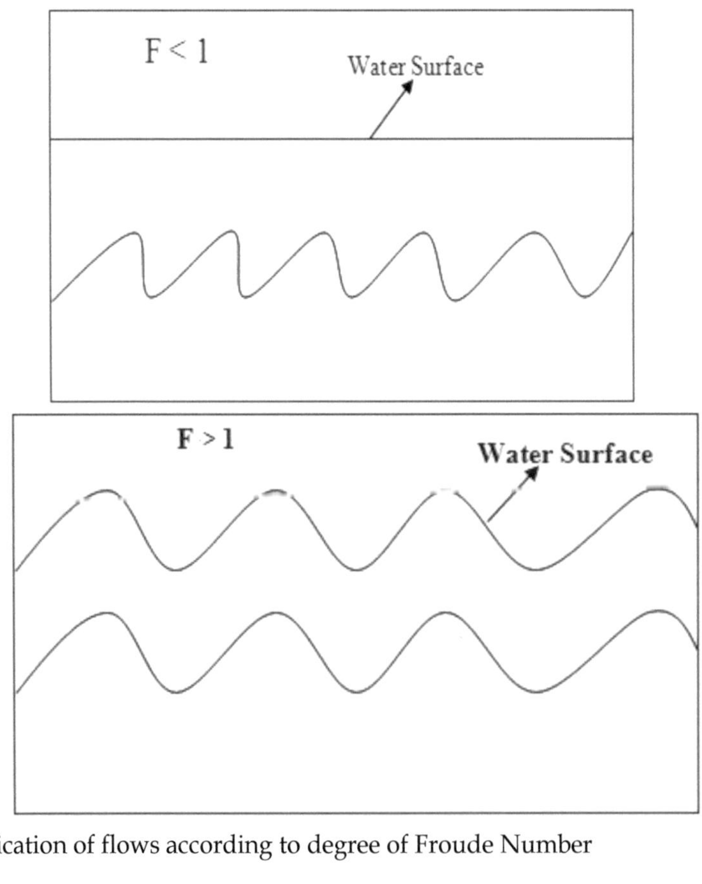

Fig. 7. Classification of flows according to degree of Froude Number

3.2 Velocity distribution, in turbulent flows

Earlier we saw that for laminar flows the velocity distribution could be determined from Eq. (4). Eq. (8). Fig. 7 shows the turbulent flows and the corresponding two regions. As per the Law of the Wall for rough boundaries under turbulent flow depth, the shear velocity are known along with the bed roughness, and in such cases Eq. (10) can be used to determine the velocity at any height *y* above the boundary.

3.3 Subdivisions of turbulent flows

Turbulent flows can be divided into three layers: **(i)** *Viscous Sub layer* is the region near the boundary that is dominated by viscous shear and quasi-laminar flow which is also referred to, inaccurately, as the laminar layer, (ii) *Transition Layer* lies intermediate between quasi-laminar and fully turbulent flow, and (iii) *Outer Layer* which is fully turbulent and momentum transfer is dominated by turbulent shear.

3.4 Viscous sub layer (VSL)

The thickness of the VSL (δ) is known from experiments to be related to the kinematic viscosity and the shear velocity of the flow by:

$$\delta = \frac{12\nu}{U_*} \tag{12}$$

It ranges from a fraction of a millimetre to several millimetres thick, and the thickness of the VSL particularly important in comparison to size of grains (d) on the bed. Next it shall be discussed about the forces that act on the grains and the variation of these relationships. The *Boundary Reynolds' Number* (**R***) is used to determine the relationship between δ and d:

$$R_* = \frac{U_* D}{\nu} \tag{13}$$

A key question is *at what value of R* is the diameter of the grains on the bed equal to the thickness of the VSL?

Given that $\delta = \frac{12\nu}{U_*}$, the condition exists when δ = d, and by substituting this relationship in R_*

$R_* = \frac{U_* D}{\nu}$ =12, thus

R* < 12 δ > d
R* = 12 d = d
R* > 12 δ < d

Turbulent boundaries are classified on the basis of the relationship between thickness of the VSL and the size of the bed material. Given that there is normally a range in grain size on the boundary, the following shows the classification (Fig. 8):

At the boundary of a turbulent flow the average boundary shear stress (τ_o) can be determined using the same relationship, as for a laminar flow. In the viscous sub layer viscous shear predominates so that the same relationship exists, as given in Eqs. (3a, 8 and 9) that applies to steady, uniform turbulent flows.

Boundary shear stress governs the power of the current to move sediment; specifically, erosion and deposition depend on the change in boundary shear stress in the downstream direction. In general, sediment transport rate (q_s) is the amount of sediment that is moved by a current that increases with increasing boundary shear stress. When τ_o increases downstream, so does the sediment transport rate; this leads to erosion of the bed providing that a τ_o that is sufficient to move the sediment. When τ_o decreases along downstream, so does the sediment transport rate; this leads to deposition of sediment on the bed. Variation in τ_o along the flow due to turbulence leads to a pattern of erosion and deposition on the bed of a mobile sediment. This phenomena is given in Fig. 9.

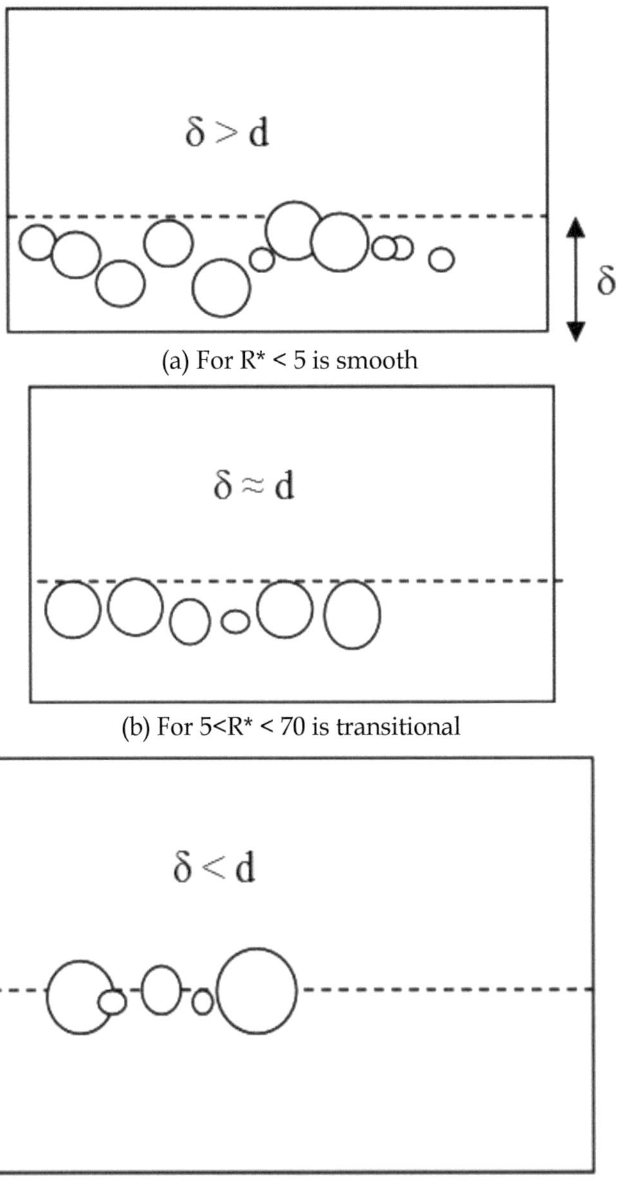

(c) For R* > 70 is Rough

Fig. 8. Classification of flows according to degree of Boundary Reynolds' Number

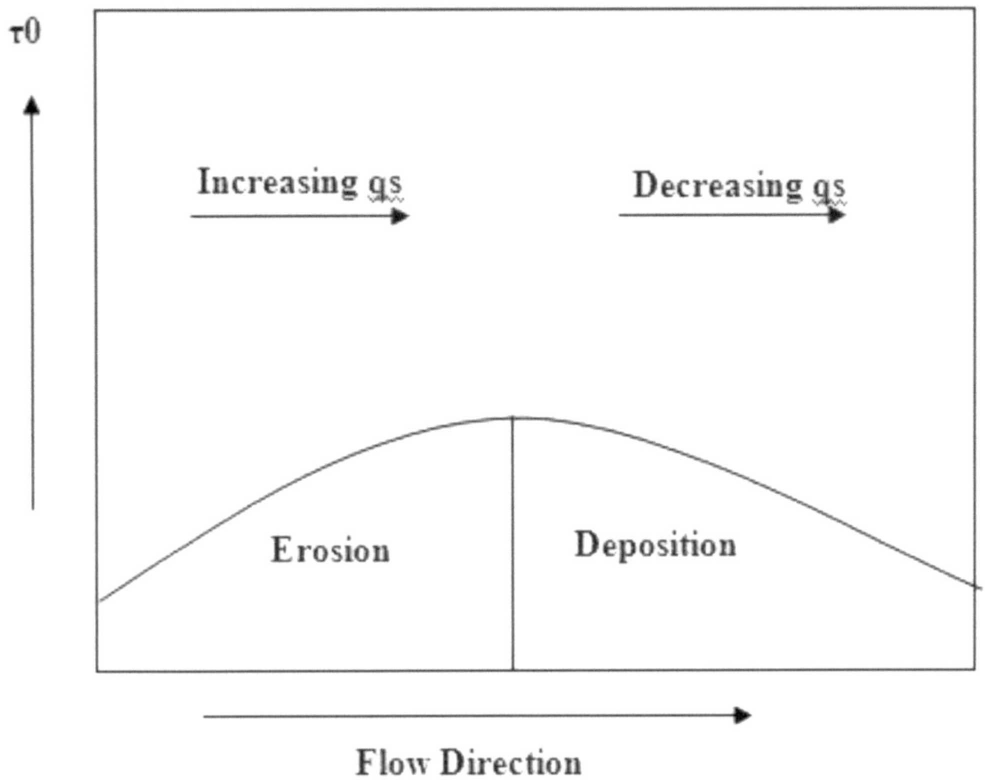

Fig. 9. Pattern of bed erosion and deposition according to variation of shear stress.

3.4.1 Large scale structures of the outer layer

Secondary flows involves a rotating component of the motion of fluid about an axis that is parallel to the mean flow direction. Commonly there are two or more such rotating structures extending parallel to each other.

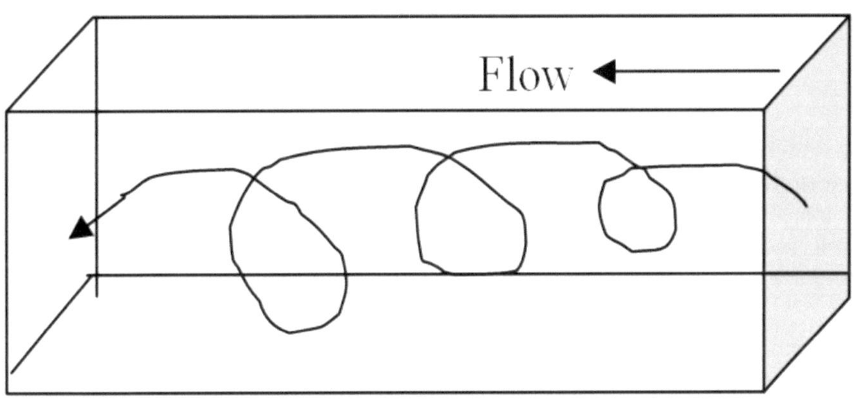

Fig. 10. Eddies about the axes perpendicular to the flow direction.

In meandering channels, characterized by a sinusoidal channel form, counter-rotating spiral cells alternate from side to side along the channel. *Eddies* are components of turbulence that rotate about axes that are perpendicular to the mean flow direction. Smaller scale than secondary flows moves downstream with the current at a speed of approximately 80% of the water surface velocity (U_∞). Eddies move up and down within the flow as the travel downstream, and this lead to variation in boundary shear stress over time and along the flow direction. Some eddies are created by the topography of the bed. In the lee of a negative step on the bed (see figure below) the flow separates from the boundary ("s" in the figure) and reattaches downstream ("a" in the figure). A *roller eddy* develops between the point of separation and the point of attachment. Asymmetric bed forms (see next chapter) develop similar eddies.

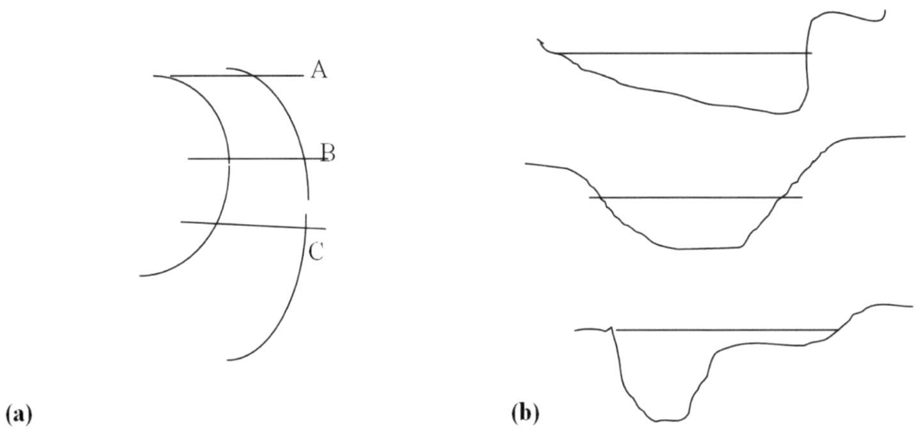

Fig. 11. Asymmetric bed forms

3.4.2 Small scale structures of the viscous sub layer

Alternating lanes of high and low speed fluid within the VSL are termed as *streaks* associated with counter-rotating, flow parallel vortices within the VSL. Streak spacing (λ) varies with the shear velocity (U_*) and the kinematic viscosity (ν)of the fluid; λ ranges from millimetres to centimetres. The relationship is as follows:

$$\lambda = \frac{100\nu}{U_*} \tag{14}$$

λ increases when sediment is present. Due to fluid speed, a bursting cycle is referred as:
Burst: ejection of low speed fluid from the VSL into the outer layer.
Sweep: injection of **high** speed fluid from the outer layer into the VSL.
Often referred to as the *bursting cycle* but not every sweep causes a burst and vise versa, however, the frequency of bursting and sweeps are approximately equal.

3.5 Sediment transport under unidirectional flows

The sediment that is transported by a current comes under two main classes:
Wash load: silt and clay size material that remains in suspension even during low flow events in a river.

Bed material load: sediment (sand and gravel size) that resides in the bed but goes into transport during high flow events e.g., floods.
Bed material load makes up many arsenates and ratites in the geological record. Three main components of bed material load are: *Contact load*: particles that move in contact with the bed by sliding or rolling over it. *Saltation load*: movement as a series of *hops* along the bed, each hop following a ballistic trajectory.

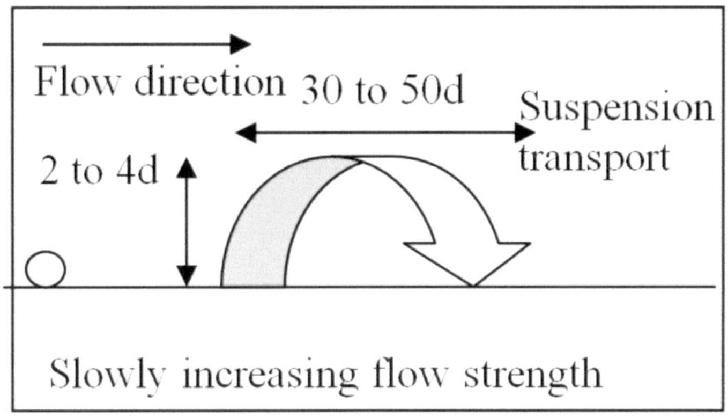

Fig. 12. The ballistic trajectory in the flow

When the ballistic trajectory is disturbed by turbulence, the motion is referred to as *Suspensive saltation.*
Intermittent suspension load: carried in suspension by turbulence in the flow. *Intermittent* because it is in suspension only during high flow events, and otherwise, resides in the deposits of the bed. Bursting is an important process in initiating suspension transport.

3.6 Hydraulic interpretation of grain size distributions

In the section on grain size distributions we saw that some sands are made up of several normally distributed sub-populations. These sub-populations can be interpreted in terms of the modes of transport that they underwent prior to deposition. The finest sub-population represents the wash load. Only a very small amount of wash load is ever stored within the bed material so that it makes up a very small proportion of these deposits. The coarsest sub-population represents, the contact and saltation loads. In some cases they make up two sub-populations (only one is shown in the Fig.13).
The remainder of the distribution, normally making up the largest proportion, is the intermittent suspension load. This interpretation of the subpopulations gives us two bases for quantitatively determining the strength of the currents that transported the deposits. The grain size X is the coarsest sediment that the currents could move on the bed. In this case, $X = -1.5\ \phi$ or approximately 2.8 mm. If the currents were weaker, that grain size would not be present. And, if the currents were stronger, coarser material would be present. This assumes that there are no limitations to the size of grains available in the system. The grain size Y is the coarsest sediment that the currents could take into suspension. In this case, $Y = 1.3$ f or

approximately 0.41 mm, therefore the currents must have been just powerful enough to take the 0.41 mm particles into suspension. If the currents were stronger the coarsest grain size would be larger. This follows the above assumption of limitations to the size of grains size in a system.

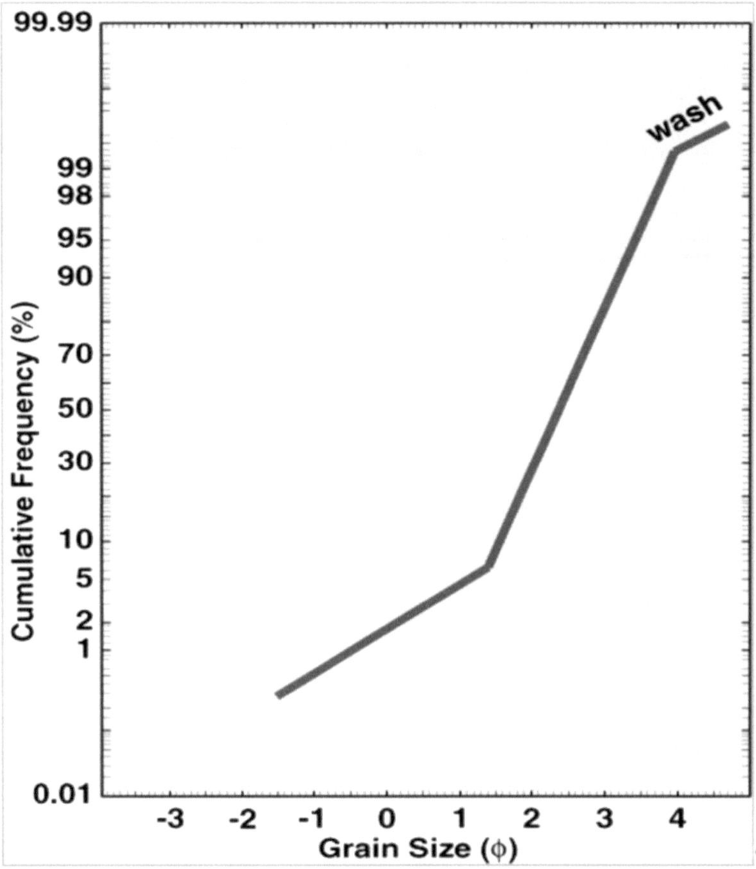

Fig. 13. The grain size frequency distribution

To quantitatively interpret *X*, we need to know the hydraulic conditions needed to just begin to move of that size. This condition is the *threshold for sediment movement.* To quantitatively interpret *Y* we need to know the hydraulic conditions needed to just begin carry that grain size in suspension. This condition is the *threshold for suspension.*

3.7 The threshold for grain movement on the bed

Grain size *X* can be interpreted, if we know what flow strength is required to just move a particle of that size. That flow strength will have transported sediment with that maximum grain size. Several approaches have been taken to determine the critical flow strength to initiate motion on the bed.

Hjulstrom's Diagram shows the diagram of the critical velocity that is required to just begin to move sediment of a given size i.e. the top of the mud region. It also shows the critical velocity for deposition of sediment of a given size at the bottom of the field. The experiment is based on a series of experiments using unidirectional currents with a flow depth of 1 m. It can be noted here that for grain sizes coarser than 0.5 mm the velocity that is required for transport increases with grain size; the larger the particles the higher velocity the is required for transport. For finer grain sizes (with cohesive clay minerals), the greater the critical velocity for transport. This is because the more mud is present means that the cohesion is greater, and the resistance to erosion increases, despite the finer grain size. In our example, the coarsest grain size was 2.8 mm. According to Hjulstron's diagram that grain size would require a flow with a velocity of approximately 0.65m/s. Therefore, the sediment shown in the cumulative frequency curve, was transported by currents at 0.65 m/s.

The problem is that the forces that are required to move sediment, are not only related to flow velocity, but also the boundary shear stress that is a significant force. Boundary shear stress varies with flow depth, as shown the relationship earlier given in Eq. (9) as $\tau_0 = \rho g D Sin(\theta)$. Therefore, Hjulstrom's diagram is reasonably accurate only for sediment that has been deposited under flow depths of 1 m.

3.8 Shield's criterion for the initiation of motion

Based on a large number of experiments Shield's criterion considers the problem in terms of the forces that act to move a particle. The criterion applies to beds of spherical particles of uniform grain size. Forces that are important to initial motion are as follows:

1. The submerged weight of the particle can be taken as $\pi/6\,(\rho_s - \rho)\,g\,d^3$ which resists motion.
2. To which causes a drag force that acts to move the particle down current
3. Lift force (L) that reduces the effective submerged weight.

The flow velocity that is felt by the particle varies from approximately zero at its base to some higher velocity at its highest point.

Pressure specifically *dynamic pressure* in contrast to static pressure is also imposed on the particle and the magnitude of the dynamic pressure varies inversely with the velocity. For, higher velocity, lower dynamic pressure, and maximum dynamic pressure is exerted at the base of the particle and minimum pressure at its highest point. The dynamic pressure on the particle varies symmetrically from a minimum at the top to a maximum at the base of the particle. As shown in Fig. 14, this distribution of dynamic pressure results in a net pressure force that acts upwards. Thus, the net pressure force known as the Lift Force acts opposite to the weight of the particle reducing its effective weight. This makes it easier for the flow to roll the particle along the bed. The lift force reduces the drag force that is required to move the particle. If the particle remains immobile to the flow and the velocity gradient is large enough so that the Lift force exceeds the particle's weight, it will jump straight upwards away from the bed. Once off the bed, the pressure difference from top to bottom of the particle is lost and it is carried down current as it falls back to the bed following the ballistic trajectory of saltation.

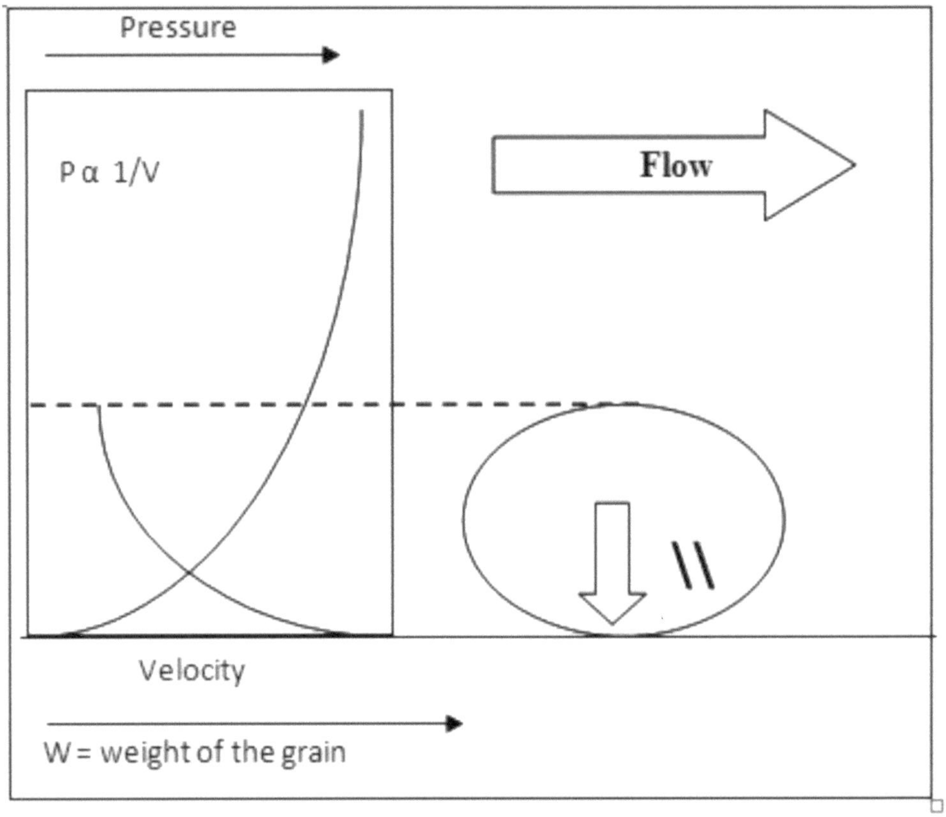

Fig. 14. Simplified ray diagram showing the forces required for initial motion

Shield's experiments involved determining the critical boundary shear stress required to move spherical particles of various size and density over a bed of grains with the same properties (uniform spheres). He produced a diagram that allows the determination of the critical shear stress required for the initiation of motion. A bivariate plot of "Shield's Beta" versus Boundary Reynolds' Number

$$\beta = \frac{\tau_0}{(\rho_s - \rho)gd} = \text{(Force acting to move the particle excluding lift) / (Force resisting movement)} \quad (15)$$

τ_0 is the critical shear stress for motion, and the denominator gives the submerged weight of grains per unit area on the bed. As the lift the force increases β will decrease that shall lower required for movement. Reflects $R_* = \frac{U_* d}{\upsilon}$ something of the lift force (related to the velocity gradient across the particle).

For low boundary Reynold's numbers Shield's β decreases with increasing R* (Fig. 15). For high boundary Reynold's numbers Shield's β increases with increasing R*. The change takes place at R* ≈ 12.

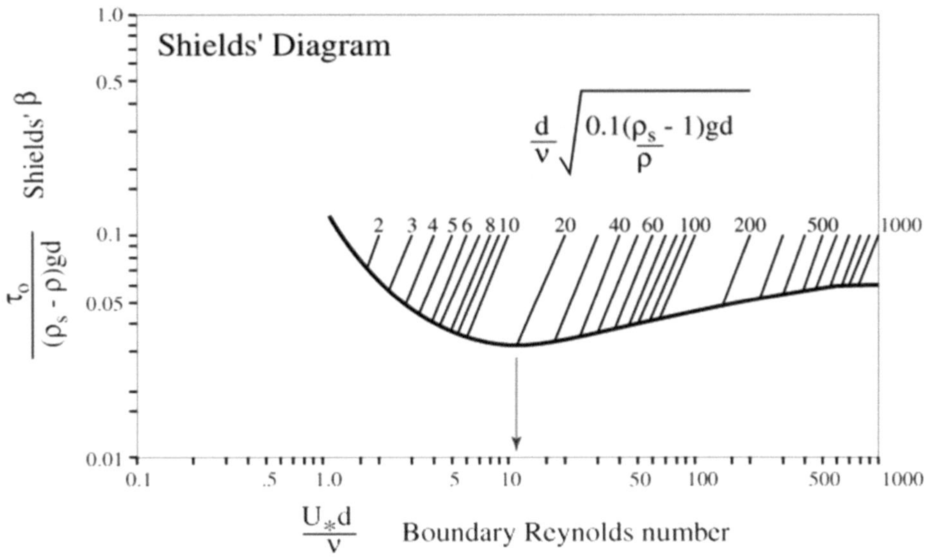

Fig. 15. Shield's Diagram

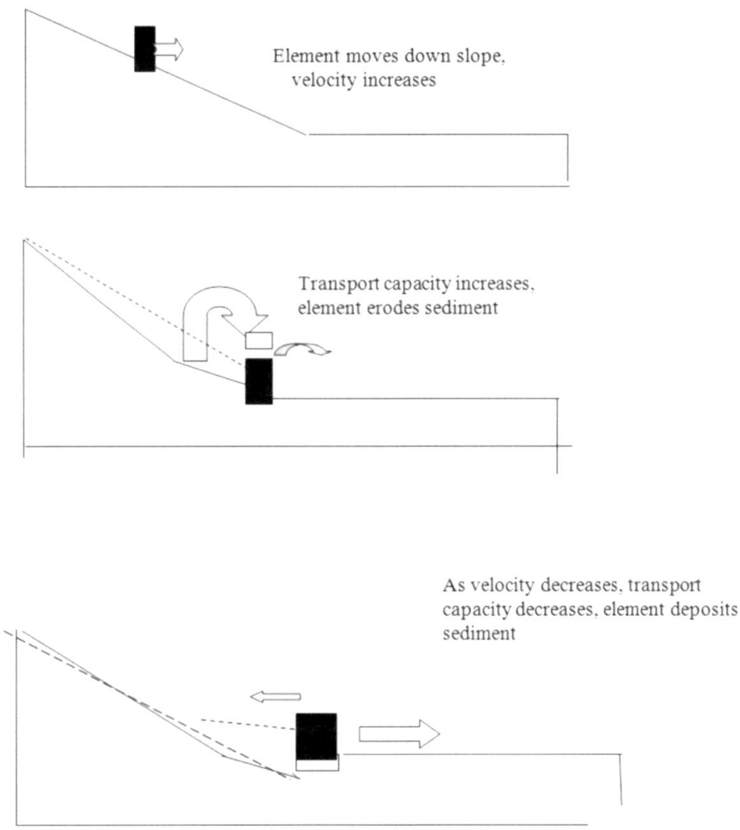

Fig. 16. Two dimensional flow simulation with flow depth

The upstream boundary condition needed to route sediment through a network of stream channels, there is no established method exists for a specific watershed. An example is illustrated in Fig. 17.

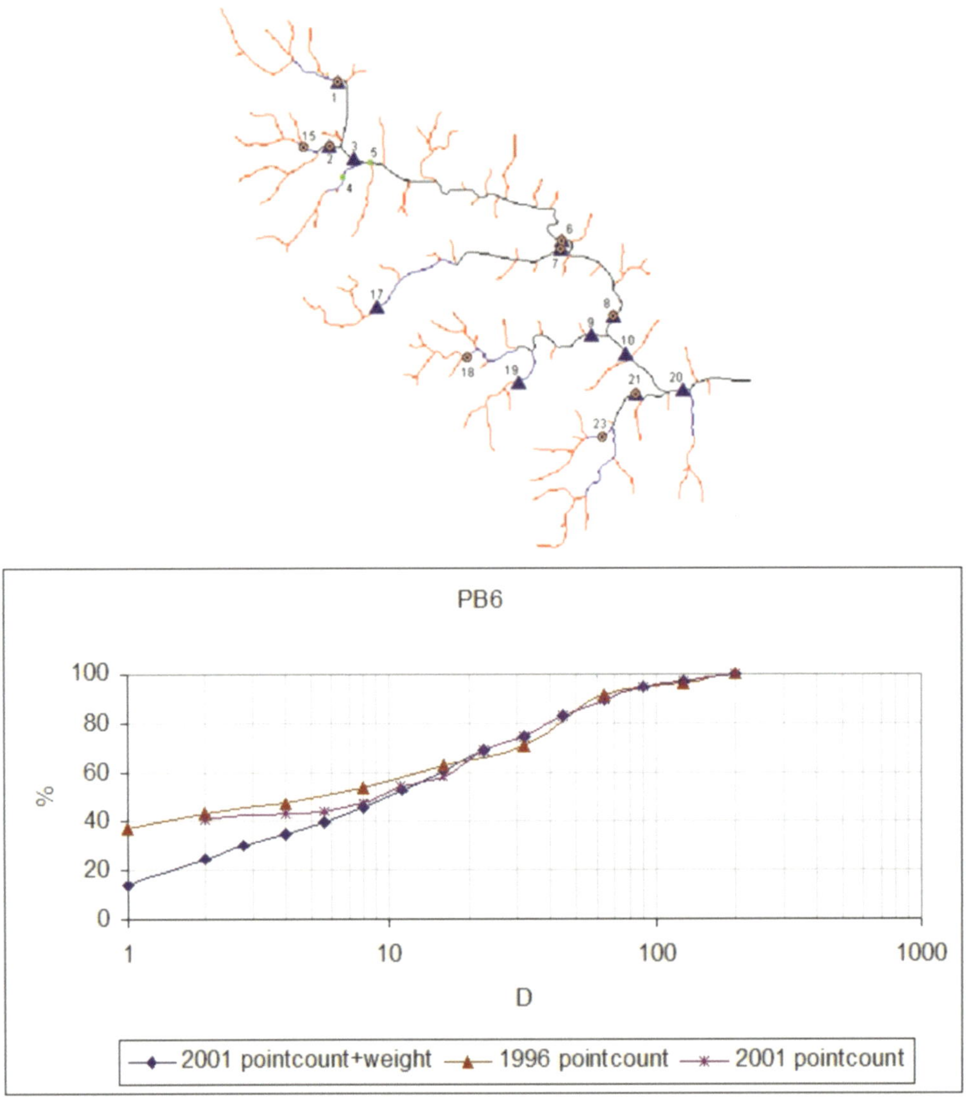

Fig. 17. Regression equations relating sediment grain size distribution of the bed and bank sediment throughout a % of the basin over decadal timescales

4. Sediment transport

This is the movement of solid particles and sediment is naturally-occurring material that is broken down by processes of weathering and erosion, and is subsequently transported by the action of fluids such as wind, water, or ice and/or by the force of gravity acting on the particle itself , typically due to a combination of the force of gravity acting on the sediment and/or the movement of the fluid. A fluid is a substance that continually deforms under an applied shear stress, no matter how small it is. In general, fluids are a subset of the phases of matter and include liquids, gases, plasmas and, to some extent, plastic solids in which the sediment is entrained. An understanding of sediment transport is typically used in natural systems, where the particles are elastic rocks.

The estimation of sediment yield is needed for studies of reservoir sedimentation, river morphology, and soil and water conservation planning. However, sediment yield estimate of a watershed is difficult as it results due to a complex interaction between topographical, geological, and soil characteristics. Sediment graph provides useful information to estimate sediment yield to study transport of pollutants attached to the sediment. To determine these sediment graphs, simple conceptual models are used, which are based on spatially lumped form of continuity and linear storage-discharge equations. Here a watershed is represented by storage systems that include the catchment processes, without including the specific details of process interactions. Examples of few conceptual models are given by (Rendon-Herrero, 1978; Williams, 1978; Singh et al., 1982; Chen and Kuo, 1984; Kumar and Rastogi, 1987; and Lee and Singh, 2005). Rendon-Herrero, (1978) defined the unit sediment graph (USG) resulting due to one unit of mobilized sediment for a given duration uniformly distributed over a watershed. Similarly, Williams (1978} model is based on the instantaneous unit sediment graph (IUSG) concept, where IUSG was defined as the product of the IUH and the sediment concentration distribution (SCD), which was assumed to be an exponential function for each event and was correlated with the effective rainfall characteristics. In Chen and Kuo (1984) model the mobilized sediment was related regressionally with effective-rainfall, and rainfall records and watershed characteristics are to be known necessarily. A similar regression approach was followed by Kumar and Rastogi (1987), Raghuwanshi et al. (1994, 1996), and Sharma and Murthy (1996) to derive sediment graph and peak sediment flow rates from a watershed to reflect the respective changes due to land management practices. However, this routine procedure of regression between mobilized sediment and effective-rainfall always does not produce satisfactory results (Raghuwanshi et al., 1994, 1996). Moreover, the IUSG models utilizing the regression relationship for sediment graph derivation does not explicitly consider the major runoff and sediment producing characteristics of watershed i.e. soil, land use, vegetation and hydrologic condition in their formulation.

In addition to the above approaches discussed so far, the Soil Conservation Service Curve number (SCS-CN) method has also been used for sediment yield modeling (Mishra et al. 2006). Since the method is simple and well established in hydrologic, agriculture and environmental engineering, and is discussed here as it considers the effects of soil type, land use/treatment, surface condition, and antecedent condition. In a recent book by Singh and Frevert (2002), at least six of the twenty-two chapters present mathematical models of watershed hydrology that use the SCS-CN approach, and it shows a lot about the robustness of the SCS-CN methodology and its lasting popularity. Recently Mishra et al. (2006) developed sediment yield models using SCS-CN method, delivery ratio (D_R) concept, and USLE. The models take care of various elements of rainfall-runoff process such as initial abstraction; initial soil moisture; and initial flush. However, the developed models are not applicable for estimation of sediment graphs (sediment flow rate versus time).

With the above back ground, the following sections discuss a simple sediment yield model based on SCS-CN method, Power law (Novotony and Olem, 1994), and utilizes linear reservoir concept similar to Nash (1960) to estimate sediment flow rates and total sediment yield as well. Briefly the model comprises of (i) the mobilized sediment estimation by SCS-CN method and Power law (Novotony and Olem, 1994), instead of relating mobilized sediment and effective-rainfall regressionally; and (ii) the mobilized sediment is then routed through cascade of linear reservoirs similar to Nash (1960). The shape and scale parameters of the IUSG are determined from available storm sediment graphs and then direct sediment graphs are computed by convolution of the IUSG with mobilized sediment. It is noteworthy here that the model does not explicitly account for the geometric configuration of a given watershed.

4.1 Mathematical formulation of proposed model

The suspended sediment dynamics for a linear reservoir can be represented by a spatially lumped form of continuity equation and a linear-storage discharge relationship, as follows:
First linear reservoir:

$$I_{s1}(t) - Q_{s1}(t) = dS_{s1}(t) / dt \tag{16}$$

$$S_{s1}(t) = K_s Q_{s1}(t) \tag{17}$$

where $I_{s1}(t)$ is the sediment inflow rate to the first reservoir [MT^{-1}], and specified in units of (Tons/hr), $Q_{s1}(t)$ is the sediment outflow rate [MT^{-1}] in units of (Tons/hr), $S_{s1}(t)$ is the sediment storage within the reservoir specified in Tons, and K_s is sediment storage coefficient in hours..
For an instantaneous inflow i.e. $I_{s1}(t)$ = 0, Eq. (16) converts to

$$0 - Q_{s1}(t) = dS_{s1}(t) / dt \tag{18}$$

Substituting the value of $S_{s1}(t)$ from Eq. (17) in Eq. (18), a simplified form of Eq. (18) is deduced as follows:

$$0 - Q_{s1}(t) = d(K_s Q_{s1}(t)) \tag{19}$$

On rearranging Eq. (19) and performing integration operation one gets

$$\int dQ_{s1}(t) / Q_{s1}(t) = -(1 / K_s) \int dt \tag{20}$$

or

$$-t / K_s + C_1 = \ln Q_{s1}(t) \tag{21}$$

where C_1 is the constant of integration. C_1 can be estimated by putting t = 0 in Eq. (21) to get $C_1 = -\ln Q_{s1}(0)$, which on substituting in Eq. (21) and on rearranging gives

$$Q_{s1}(t) = Q_{s1}(0) e^{-t / K_s} \tag{22}$$

For t = 0, Eq. (17) reduces to

$$S_{s1}(0) = K_s Q_{s1}(0) \tag{23}$$

Defining A_c as the watershed area in Km^2 and Y as mobilized sediment per storm in $Tons/km^2$, the total amount of mobilized sediment Y_T = A_c Y Tons. If this much amount occurs instantaneously for one unit, i.e., $S_{s1}(0) = A_c Y = 1$, Eq. (23) simplified to the following form

$$1 = K_s Q_{s1}(0) \tag{24}$$

Coupling of Eqs. (22) & (23) results,

$$Q_{s1(t)} = (1 / K_s) e^{-t/K_s} \tag{25}$$

Eq. (25) gives nothing but the rate of sediment output from the first reservoir. This output forms the input to second reservoir and if it goes on up to n^{th} reservoir, then the resultant output from the nth reservoir can be derived as:

$$Q_{sn}(t) = [(t / K_s)^{n-1} e^{-t/K_s}] / K_s \Gamma(n) \tag{26}$$

where Γ() is the Gamma function. Eq. (26) represents the IUSG ordinates at time t (hr^{-1}). For the condition, at t = t_p or $dQ_{sn}(t)/dt = 0$, yields

$$K_s = t_p / (n-1) \tag{27}$$

Coupling of Eqs. (26) & (27) yields

$$Q_{sn}(t) = (n-1)^n / t_p \Gamma(n) [(t / t_p) e^{-(t/t_p)}]^{n-1} \tag{28}$$

Eq. (28) gives the output of the nth linear reservoir.
The SCS-CN method is based on the water balance equation and two fundamental hypotheses, which can be expressed mathematically, respectively, as:

$$P = I_a + F + Q \tag{29}$$

$$Q / P - I_a = F / S \tag{30}$$

$$I_a = \lambda S \tag{31}$$

where, P is total precipitation, I_a initial abstraction, F cumulative infiltration, Q direct runoff, S potential maximum retention, and λ initial abstraction coefficient. Combination of Eqs. (29) and (30) leads to the popular form of SCS-CN method, expressible as:

$$Q = (P - I_a)^2 / P - I_a + S \quad \text{for } P > I_a \tag{32}$$

$$= 0 \quad \text{otherwise}$$

Alternatively, for I_a = 0, Eq. (32) reduces to

$$Q = P^2 / P + S \quad \text{for } P > 0 \tag{33}$$

$$= 0 \quad \text{otherwise}$$

Following Mishra and Singh (2003) for the condition, $f_c = 0$, the Horton's method (Horton, 1938) can be expressed mathematically as:

$$f = f_0 e^{-kt} \tag{34}$$

where f is the infiltration rate (L T^{-1}) at time t, f_o is the initial infiltration rate (LT^{-1}) at time t=0, k is the decay constant (T^{-1}), and f_c is the final infiltration rate (LT^{-1}). The cumulative infiltration F can be derived on integrating Eq. (34) as:

$$F = f_0(1 - e^{-kt}) / k \tag{35}$$

It can be observed from Eq. (35) that as $F \rightarrow f_o/k$, as $t \rightarrow \infty$,. Similarly, for Eq. (30) as $Q \rightarrow (P-I_a)$, $F \rightarrow S$, and time $t \rightarrow \infty$, therefore the similarity between the two yields

$$S = f_o / k \tag{36}$$

On the basis of infiltration tests, Mein and Larson, (1971) got $f_o = i_o$, where i_o is the uniform rainfall intensity when t = 0. Substituting this into Eq. (36) yields

$$f_0 = i_0 = kS \tag{37}$$

Eq. (37) describes the relationship among the three parameters f_o, k, and S. Thus Eq. (37) shows that k depends on the magnitude of the rainfall intensity and soil type, land use, hydrologic condition, and antecedent moisture that affect S and the results are consistent as reported by Mein and Larson (1971). An assumption that rainfall P linearly increases with time t leads to

$$P = i_0 t \tag{38}$$

which is a valid and reasonable assumption for infiltration rate computation in experimental tests (Mishra and Singh, 2004). Coupling of Eqs. (37) & (38) gives,

$$P = k\,S\,t \tag{39}$$

The Power law proposed by Novotony and Olem (1994) can be expressed as

$$D_R = \alpha C_r^{\ \beta} \tag{40}$$

where C_r = runoff coefficient; D_R = sediment delivery ratio; α and β = the coefficient and exponent of power relationship. The ratio, D_R, is dimensionless and is expressed in terms of Sediment yield Y and Potential maximum erosion A as follows:

$$D_R = Y/A \tag{41}$$

The coefficient, C is also dimensionless, and expressed in terms of Q and P, as:

$$C_r = Q/P \tag{42}$$

Substituting the expressions of D_R and C_r in Eq. (40) one gets

$$Y = \alpha A (Q / P)^{\beta} \tag{43}$$

In general, the potential maximum erosion (A) for storm based applications is computed by MUSLE (Williams, 1975a) as:

$$A = 11.8(V_Q Q_P)^{0.56} K(LS)CP \tag{44}$$

where V_Q is the volume of runoff in m^3, Q_P is the peak flow rate in m^3/s, K is the soil erodibility factor, LS is the topographic factor, C is the cover and management factor and P is the support practice factor.
For the condition $I_a = 0$, equating Eqs. (30) & (32) reduces to

$$Q / P = P / (P + S) = F / S \tag{45}$$

Substituting the equality $Q / P = P / (P + S)$ (Eq. 45) in Eq. (43) results

$$Y = \alpha A[P / (P + S)]^{\beta} \tag{46}$$

Similarly, the coupling of Eqs. (24) & (31) yields

$$Y = \alpha A[kt / (1 + kt)]^{\beta} \tag{47}$$

Thus, Eq. (47) gives the expression for mobilized sediment due to an isolated storm event occurring uniformly over the watershed. Hence, total amount of mobilized sediment is expressed as:

$$Y_T = \alpha A A_c [kt / (1 + kt)]^{\beta} \tag{48}$$

Finally, coupling of Eq. (48) results as follows:

$$Q_{ns}(t) = \left[\alpha A A_c [kt / (1 + kt)]^{\beta} (n-1)^{n} / t_p \Gamma(n) [(t / t_p) e^{-(t/t_p)}]^{n-1} \right] \tag{49}$$

The expression given by Eq. (49) is the proposed model for computations of sediment graphs. The proposed model has four parameters α , β , k, and n.

4.2 Application

The workability of the proposed model is tested using the published data of Chaukhutia watershed of Ramganga Reservoir catchment (Kumar and Rastogi, 1987, Raghuwanshi et al., 1994, 1996), a schematic map of the watershed is given in Fig. 18. The basic characteristics of sediment graph data are given in Table 1.

4.3 Parameter estimation

The shape parameter (n_s) was estimated by the relationship given by Bhunya et al. (2003) as:

$$n_s = 5.53\beta_s^{1.75} + 1.04 \qquad \text{for } 0.01<\beta_s<0.35$$

$$n_s = 6.29\beta_s^{1.998} + 1.157 \qquad \text{for } \beta_s \geq 0.35 \tag{50}$$

where β_s is a non dimensional parameter defined as the product of peak sediment flow rate (q_{ps}) [Tons/hr/Tons] and time to peak sediment flow rate (t_{ps}) [hr]. The rest of the parameters were estimated by using the non-linear Marquardt algorithm (Marquardt, 1963) of the least squares procedure. In the present application, potential maximum erosion A is also taken as a parameter due to lack of their observations. The estimated parameters along with storm event values are given in Table 1 and 2.

Date of Event	q_s (Tons/hr/Tons)	t_{ps} (hr)	β_s	$Q_{s(o)}$ (Tons)	$Q_{ps(o)}$ (Tons/hr)
July 17, 1983	0.38	2	0.76	2739	1025
August 21/22, 1983	0.418	2	0.836	2070	875
July 15, 1984	0.397	2	0.794	3145	1043
August 18/19, 1984	0.404	2	0.81	2105	743
September 1/2, 1984	0.39	2	0.78	1205	475
September 17/18, 1984	0.41	2	0.82	963	392

Table 1. Characteristics of storm events

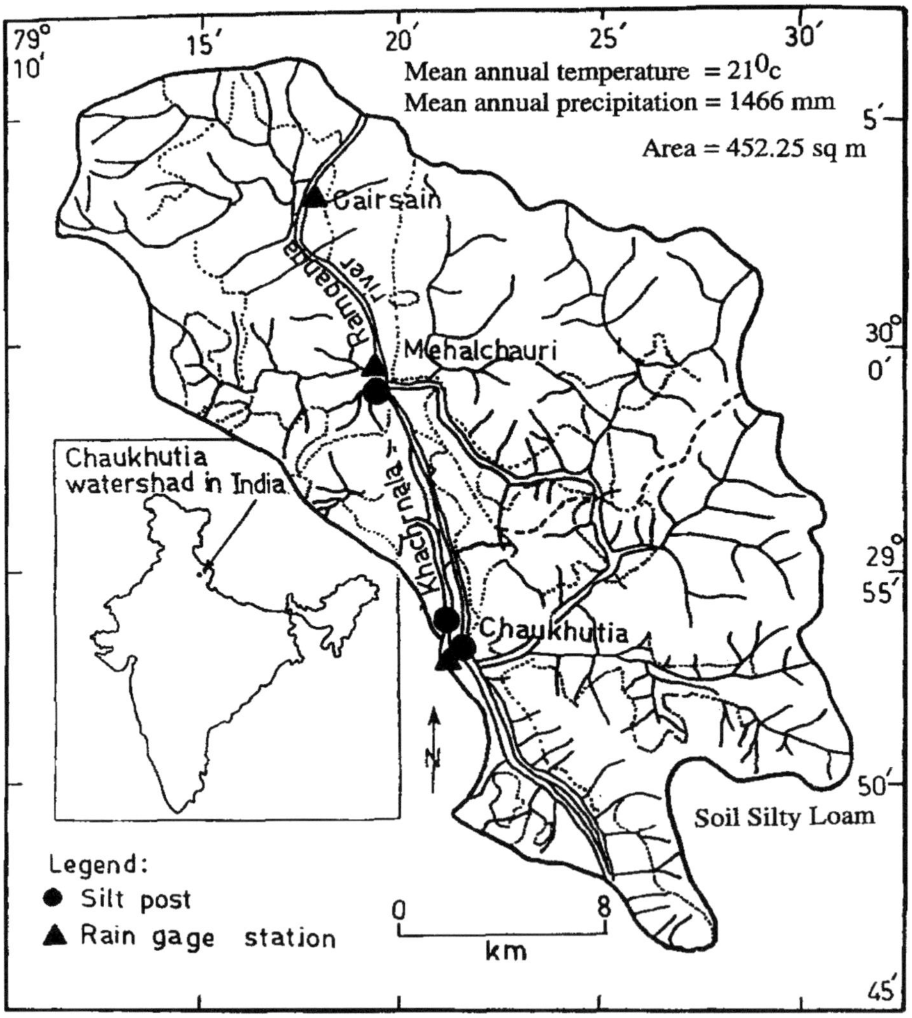

Fig. 18. Location of Chaukhutia watershed in Ramganga reservoir catchment (Source: Raghuwanshi et al. 1994)

Date of Event	Model parameters				
	n_s	α	β	k	A (Tons/Km²)
July 17, 1983	4.79	0.530	0.351	0.029	26.66
August 21/22, 1983	5.55	0.727	0.701	0.030	40.78
July 15, 1984	5.12	0.735	0.721	0.030	62.69
August 18/19, 1984	5.27	0.714	0.663	0.030	38.14
September 1/2, 1984	4.99	0.388	0.425	0.030	19.64
September 17/18, 1984	5.39	0.587	0.781	0.030	29.34

Table 2. Optimized parameter values for Chaukhutia watershed

4.4 Performance of the proposed model

The performance of the proposed sediment graph model was evaluated on the basis of their (i) closeness of the observed and computed sediment graphs visually; and (ii) goodness of fit (GOF) in terms of model efficiency (ME) and relative error (RE) of the results defined as:

$$ME = 1 - \frac{\sum (Q_{so} - Q_{sc})^2}{\sum (Q_{so} - \overline{Q_{so}})^2}\,;\; RE(Qs) = \frac{Q_{s(o)} - Q_{s(c)}}{Q_{s(o)}} \times 100\,;\; RE(Qps) = \frac{Q_{ps(o)} - Q_{ps(c)}}{Q_{ps(o)}} \times 100 \qquad (51)$$

where $Q_{s(o)}$ and $Q_{s(c)}$ are observed and computed total sediment outflow, respectively $RE_{(Qs)}$ and $RE_{(Qps)}$ are relative errors in total sediment outflow and peak sediment flow rates, respectively.

For visual appraisal, the sediment graph computed using the proposed model is compared with the observed values using the data of August 18-19, 1984 event (Fig. 19). From the figure, it is observed that the computed sediment graph exhibits fair agreement with the observed graph. Similar results were also obtained for rest of the storm events that are not reported here. However, Fig. 20 & 21 shows the comparison between computed and observed total sediment outflow and peak sediment outflow rates for all the storm events. The closeness of data points in terms of a best fit line and a value of $r^2 \approx 1.000$ indicate a satisfactory model performance for the assigned Job.

Further the results of GOF criteria given by Eq. (51) for all the events are shown in Table 3. The results indicate that the RE for total sediment outflow and peak sediment flow rate estimates vary from 2.49 to 10.04% and 12.59 to 16.56%, respectively. Though error in case of peak sediment flow rate estimation is on higher side, this may be taken safely because even the more elaborate process-based soil erosion models are found to produce results with still larger errors (Vanoni 1975; Foster 1982; Hadley et al. 1985; Wu et al. 1993; Wicks and Bathurst 1996; Jain et al. 2005). Table 3 also shows the GOF in terms of ME for the storm events considered in the application. It is observed that ME varies from 90.52 to 95.41%, indicating a satisfactory performance of the model for sediment graph computations.

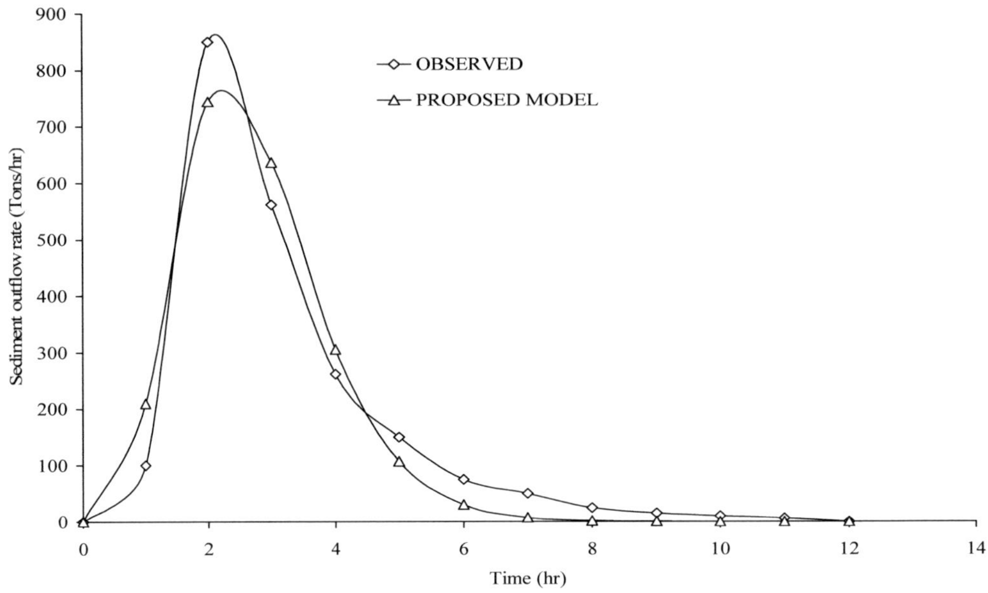

Fig. 19. Comparison of observed and computed sediment graphs for the storm of August, 18-19, 1984.

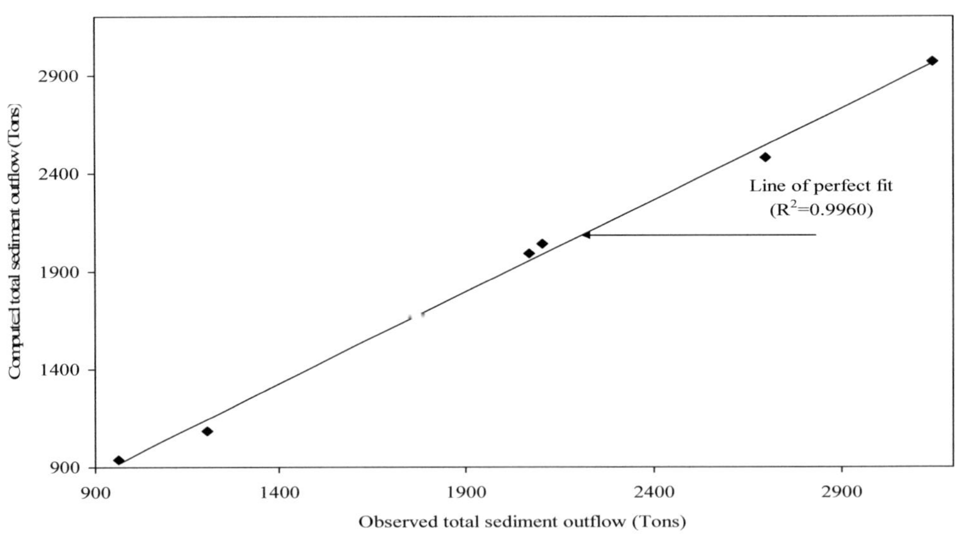

Fig. 20. Comparison between observed and computed total sediment outflow using proposed model for all storm events

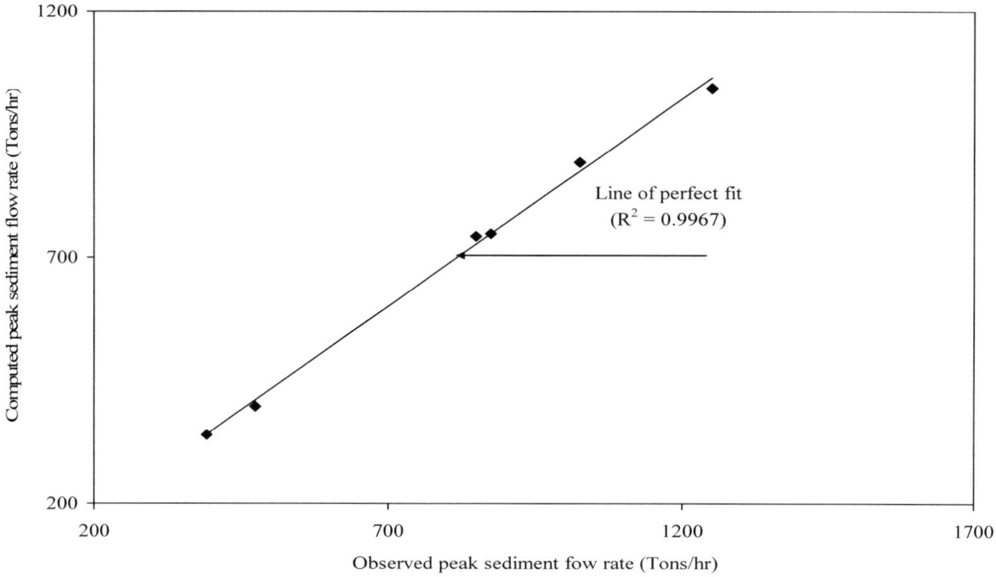

Fig. 21. Comparison between observed and computed peak sediment flow rates using proposed model for all storm events

Date of Event	RE $_{(QS)}$	RE$_{(Qps)}$	Efficiency
July 17, 1983	8.04	12.88	92.91
August 21/22, 1983	3.77	14.51	93.48
July 15, 1984	5.56	16.56	90.52
August 18/19, 1984	3.04	12.59	95.34
September 1/2, 1984	10.04	16.42	93.65
September 17/18, 1984	2.49	13.52	95.41

Table 3. Goodness of fit Statistics

4.5 Sensitivity analysis

From the results so far, it is imperative to analyze the sensitivity of different parameters of the proposed model for their effect on overall output. Here, the conventional analysis for sensitivity similar to the work of McCuen and Snyder (1986) and Mishra and Singh (2003) is followed as discussed in the following section.

It is evident form Eq. (49) that is a function of α, β, k, n and Λ i.e. $Q_s(t) - f(\alpha, \beta, k, n, A)$. Therefore, the total derivative of C can be given as

$$dQ_s(t) = \frac{\partial Q_s(t)}{\partial \alpha} d\alpha + \frac{\partial Q_s(t)}{\partial \beta} d\beta + \frac{\partial Q_s(t)}{\partial k} dk + \frac{\partial Q_s(t)}{\partial n} dn \tag{52}$$

where $\frac{\partial Q_s(t)}{\partial \alpha}$, $\frac{\partial Q_s(t)}{\partial \beta}$, $\frac{\partial Q_s(t)}{\partial k}$ and $\frac{\partial Q_s(t)}{\partial n}$ are the partial derivatives of $Q_s(t)$ with respect to α, β, k, n respectively. The total derivative, $dQ_s(t)$, corresponding to the increments dα,

dβ, dk and dn can be physically interpreted as the total variation of $Q_s(t)$ due to the variation of α, β, k and n at any point in the (α, β, k, n) domain. The variation of $Q_s(t)$ with respect to the variable under consideration can be derived from Eq. (49).
A more useful form of Eq. (52) can be given as

$$\frac{dQ_s(t)}{Q_s(t)} = \left(\frac{\partial Q_s(t)}{\partial \alpha}\frac{\alpha}{Q_s(t)}\right)\frac{d\alpha}{\alpha} + \left(\frac{\partial Q_s(t)}{\partial \beta}\frac{\beta}{Q_s(t)}\right)\frac{d\beta}{\beta} + \left(\frac{\partial Q_s(t)}{\partial k}\frac{k}{Q_s(t)}\right)\frac{dk}{k} + \left(\frac{\partial Q_s(t)}{\partial n}\frac{n}{Q_s(t)}\right)\frac{dn}{n} \quad (53)$$

where $\left(\frac{\partial Q_s(t)}{\partial \alpha}\frac{\alpha}{Q_s(t)}\right)$, $\left(\frac{\partial Q_s(t)}{\partial \beta}\frac{\beta}{Q_s(t)}\right)$, $\left(\frac{\partial Q_s(t)}{\partial k}\frac{k}{Q_s(t)}\right)$ and $\left(\frac{\partial Q_s(t)}{\partial n}\frac{n}{Q_s(t)}\right)$ are referred to as the ratio of the error in the sediment flow rate ($dQ_s(t)/Q_s(t)$) to the error in α (dα/ α), to the error in β (dβ/ β), to the error in k (dk/k), and to the error in n (dn/n). Now, individual ratio terms corresponding to each parameter can be derived from Eq. (49) as follows:

$$\frac{\partial Q_s(t)}{\partial \alpha}\frac{\alpha}{Q_s(t)} = 1. \quad (54)$$

A similar error ratio term for parameter 'A' $\left(\frac{\partial Q_s(t)}{\partial A}\frac{A}{Q_s(t)}\right)$ can be obtained as well.

Similarly, for rest of the parameters, the error ratio terms are derived as

$$\frac{\partial Q_s(t)}{\partial \beta}\frac{\beta}{Q_s(t)} = \beta \ln\left(\frac{kt}{1+kt}\right) \quad (55)$$

$$\frac{\partial Q_s(t)}{\partial k}\frac{k}{Q_s(t)} = \frac{\beta}{t(1+kt)} \quad (56)$$

$$\frac{\partial Q_s(t)}{\partial n}\frac{n}{Q_s(t)} = \frac{\left[(n-1)(2-n)\ln c - (6-3.5n\right]n}{(n-1)^{2.5}}, \text{ where } c = (t/t_p)\exp(-t/t_p), \; n>1 \quad (57)$$

Eq. (57) is based on the expansion of exponential term up to first order only.

Sensitivity to α

In order to analyze the model sensitivity to parameter α the terms pertaining to β, k and n are eliminated from Eq. (53) and the resulting expression reduces to

$$\frac{dQ_s(t)}{Q_s(t)} = \left(\frac{\partial Q_s(t)}{\partial \alpha}\frac{\alpha}{Q_s(t)}\right)\frac{d\alpha}{\alpha} \quad (58)$$

Coupling of Eqs. (58) and (54) results

$$\frac{dQ_s(t)}{Q_s(t)} = \frac{d\alpha}{\alpha}, \text{ or } \frac{dQs(t)/Qs(t)}{d\alpha/\alpha} = 1 \quad (59)$$

From Eq. (59) it can be inferred that the ratio of the error in $Q_s(t)$ to the error in α is 1. This indicate that the any variation (increase or decrease) in α estimates will cause a same amount of variation (increase or decrease) in $Q_s(t)$, as depicted in Fig. 22. Similar pattern can be observed for parameter A also.

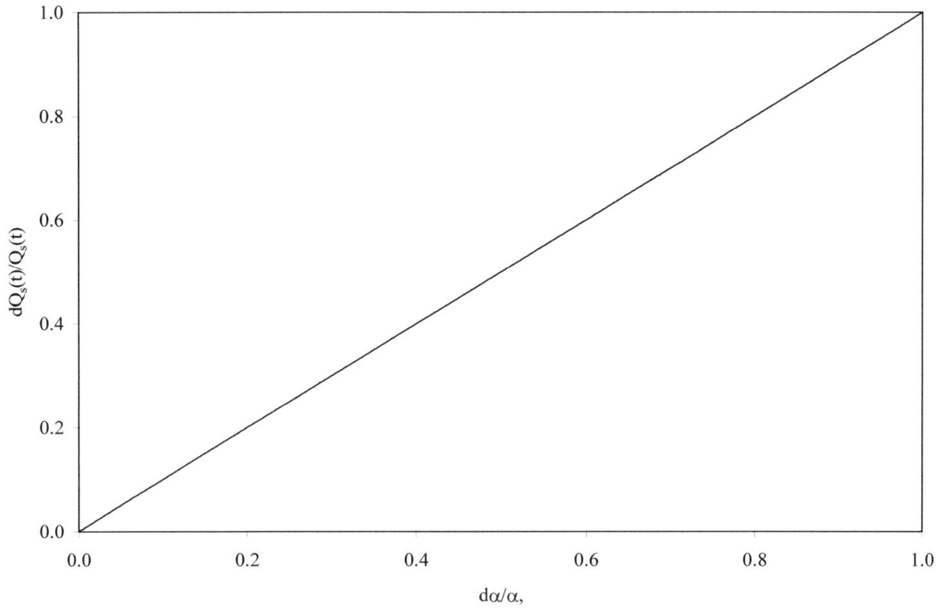

Fig. 22. Sensitivity of sediment outflow rate to α

Sensitivity to β

Similar to the above, the variation of β only is considered after ignoring the impact of α, k, and n, Eq. (38) in such case reduces to the following form

$$\frac{dQ_s(t)}{Q_s(t)} = \left(\frac{\partial Q_s(t)}{\partial \beta} \frac{\beta}{Q_s(t)} \right) \frac{d\beta}{\beta} \tag{60}$$

or

$$\frac{dQ_s(t)/Qs(t)}{d\beta / \beta} = \left(\frac{\partial Q_s(t)}{\partial \beta} \frac{\beta}{Q_s(t)} \right) \tag{61}$$

Equating Eqs. (61) and (55) one gets

$$\frac{dQ_s(t)/Qs(t)}{d\beta / \beta} = \beta \ln \left(\frac{kt}{1+kt} \right) \tag{62}$$

Analogous to the previous analysis, the left hand side of Eq. (62) represents the ratio of error in $Q_s(t)$ to the error in β, and the same is shown in Fig. 23 . It is apparent from Fig. 23 that any variation (increase) in β for a given t and k causes $Q_s(t)$ to decrease.

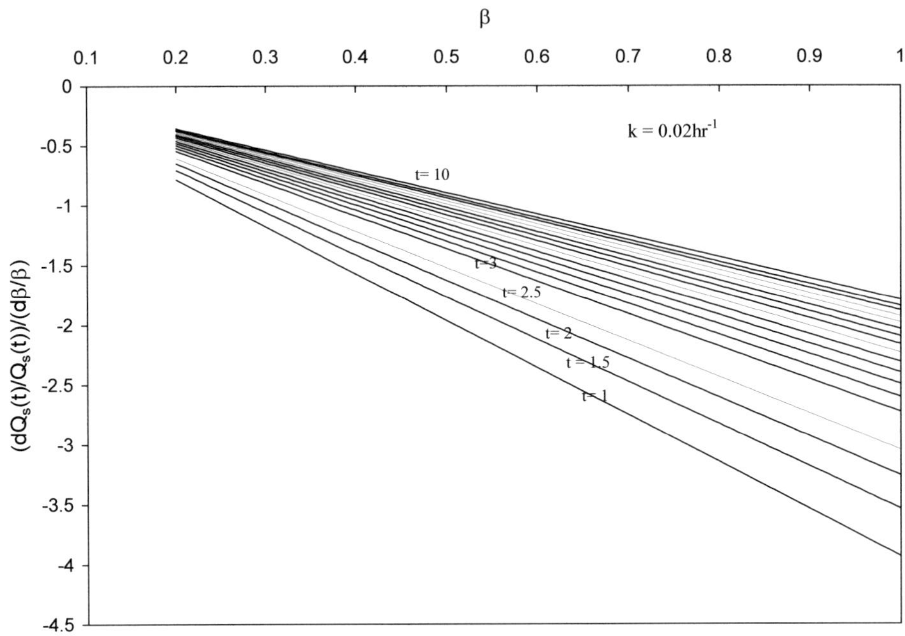

Fig. 23. Sensitivity of sediment outflow rate to β

Sensitivity to k

As above, considering the variation of k only reduces Eq. (53) reduces to the following form.

$$\frac{dQ_s(t)}{Q_s(t)} = \left(\frac{\partial Q_s(t)}{\partial k} \frac{k}{Q_s(t)} \right) \frac{dk}{k} \tag{63}$$

Alternatively, Eq. (48) can be expressed as

$$\frac{dQ_s(t)/Qs(t)}{dk\ /\ k} = \left(\frac{\partial Q_s(t)}{\partial k} \frac{k}{Q_s(t)} \right) \tag{64}$$

Equating Eqs. (64) and (56) one gets

$$\frac{dQ_s(t)/Qs(t)}{dk\ /\ k} = \frac{\beta}{t(1+kt)} \tag{65}$$

As expressed in Eq. (65) and shown in Fig. 24, for any increase in k the ratio of errors tends to decrease, implying the Q_s (t) to increase and vice versa.

Sensitivity to n

Similar to the preceding analysis, if the variation of only n is considered ignoring the impact of α, β, k, Eq. (53) reduces to

$$\frac{dQ_s(t)}{Q_s(t)} = \left(\frac{\partial Q_s(t)}{\partial n} \frac{n}{Q_s(t)} \right) \frac{dn}{n} \tag{66}$$

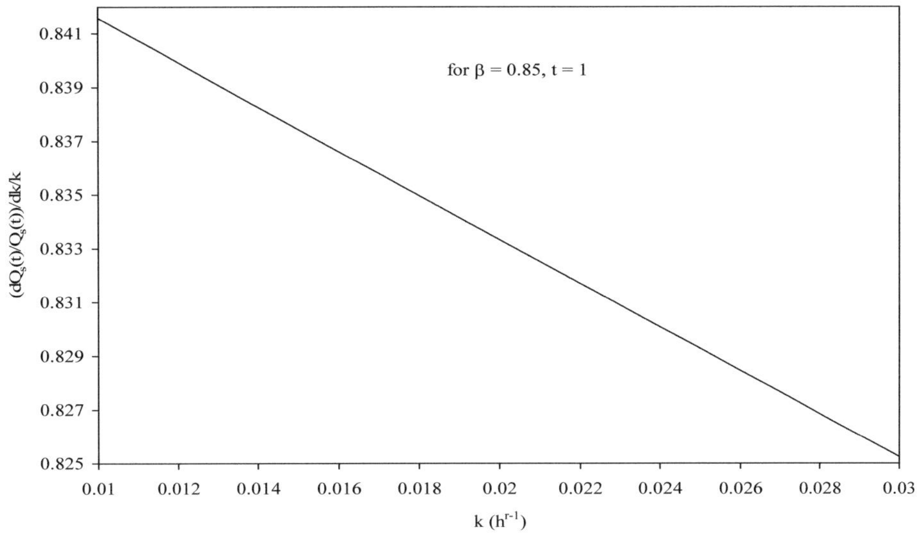

Fig. 24. Sensitivity of sediment outflow rate to k

Equating Eqs. (66) & (57) results

$$\frac{dQ_s(t)/Qs(t)}{dn/n} = \frac{[(n-1)(2-n)\ln c - (6-3.5n]n}{(n-1)^{2.5}}, \text{ where c} = (t/t_p)\exp(-t/t_p), \text{ n>1} \quad (67)$$

Analogous to the previous analysis, the left hand side of Eq. (67) represents the ratio of error in $Q_s(t)$ to the error in n. It is apparent from Fig. 25 that any variation (increase) in n for a given t/t_p causes the ratio to increase, implying $Q_s(t)$ to increase.

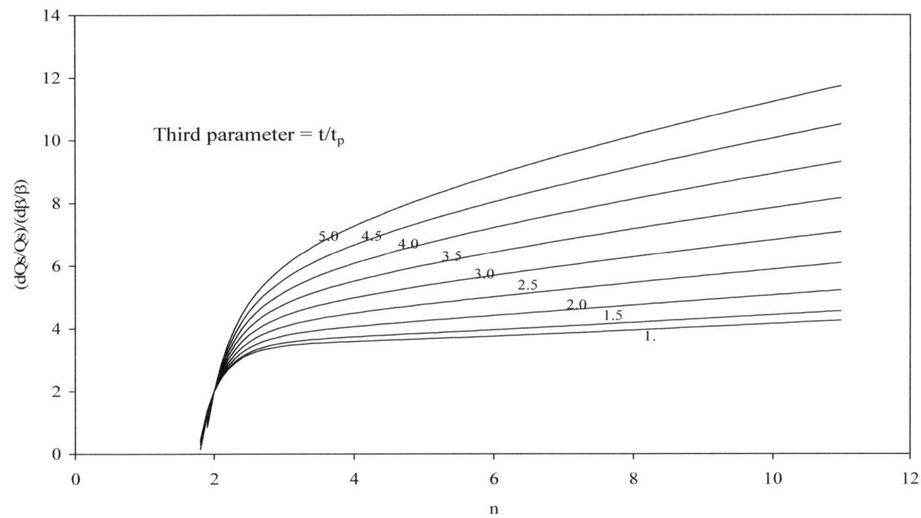

Fig. 25. Sensitivity of sediment outflow rate to n

A comparison of the sediment graphs computed by the proposed model and IUSG model (Raghuwanshi et al., 1994) with the observed sediment graph is shown in Fig. 26. It is observed that the proposed model compares more closely than IUSG with the observed sediment graph.

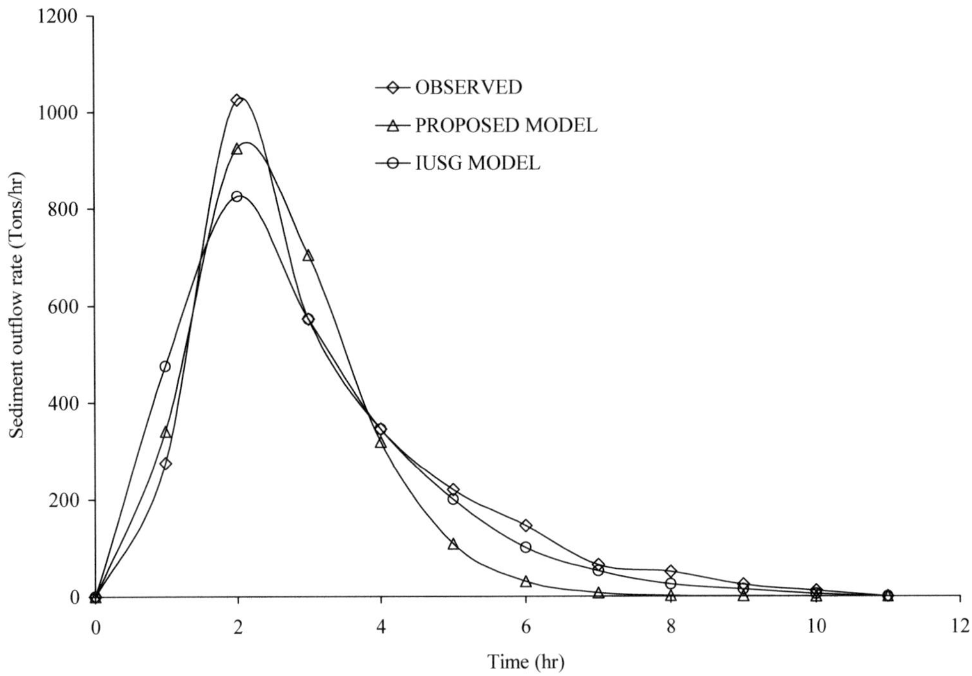

Fig. 26. Comparison of observed and computed sediment graphs for the storm of July 17, 1983.

5. References

Aksoy, H., and Kavvas, M. L., 2005. A review of hillslope and watershed scale erosion and sediment transport models. CATENA 64, 247-271.

Bhunya, P. K., Mishra, S. K., and Berndtsson, R., 2003. Simplified two-parameter gamma distribution for derivation of synthetic unit hydrograph. ASCE J. Hydrol. Eng. 8 (4), 226-230.

Bogardi, I., Bardossy, A., Fogel, M., and Duckstein, L., 1986. Sediment yield from agricultural watersheds. ASCE J. Hydraul. Eng. 112 (1), 64-70.

Chen, V. J., and Kuo, C. Y., 1986. A study of synthetic sediment graphs for ungauged watersheds. J. Hydrol. 84, 35–54.

Chow, V. T., 1964. Applied Hydrology. McGraw-Hill, New York.

Foster, G. R., 1982. Modeling the erosion processes. In: C.T. Hann, Johnson, H., and Brakensiek, D.L. (Eds.), Hydrological modelling of small watersheds, ASAE Monograph No. 5, ASAE, St. Joseph, Mich.., 297–380.

Hadley, R. F., Lal, R., Onstad, C. A., Walling, D. E., and Yair, A., 1985. Recent developments in erosion and sediment yield studies. IHP-II Project A.1.3.1, UNESCO, Paris.

Hawkins, R. H., Woodward, D. E., and Jiang, R., 2001. Investigation of the runoff curve number abstraction ratio. Paper presented at USDA-NRCS Hydraulic Engineering Workshop, Tucson, Arizona.

Horton, R. I., (1938) The interpretation and application of runoff plot experiments with reference to soil erosion problems. Proc. Soil Science Society of America, 3, 340-349.

Jain, M. K., Kothyari, U.C., and RangaRaju, K. G., 2005. GIS based distributed model for soil erosion and rate of sediment outflow from catchments. ASCE J. Hydraul. Eng. 131 (9), 755-769.

Kavvas, M. L., Chen, Z. Q., Dogrul, C., Yoon, J. Y., Ohara, N., Liang, L., Aksoy, H., Anderson, M. L., Yoshitani, J., Fukami, K., and Matsuura, T., 2004. Watershed environmental hydrology (WEHY) model based on upscaled conservation equations: hydrologic module. ASCE J. Hydrol. Eng. 9 (6), 450-464.

Kavvas, M. L., Yoon, J. Y., Chen, Z. Q., Liang, L., Dogrul, C., Ohara, N., Aksoy, H., Anderson, M. L., Reuters, J., and Hackley, S., 2006. Watershed environmental hydrology model: Environmental module and its application to a California watershed. ASCE J. Hydrol. Eng. 11 (3), 261-272.

Kothyari, U. C., Tiwari, A. K., and Singh, R., 1996. Temporal variation of sediment yield. ASCE J. Hydrol. Eng. 1 (4), 169-176.

Kumar, S., and Rastogi, R. A., 1987. Conceptual catchment model for estimation of suspended sediment flow. J. Hydrol. 95, 155-163.

Lee, Y. H., and Singh, V. P., 1999. Prediction of sediment yield by coupling Kalman filter with instantaneous unit sediment graph. Hydrol. Process. 13, 2861-2875.

Lee, Y. H., and Singh, V. P., 2005. Tank model for sediment yield. Water Resour. Mgmt. J. 19, 349-362.

Marquardt, D.W., 1963. An algorithm for least-square estimation of non-linear parameters. J. Soc. Indust. Appl. Math. 11 (2), 431-441.

McCuen, R.H., 2002. Approach to confidence interval estimation for curve numbers. ASCE J. Hydrol. Eng. 7 (1), 43-48.

Mein, R. G., and Larson, C. L., 1971. Modeling the infiltration component of the rainfall-runoff process. Water Resources Research Center, University of Minnesota, Graduate School, Minneapolis, Minnesota.

Merritt, W.S., Letcher, R.A., and Jakeman, A.J., 2003. A review of erosion and sediment transport models. Environ. Model. Software 18, 761–799.

Michel, C., Andreassian, V., and Perrin, C., 2005. Soil Conservation Service Curve Number method: how to mend a wrong soil moisture accounting procedure. Water Resour. Res. 41.

Mishra, S. K., and Singh, V. P., 1999. Another look at SCS-CN method. ASCE J. Hydrol. Eng. 4 (3), 257-264.

Mishra, S. K., and Singh, V. P., 2003. Soil Conservation Service Curve Number (SCS-CN) Methodology. Kluwer Academic Publishers, P.O. Box 17, 3300 AA, Dordrecht, The Netherlands.

Mishra, S. K., and Singh, V. P., 2004. Validity and extension of the SCS-CN method for computing infiltration and rainfall-excess rates. Hydrol. Process. 18 (17), 3323-3345.

Mishra, S. K., Tyagi, J.V., Singh, V. P., and Singh, R., 2006. SCS-CN based modeling of sediment yield. J. Hydrol. 324, 301-322.

Nash, J. E., 1957. The form of instantaneous unit hydrograph. Bull. Int. Assoc. Sci. Hydrol. 3, 114–121.

Nash, J.E., and Sutcliffe, J.V., 1970 River flow forecasting through conceptual models. Part I - A discussion of principles. J. Hydrol. 10, 282-290.

Novotny, V., and Olem, H., 1994. Water Quality: Prevention, Identification, and Management of Diffuse Pollution. John Wiley & Sons, New York, N.Y.

Ponce, V.M., and Hawkins, R. H., 1996. Runoff curve number: has it reached maturity?. ASCE J. Hydrol. Eng. 1 (1), 11-19.

Raghuwanshi, N. S., Rastogi, R. A., and Kumar S., 1994. Instantaneous unit sediment graph. ASCE J. Hydraul. Eng. 120 (4) 495-503.

Rendon-Herrero, O., 1978. Unit sediment graph. Water Resour. Res. 14(5), 889-901.

SCS, 1956. National Engineering Handbook, Hydrology, Section 4, Soil Conservation Service, USDA, Washington DC.

Sharma, K. D., and Murthy, J. S. R., 1996. A conceptual sediment transport model for arid regions. J. Arid Environ. 33, 281-290.

Soil and Water Conservation Division (SWC&D), 1991–1996. Evaluation of hydrological data, Vols. I and II, Ministry of Agriculture, Government of India, New Delhi, India.

Vanoni, V. A., ed. 1975. Sedimentation Engineering, ASCE, New York.

Wicks, J. M., and Bathurst, J. C. 1996. SHESED: A physically based, distributed erosion and sediment yield component for the SHE hydrological modeling system. J. Hydrol. 175, 213–238.

Williams, J. R., 1975. Sediment routing for agriculture watersheds. Water Resour. Bull. 11 (5), 965–974.

Williams, J. R., 1978. A sediment graph model based on an instantaneous unit sediment graph. Water Resour. Res. 14(4), 659–664.

Wilson, B. N., Barfield, B. J., Moore, I. D., and Warner, R. C., 1984. A hydrology and sedimentology watershed model. II: Sedimentology component." Transaction ASAE, 17, 1378-1384.

Wischmeier, W.H. and Smith, D.D., 1978. Predicting rainfall-erosion losses-A guide to conservation planning. Agricultural Handbook No. 537, Science and Education Administration, USDA, Washington DC.

Wu, T. H., Hall, J. A., and Bonta, J. V., 1993. Evaluation of runoff and erosion models. ASCE J. Irrig. and Drain. Eng. 119(4), 364-382.

Young, R. A., Onstad, C. A., Bosch, D.D., and Anderson, W.P., 1987. AGNPS: An agricultural non-point source pollution model: a large water analysis model, U. S. Dept. of Agriculture, Cons. Res. Report No.35, p.77.

2

Bed Forms and Flow Mechanisms Associated with Dunes

Ram Balachandar and H. Prashanth Reddy
Department of Civil and Environmental Engineering
University of Windsor
Canada

1. Introduction

This section will focus on the general features of flow in open channels with a brief analysis of rough open channel flow, shallow open channel flow and the latest developments. This section will serve as the background material to enable the reader to develop an appreciation of the important features that lead to the formation of dunes.

1.1 General description of flow in open channels

Open channel flow (OCF) occurs in canals, rivers and various types of waterways. Open channel flow is mainly driven by gravity unlike flows in closed conduits where the flow is mainly driven by a pressure gradient. An important feature of an open channel flow is the presence of the free surface (zero-shear state). The friction at the bed and the side walls acts as resistance to the flow which is mostly turbulent. The type of bottom and side walls are dependent on the material in which the flow is taking place. Lined canals and canals in rocky strata can be classified as fixed bed channels, and canals in erodible medium are termed as movable bed channels. The shear stress distribution and roughness characteristics of the boundary do not vary with time for a given flow rate in a fixed bed open channel. A movable bed generally accompanies the flow in an alluvium soil. The side slopes and bottom slope of erodible channels vary with time. The flow in open channels can be classified depending on the velocity changes in space and time. The flow is steady if $dU/dt = 0$ at any point in the flow field and it is unsteady if $dU/dt \neq 0$. Though the turbulent flow accompanied by the formation of ripples and dunes on the bed is strictly unsteady, in most practical situations, the time-averaged velocity acquired over long time intervals result in a quasi-steady flow. Flood flows in rivers and surges in power canals are examples of unsteady flows in open channels. Unsteady flows are more difficult to solve because the governing equations are hyperbolic (continuity and momentum). The flow is uniform if $dU/dx = 0$ and non-uniform if $dU/dx \neq 0$. Uniform flow can only occur in long prismatic channels or generated in laboratory flumes. Non-uniform flow can be classified into gradually varied flow (dy/dx is very small) and rapidly varied flow ($dy/dx >> 0$) depending on the amount of change in the velocity over space. Furthermore, most practical open channel flows are turbulent ($Re \geq 2000$) and accompanied by rapid changes in velocities and pressures, in space and time. Eddies and swirls are present in turbulent flow leading to lateral mixing. Figure 1 illustrates the presence of dunes in the Rhine River in The Netherlands.

1.2 Rough open channel flows

Turbulent wall-bounded flows on rough surfaces, including that in open channels, have been studied extensively. One should note that smooth open channel flows occur only in laboratory flumes. In rough open channel flows the non-dimensional equivalent sand roughness, $k_s^+ = ku_*/\nu$ is greater than 70, and the roughness elements completely penetrate the fully turbulent logarithmic layer. In fully developed open channel flow, the height of the turbulent boundary layer occupies the entire flow depth (d) and thus $\delta \approx d$. Recent particle image velocimetry (PIV) and planar light induced fluorescence measurements by Djenidi et al. (2008), Manes et al., (2007) and numerical results of Krogstad et al., (2005) document an increase in the Reynolds stresses in the outer layer which was different from that obtained on the smooth wall. Nikora et al. (2001) developed a comprehensive classification of rough open channel flows based on the value of the relative submergence (d/k). Krogstad et al. (2005) and Volino et al. (2009) speculated that in 3-D rough wall flow, the effect of the surface roughness on the outer layer may be dependent not only on the surface conditions but also on the type of flow: internal or external flow. Balachandar and Patel (2002) have shown that some types of 3-D roughness might produce very thin roughness sub-layers at low submergence, which will allow for development of the logarithmic layer. Roussinova and Balachandar (2011) investigated the effect of depth on turbulence characteristics in the fully rough regime of train of ribs and found that turbulence structures are affected by both roughness and depth.

1.3 Shallow open channel flow

Open channel flows can be classified as shallow when the vertical length scale of the flow (usually the depth, d) is significantly smaller than the width of the flow (Jirka and Uijttewaal, 2004). Fully turbulent shallow channel flow is highly chaotic with the presence of coherent motions. The key feature to understand shallow open channel flow may lie in analyzing the nature of the interactions of events occurring near the bed and near the free surface in greater detail. Ejection and sweep events, hairpin vortices of a length scale of the order of two to three wall units (Blackwelder and Kovasznay, 1972, Theodorsen 1955) and bursting phenomenon (Kline and Robinson 1989) have been observed in the shallow flows. Wu and Christensen (2006), Lin et al. (2003), and Roussinova et al. (2010) used PIV data to expose the signature of the horseshoe vortices, sweep and ejections for a typical shallow channel flow. Experimental evidence put forward by Balachandar and Patel (2005) shows that for a low turbulent intensity flow, a change in both the Cole's boundary layer wake parameter as well as in the rate of production of turbulent kinetic energy due to the presence of free surface disturbance is observed.

2. Bed forms

In alluvial streams, bed load and contact load movement results in the formation of bed forms, which vary geometrically in size and shape. Generally, all of them are grouped into "sand waves" because of the similarity of appearance to sea waves. The size and shape of bed forms depend on the flow velocity, flow depth, Froude number, stream power (defined as the product of mean bed shear stress and mean velocity of flow), sediment properties and fluid properties. In practice, a few principal types of bed forms are usually distinguished: ripples, dunes, plane bed, antidunes, standing waves and chutes and pools (Figure 2).

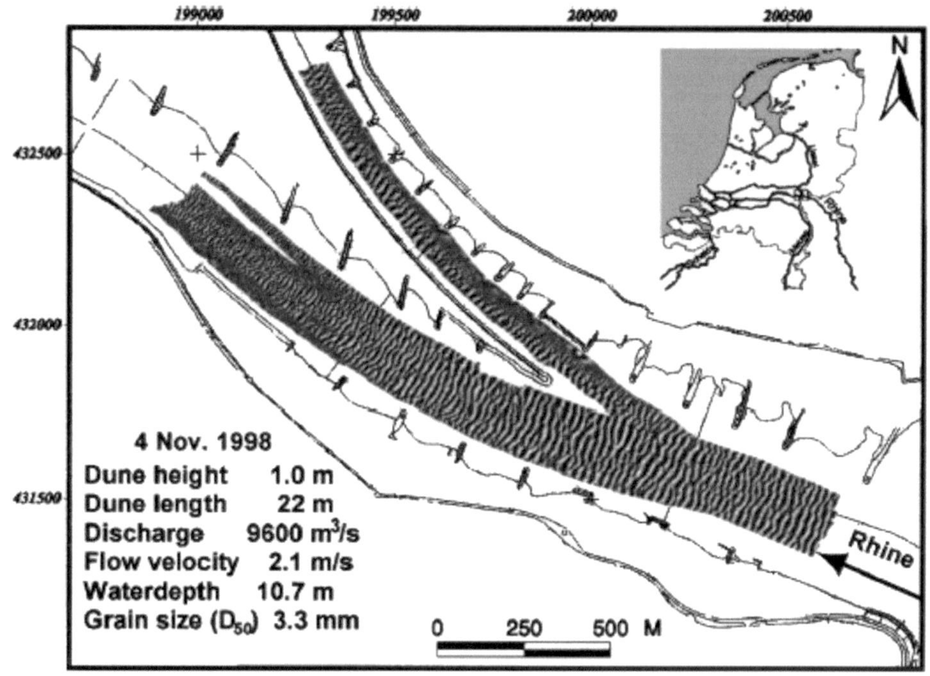

Fig. 1. Dunes around a bifurcation of the Rhine in The Netherlands at peak flood stage in 1998, mapped using data of a multibeam echosounder (from ASCE task committee 2002, copyright permission of ASCE)

Ripples are transverse bed forms, which normally have heights of less than 0.04 m and lengths below 0.6 m. Ripples are formed under hydraulically smooth flow conditions. If the flow velocity is great enough to move the individual sand grains but less than another limiting value, the bed is spontaneously deformed into irregular features called dunes. Dunes are formed under hydraulically rough conditions. Dunes have much larger dimensions and may have heights up to several meters and wavelength up to hundreds of meters. Observations and measurements suggest that lengths of dunes are about 3 to 8 times the water depth (Yalin 1977). When the velocity of flow is increased in beds composed of fine sediments (< 0.4 mm), a situation will be reached when dunes can no longer be sustained, and flat bed is formed in which the bed and free surface become flat. With increase in velocity and the Froude number, the water surface becomes unstable, and even small disturbances give rise to stationary surface waves thereby forming trains of long, sinusoidal-shaped waves of sand that are in phase with the surface waves and usually move slowly upstream. These features are called antidunes, and surface waves accompanying them are referred to as stationary waves. At higher velocities and stream powers, the bed reorients itself to create a series of hydraulic jumps for energy dissipation and these bed forms are called as chutes and pools. These high Froude number flows may occur when flash floods sweep down steep gullies of rocky areas. Large quantities of suspended sediment are carried by resulting breaking waves.

2.1 Dune geometry

The upstream side of a dune is commonly referred to as the stoss side (for example, see Figures 3 and 4). The downstream side is usually referred to as the lee face or slipface. Dune height h (Figure 4) is most often defined as the difference in elevation between the dune crest and corresponding downstream trough (Gabel 1993, Julien and Klaassen 1995): although sometimes dune height is measured from upstream trough to downstream crest (e.g., Nordin 1971). Dune length, λ is the longitudinal distance between subsequent crests or troughs. Dune steepness, $\delta = h/\lambda$, is the ratio of dune height to length and its value varies between 0.1 to 0.03 (Yalin, 1964). If the aspect ratio of the channel is small, dunes tend to be approximately triangular with straight crests and extend nearly across the width of the channel (Klaassen et al. 1986). On the other hand, when the aspect ratio is large, dunes will be three-dimensional with sinusoidal crests or several crests stretching across the flume.

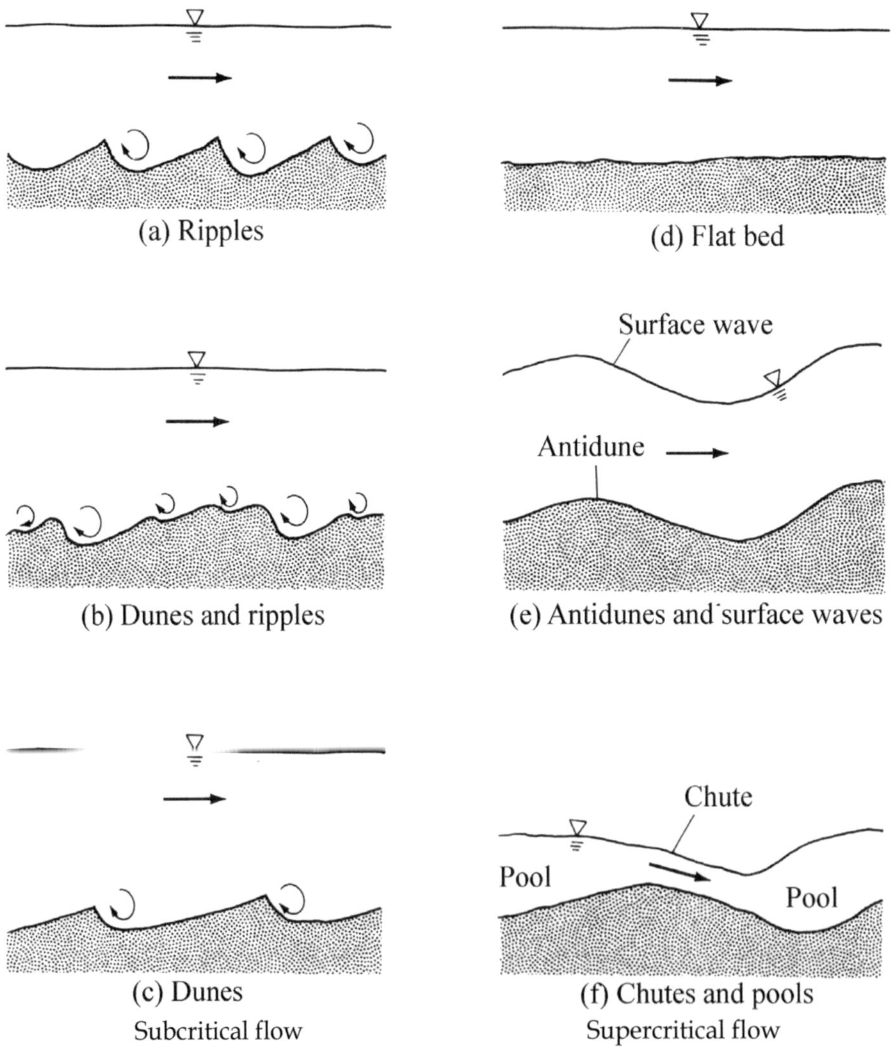

Fig. 2. Bed forms in alluvial channel (from Vanoni, 1975, copyright permission of ASCE)

2.2 Fully developed (equilibrium) bed forms

The simplification with equilibrium bed forms is that uniformly sized periodic dunes vary only in the streamwise direction and do not change with time. As a first step towards developing a comprehensive understanding of flow in natural channels, researchers have studied the flow field over a train of fixed, well developed dunes. Best et al. (2001) opined that a fixed dune approach doesn't capture the effect of sediment transport on velocity profiles and despite this limitation the simplification of dune morphology to a fixed two-dimensional profile does allow investigation of the major features of flow, as has been successfully demonstrated in previous studies of flow over two-dimensional bed forms (e.g. Bennett and Best, 1995; Lyn, 1993; Nelson et al., 1993; Raudkivi, 1966). Such studies also provide a basis for a better understanding of the effects of dune three-dimensionality. The laser-Doppler (LDV) technique has been used in several laboratory fixed sediment bed form investigations (e.g., Balachandar et al., 2007; Balachandar and Patel 2008; Bennett and Best 1995; Coleman et al. 2006; Kadota and Nezu 1999; Lyn 1993; McLean et al. 1994 and 1996; Nelson et al. 1993; Nelson et al. 1995; Van Mierlo and de Ruiter 1988; Venditti and Bennett 2000). These studies have suggested that near-bed turbulence over much of the stoss side of a dune deviates markedly from either classical boundary-layer or wake turbulence. Cellino and Graf (2000) carried out detailed study of sediment-transport flows over mobile bed forms using acoustic Doppler technique. Mendoza and Shen (1990), Johns et al. (1993), Yoon and Patel (1996), Yue et al. (2006), Stoesser et al. (2008) and Noguchi et al., (2009) performed numerical simulations of flow over fixed bed forms by applying sophisticated turbulence models. Detailed flow characteristics will be discussed in the upcoming sections.

2.3 Transient bed forms

The shape and size of bed forms and flow conditions are inter-related. Significant progress has been made to understand the role played by topography on flow characteristics, separated flow dynamics, internal boundary layer development, and turbulence structures (Smith 1996). Raudkivi (1997) studied the formation and development of ripples. Bennett and Best (1996) and Lopez et al. (2000) studied the ripple to dune transition. Recent experimental and theoretical studies carried out by Coleman and Melville (1996) and Coleman and Fenton (2000) have focused on the bed form initiation process. Coleman and Eling (2000) observed that turbulence may not be an essential feature of the initial instability of a sediment bed.

Robert and Uhlman (2001) carried out experiments on different bed stages across the ripple-dune transition and found that turbulence intensity and Reynolds stresses gradually increase throughout the transition. Coleman et al. (2006) studied developing dunes using an innovative experimental approach and double averaging methodology to advance our understanding of the structure of the rough flow near the bed region. They found that flow structure doesn't change as dunes develop in time for a steady flow, even though overall flow structure must change as dunes develop and traditional boundary layer type features can be potentially destroyed. Interesting findings of this work include: (a) friction factor was noted to increase with bed form growth; (b) the location of the separated shear layer reattachment point was determined as approximately 4h downstream of the crest for time averaged flow fields over the developing dunes as compared to 4.2h of previous flows over fixed dunes (Bennett and Best 1995; Kadota and Nezu 1999; Lyn 1993; McLean et al. 1999;); (c) negligible variation of the vertical velocity distribution, spatial fields of Reynolds stresses , form induced stress, skin friction, form drag, bed stress and overall momentum flux as dunes grow and the overall form of the velocity profiles display an approximately linear distribution below roughness tops and a potentially logarithmic distribution above roughness tops; (d) peak Reynolds stresses occur in the shear layer associated with the separation zone consistent with shear layer instabilities and vortex shedding off this layer.

3. Temporal and spatial flow over dunes

This section presents a discussion of results related to temporal and spatial velocity characteristics and flow resistance of open channels with bed forms

3.1 Temporal and spatial flow over dunes

Compared to flow over a hydrodynamically smooth channel boundary, the interactions of flow with the bed formation are considerably harder to predict. Dunes disrupt the boundary layer type flow and generate turbulence. Extensive research regarding fluid flow over dunes has been undertaken in the past forty years (Engelund and Fredsøe 1982; Lyn 1993; McLean et al. 1994; Nelson et al. 1993; Raudkivi 1966). To summarize, five zones (Figure 3) have been recognized within the dune-flow interaction region (Balachandar et al., 2007; Balachandar et al. 2008; Bennett and Best, 1995; Best 2003).

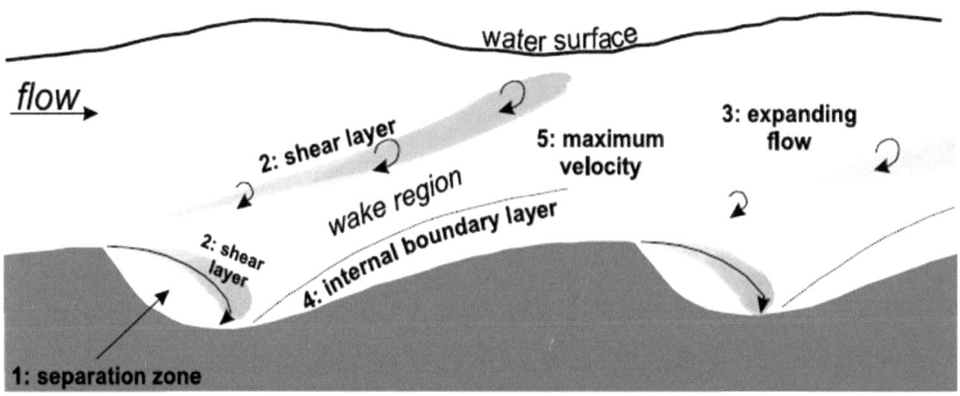

Fig. 3. Schematic diagram of the principal regions of flow over asymmetrical angle-of-repose dunes (from Best 2005, Reproduced by permission of AGU)

A typical dune and the flow over the dune are indicated in Figure 4 (Balachandar et al., 2007). The flow ahead of the crest (indicated by C in Figure 4) resembles a typical near-wall, boundary-layer like flow. Following separation at C, a typical recirculation pattern is formed. Above this separation zone lies a decelerating flow with wake-like characteristics that extends downstream beyond the reattachment point (denoted as R in Figure 4). Experimental and field investigations have documented the formation of kolks or boils, which are tilting vortices emanating from the reattachment region and rising to the free surface. Near the bottom, following reattachment, an internal boundary layer develops and interacts with the overlying wake zone. Close to the free surface and over the dune wavelength, there is an outer zone that is generally modeled as a quasi-inviscid region. Along the separating shear layer, regions of high Reynolds stresses have been observed. Many previous studies have suggested that the macroturbulence associated with dunes is important in controlling the stability of the bed forms and the entrainment and transport of the sediments. It has been suggested that this macroturbulence has its origin in Kelvin-Helmoltz instabilities in the separating shear layer (Best 2003; Kostaschuk and Church 1993; Muller and Gyr 1987). In particular, Best (2003) and Hyun et al. (2003) provide whole field quantification using particle image velocimetry. As an important step towards better understanding the flow-dune interaction, researchers have analyzed the flow field for the case of a fixed dune.

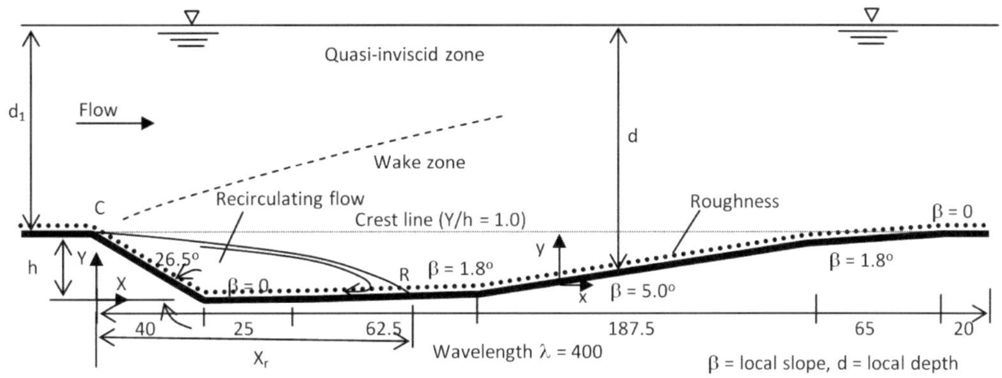

Fig. 4. Schematic of the flow field. Not to scale, all dimensions in millimeters (from Balachandar et al., 2007, copyright permission of NRC Research Press).

3.2 Time averaged flow along the dune

Using laser-Doppler measurements, Balachandar et al., (2007) analyzed the velocity data at various axial stations between two dune crests and is shown in Figure 5. In the figure, d_1 = flow depth ahead of the crest at X/h = -2 or 18; X = horizontal distance from the dune crest; h = dune height. It was observed that the flow over the dune train was periodic in space insofar as the flow pattern was the same over successive dunes, but there was a significant variation in flow properties along the wavelength of the dune. The velocity profile at X/h = -2, just ahead of the crest, indicates a log-region, although of a much reduced extent compared with a plane channel flow (see Figure 5a). A region of reverse flow can be seen at X/h = 4 and the mean reattachment point is located around X/h = 4.5. It was also observed that the flow in the outer region was not influenced by the local bed geometry, which is shown in Figure 5b. The average vertical velocity is zero at the dune crest, yet flow expansion and separation cause bed wise movement and thus negative velocities over much of the dune. This indicates downward deflection of streamlines (Figure 5c). The positive vertical velocities at X/h = 12 suggest streamlines following the upward slope of the surface. The profiles of streamwise turbulence intensity show that the turbulence reaches a high peak just below the crest line, which is also the location of maximum value of mean shear, shown in Figure 5d. A second peak at all the stations in the vicinity of y/d = 0.2, leads to the conclusion that there is a distance from the bed beyond which the turbulence is little affected by the dune shape. The peak exists at X/h = 12 and -2, and suggests the presence of remnant of the turbulence generated in the shear layer of the upstream dune. Figure 5e indicates that large values of vertical components of the turbulent intensities occur in recirculation zone and in the shear layer above it. Quite large values of shear stress (Figure

5f) can be seen measured in the separation zone. Shear stresses decrease to upstream values with increasing distance from the reattachment.

3.3 Effect of flow depth

In studying ripple-dune transition, Bennett and Best (1996) noted important changes in macroturbulence structure due to effect of depth on flow past bed forms. Wiberg and Nelson (1992) studied two flow depths over a train of ripples closely resembling the bed form shape of Nelson and Smith (1989) and noticed a higher value of bed friction at the smaller depth. Figure 6 (Balachandar et al., 2007) shows streamwise mean velocities profiles at six locations along the dune for four depth ratios (y/h), ranging from 3 to 8. At X/h = -2, in inner scaling, U^+ versus y^+ (inset of Figure 6a) collapses at all depths but the outer region is different from a simple open channel with a smooth bed. Velocity profiles at X/h = 2 to 12 show similarity in the near wall region and independent of depth in the outer region (y/d > 0.35). It is important to note that at X/h =5, the mean velocity in the near-wall region is positive for $d_1/h \geq 4$ and is negative at d_1/h = 3. This indicates the depth of flow influences the near-bed region and the length of the separation region is longer at the shallower depth. The profiles of the vertical component of turbulence reveal a systematic dependence on flow depth near the crestline and at the larger depths indicate lower values of turbulent intensity. In the recirculation region, the shear stresses are very high and increase with decreasing flow depth.

3.4 Effect of bed roughness on the flow over dunes

Balachandar and Patel (2008) studied the effect of bed roughness superimposed on a train of well-formed dunes by conducting laser-Doppler measurements at several stations between two dune crests. In these studies, rough surfaces were generated using (a) stainless steel wire mesh made of 0.72 mm wires with 6.35 mm centerline spacing and (ii) sand grain roughness created from sand grains of 1.8 mm nominal mean diameter carefully glued onto a double-sided tape to ensure a uniform distribution. Their results showed that the wire mesh provided for a higher degree of roughness as compared to sand grain roughness. It was found that with increasing streamwise distance from the crest, the dependence of the near-bed part of the profiles on the roughness becomes apparent. In the outer region (y/d > 0.2), the profiles are independent of the bed roughness (Figure 7). It is also found that outer region was clearly different from that in the simple open channel flow over a smooth surface, indicating that the entire depth of flow has been affected by the dune geometry. The effect of the near-bed roughness is more dominant in the recirculation zone. The peak value of turbulence at any station is influenced by the bed roughness and the location of the peak is farther extended into the flow away from the wall with increasing roughness. It is found that the length of the separation region is longer for the flow with larger bed roughness. The results indicate that the shape of the dunes have a major influence on the flow features, where as the effect of dune roughness is limited to the near wall region extending to a distance of about 80% of the dune height about the crest.

3.5 Relation between suspended sediment and dune characteristics

Bennett and Best, (1996), Best (2005 a, b), Lopez et al., (2000), Robert and Uhlman (2001) and Schindler and Robert (2004) have shown that suspended sediment concentration increases with increase in scale and magnitude of turbulent structures. Schindler and Robert (2005)

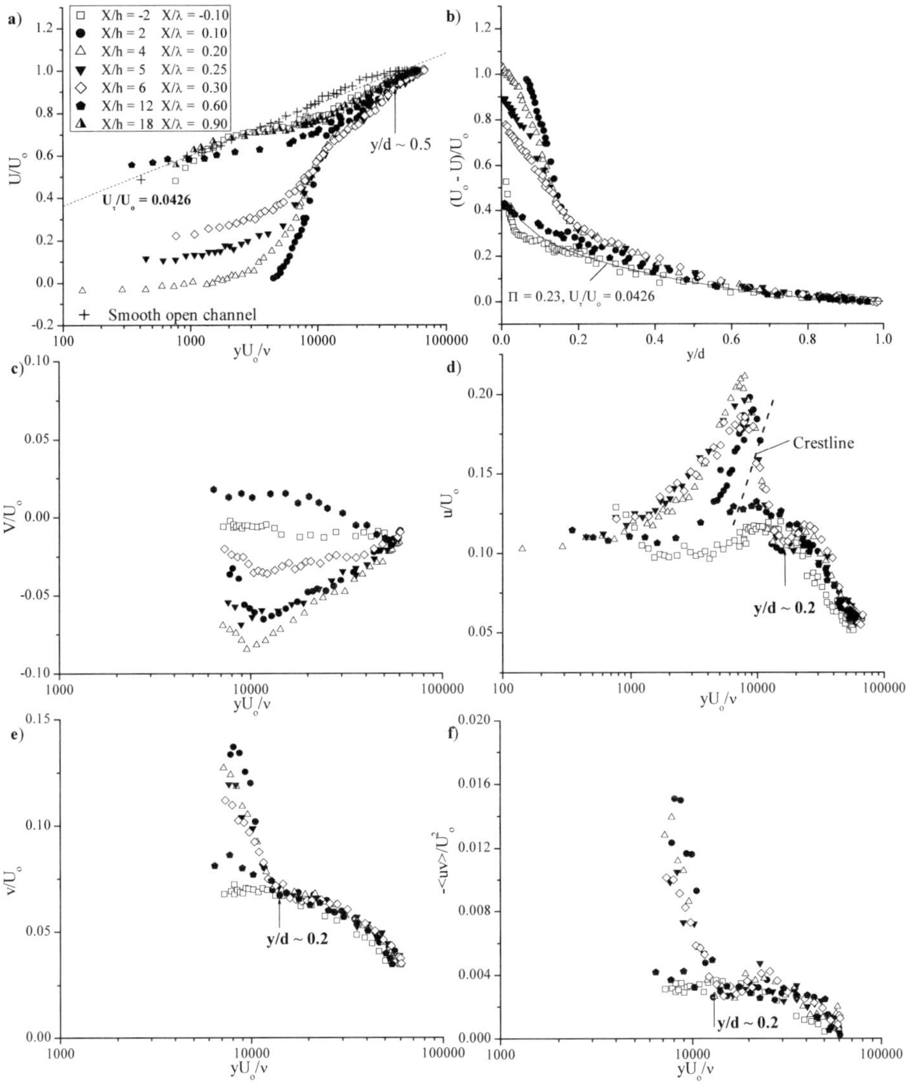

Fig. 5. Mean velocity and turbulence profiles ($d_1/h = 6$): (a) streamwise mean velocity profiles; (b) mean velocity profiles plotted in the velocity-defect format; (c) vertical component, V, of the mean velocity; (d) profiles of the streamwise root mean squared (rms) turbulence intensity (u/U_o); (e) vertical component of the rms turbulence intensity (v/U_o); (f) Reynolds shear stress profiles (from Balachandar et al., 2007, copyright permission of NRC Research Press).

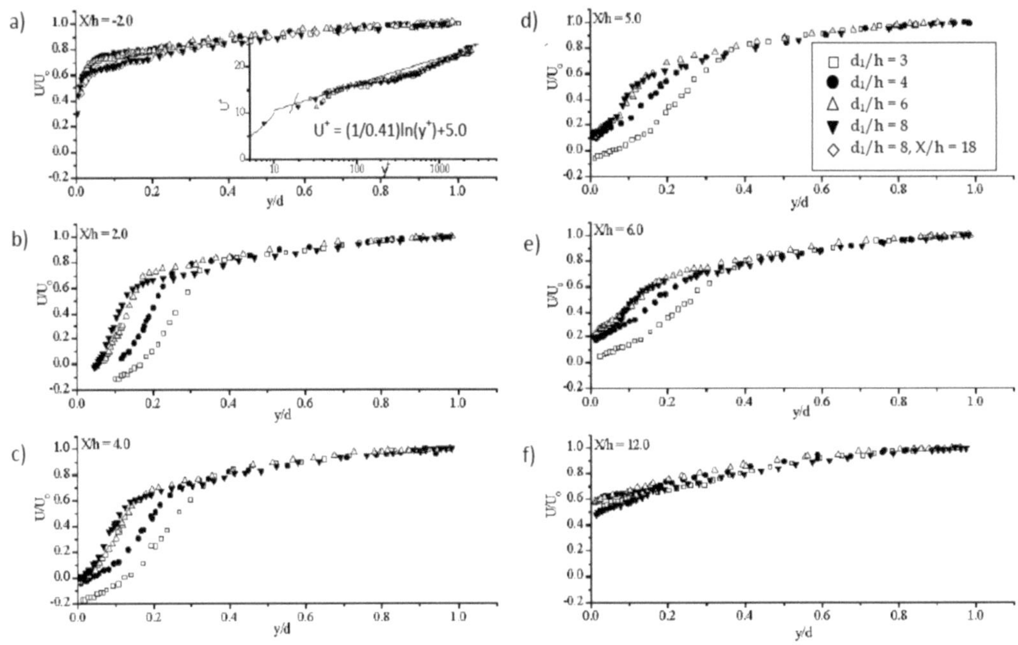

Fig. 6. Effect of depth on mean velocity profiles at various axial stations ($-2 \le X/h \le 12$) using the depth of flow, d, as the normalizing length scale (from Balachandar et al., 2007, copyright permission of NRC Research Press).

have concluded that the transition from 2-D to 3-D bed forms, result in increased sediment transport, increased turbulence and increased bed form migration rate. Venditti and Bennett (2000) found that suspended sediment concentration is highest over the dune crest and at flow reattachment. Further research is needed to spatially analyze both the fluid dynamics and the sediment transport processes over mobile bed forms.

Tevez et al. (1999) reported that spatially averaged mean velocity profiles over dunes consist of upper and lower semi log-linear segments. The upper segment reflects the total shear stress of the flow. The lower segment on symmetric dunes reflects the skin friction from sand particles, but for asymmetric dunes it is the skin friction plus the effect of form roughness from the superimposed dunes. Villard and Kostaschuk (1998) found that predictions of the Rouse equation indicated that sediment suspension is controlled by total stress for symmetric dunes, whereas for asymmetric dunes sediment suspension is related to stress associated with skin friction plus form roughness. It is reported that dunes in bed load-dominated environments are often asymmetric having low-sloping upstream side (stoss) and steep lee faces (Guy et al. 1966; Kostaschuk et al. 2004), while those in suspended load dominated environments are often more symmetric with relatively low angle lee faces (Best and Kostaschuk 2002; Kostaschuk and Villard 1996).

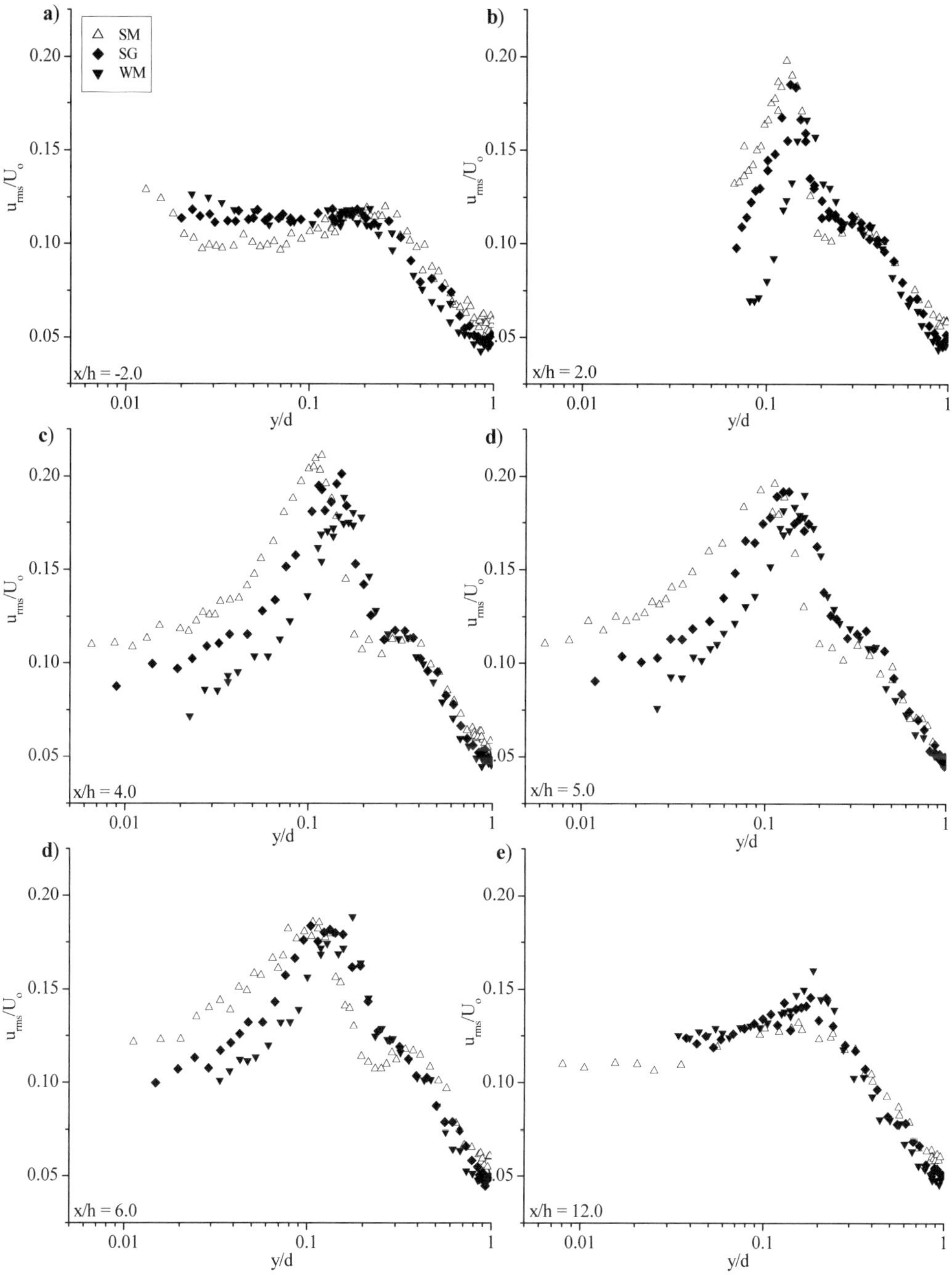

Fig. 7. Variation of streamwise turbulence intensity. SM - experiments conducted with a train of smooth bed dunes manufactured from Plexiglas, SG sand grain pasted on smooth dune; WM, wire mesh glued to dune. (from Balachandar et al., 2007, copyright permission of NRC Research Press)

3.6 Sediment heterogeneity

Existing literature mostly points to studies that focus on the case of uniform sediment while dealing with laboratory dune studies. However, sediment heterogeneity may be of practical interest because formation and development of bed forms in natural rivers happens distinctively with non-uniform sediment. It is quite challenging to study flow over dunes with sediment mixtures in a laboratory. A few studies have been conducted in this area (Parker et al., 1982, Wilcock and Southard, 1988 and Wilcock, 1992). Wilcock and Southard (1989), Klaassen et al., (1987), Ribberink, (1987) and Klaassen (1990) investigated the mutual influence of sediment gradation on ripples and dunes, on fractional transport rates and on vertical sorting. ASCE Task Committee on Flow and Transport over Dunes (2002) reported that fully developed dunes composed of highly heterogeneous sediment exhibit different geometric, flow, and transport characteristics than the more intensively studied homogenous-sediment counterparts. Klaassen (1991), Blom and Kleinhans (1999), Kleinhans (2002) and Blom and Ribberink (1999) have reported that, in the vertically sorted bimodal gravel bed streams the coarser material tends to accumulate at the base of the dunes, creating a partial barrier between the coarser substrate below and finer material in the migrating dunes above (see Figure 8). A clear armour layer was also observed at the level of the dune troughs.

Lanzoni (2000), Lanzoni and Tubino (1999), Lanzoni et al., (1994) and Lisle et al., (1991) studied the sand bar formation with bimodal mixtures of fine and coarse sediment and suggested that grain sorting associated with selective transport of graded sediment may induce a overall stabilizing effect on bottom development with appreciable reduction of bar amplitude and a shortening of bar wave lengths (see Figure 9). In the caption of Figure 9, the term FC70 is a sediment mixture of poorly sorted with strongly bimodal character, i_s is the water surface slope and MUNI is near uniform sand. In addition, these studies are designed to get a better insight on longitudinal and vertical sorting. Lanzoni (2000) opined that a suitable model for vertical sorting is required to study the presence of heterogeneous sediments.

4. Experimental studies, limitations of some of the studies, recent PIV studies

Turbulence studies in water flows over dunes were commenced with the development of hot-film anemometers and flow visualization techniques. Muller and Gyr (1982) used fluorescent dye, a light sheet and video to visualize the flow. Nezu et al. (1994) used both dye and hydrogen bubbles in conjunction with video to study turbulent structures over dunes, and Bennett and Best (1995) took long exposure photographs of small neutrally buoyant particles to view the paths of fluid particles. These studies provide valuable information about features such as point of separation, point of reattachment, shear layer development, and generation and shedding of transverse vortices. Since three decades, much more accurate measurements of water flows became feasible with the increased use of laser-Doppler anemometer or laser Doppler velocimeter (LDA or LDV).

In the last decade, quantitative flow visualization techniques such as particle tracking velocimetry (PTV) and particle image velocimetry (PIV) have become popular to study coherent eddies in space and time. The ability of PIV to yield flow-field information ensures its usage to extend further. Hyun et al. (2003) provide a discussion of the factors that determine the accuracy of PIV data, and a discussion of the physics of the flow over a dune is given in Balachandar et al., (2003).

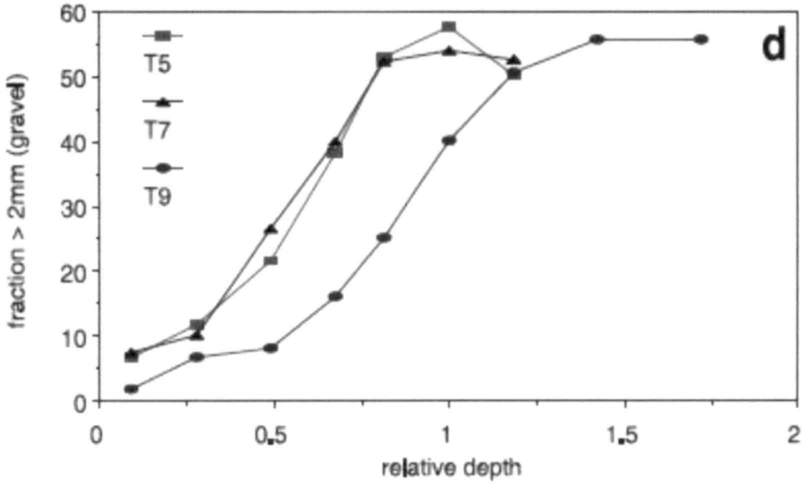

Fig. 8. Vertical sorting measured in experiments. The dune or bar top is on the left-hand side of the graphs, and the base is on the right-hand side. Vertical sorting in the dunes in Kleinhans (2002), Blom and Kleinhans (1999), experiments T5, T7 and T9. The sediment here is divided into two grain size fractions (sand and gravel), coarser and finer than 2.0 mm. The level of the dune troughs agrees with the level of the armour layer in T5 and T7, whereas in T9 the armour layer of T7 can still be observed. (From Kleinhans, 2004, copyright permission of Science Direct)

Schmeeckle et al., (1999) presented a method for the 3-D simulation of turbulence over 2-D dunes and compared the accuracy of PIV measurements and numerical simulations using a dense grid of two-dimensional laser Doppler velocimetry measurements. Balachandar et al. (2002) carried out LDV and PIV measurements and complementary LES simulation over a fixed dune. LDV and PIV provide complementary data on time-averaged and instantaneous flow structure over the dune. The time-averaged results reveal, in considerable detail, flow features such as separation and reattachment, and associated large variations in velocity and turbulence profiles. The instantaneous PIV results reveal a complex pattern of near-random but well-defined vortices. Vortices form downstream of flow separation and grow in size as they are convected along the dune. Balachandar and Patel (2008) studied the turbulent flow over a long train of fixed two-dimensional dunes, identical in size and shape by combining the complementary capabilities of LDV and PIV over a range of flow depths in a fully developed region. In this study, the points of interest were the instantaneous and mean velocity fields, the Reynolds stresses, triple-correlations, vorticity maps and analysis of events in the four quadrants.

4.1 Combined use of PIV and LDV to measure turbulence and flow over a dune

Hyun et al. (2003) assessed the relative merits of LDV and PIV to measure mean velocity and turbulence in water flow over a train of fixed two-dimensional dunes. Figure 10 shows the LDV measurement stations and PIV field-of-views. The flow field over the dune was divided into five different fields-of-view with a 20 mm overlap between the images. Though PIV is limited in the field-of-view normal to the bed, it provides instantaneous flow-field information that reveals the true complexity of the flow over dunes.

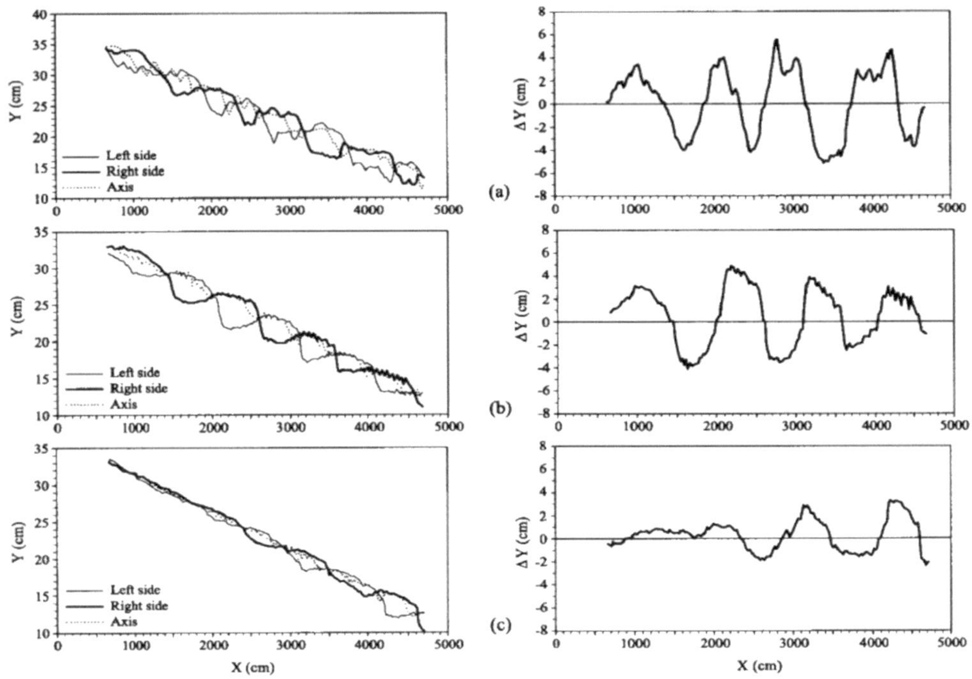

Fig. 9. Comparison between longitudinal bed profiles and differences between right-side and left-side bed elevation (Y) measured under similar hydraulic conditions during experiments with uniform (MUNI) and mixed (FC70) sediments. (a) Equilibrium phase of P2709 (MUNI) with Q = 45 L/s and i_s = 0.514%. (b) Initial phase of P1309 (FC70) with Q = 45 L/s and i_s = 0.525%. (c) Equilibrium phase of P1309 (FC70) with Q = 45 L/s and i_s = 0.525%. The data are plotted every 20 cm. (from Lanzoni, 2000, reproduced by permission of AGU)

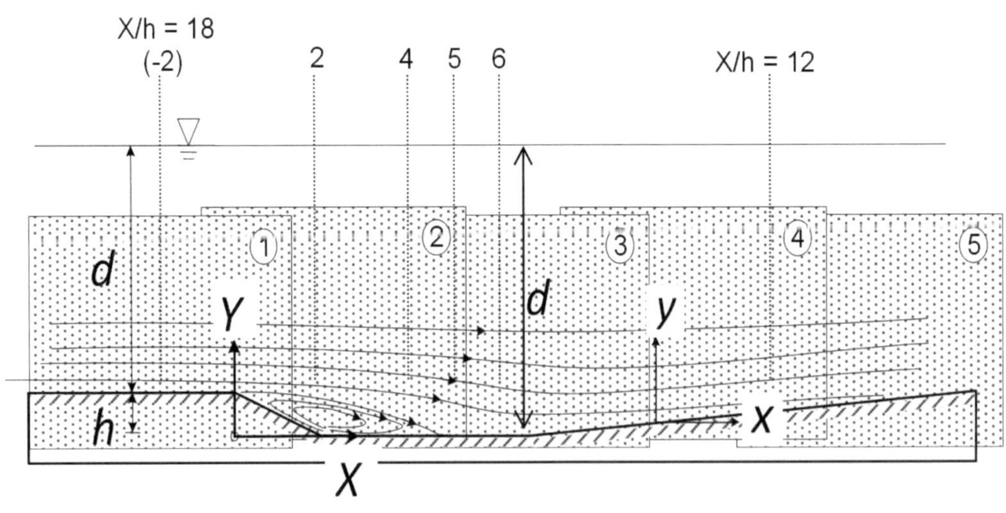

Fig. 10. Schematic of flow and notation (from Hyun et al. 2003, copyright permission of Springer)

Figure 11 (Hyun et al. 2003) shows the profiles of the x-component of mean velocity (U), the rms turbulence intensities in the x- and y-directions ($\sqrt{\overline{u'^2}}, \sqrt{\overline{v'^2}}$), and the Reynolds shear stress ($-\overline{u'v'}$) at one streamwise station (X/h = 5). All variables are made dimensionless by the maximum velocity U_o, held constant at 0.48 m/s in the experiments. The figure provides a direct comparison between PIV and LDV data at a station that is located close to the mean reattachment point. The limitation of the PIV field-of-view (and quality of light sheet) is clear from the extent of the data in the vertical direction. Also seen is the limitation of the two-component LDV in the near-wall region due to blockage of the beams by the dune. These limitations of the two systems are obvious, and present in most arrangements although their criticality will depend on the flow, or flow region of interest. In general, the PIV and LDV data show agreement within the expected uncertainties of the two systems in the mean velocity and turbulence intensities.

Hyun et al. (2003) opined that principal advantage of PIV is its capability to provide information about the instantaneous flow field. The flow is well illustrated by the two sets of successive instantaneous images shown in Figure 12. They correspond to two overlapping fields-of-view, taken at different times. In the region of the dune crest (Figure 12a), at time t = t_o, there appear to be two vortical structures, one just forming below the shear layer emanating from flow separation at the crest, and the other farther downstream. The former is small and nascent while the latter is larger and more diffused. With increasing time, the latter moves out of the field-of-view while the former grows in size and travels downstream at a velocity of about $0.4U_o$. In the last frame, this vortex is seen to again decrease in size. The subsequent evolution of this vortex could not be tracked due to limitation of the field-of-view - a limitation that is set by the PIV configuration. The second sequence of images (Figure 12b) shows the organized motions in a region downstream of the first. Here, strong ejection and sweep-type events are observed along with other flow structures. Further processing of the PIV images using some of the techniques suggested by Adrian et al. (2000) could provide more quantitative information, such as the length and time scales of the observed flow structures. The most obvious flow feature of the instantaneous flow is the formation of vortices just downstream of the crest below the shear layer, and their growth as they travel downstream.

5. Modeling and numerical simulation–RANS, LES

The RANS based turbulence models such as the k- ε model, the algebraic model and the Reynolds stress model (RSM) are based on the statistical theory of turbulence, and therefore these models cannot in principle simulate coherent structures such as bursting phenomenon (Nezu and Nakagawa 1993). Johns et al. (1993) developed a numerical scheme with a one-equation turbulence model in which a transport equation was solved for eddy viscosity and their comparison of mean flow information, such as velocity profiles, Reynolds shear stress, and bed resistance with experiments were encouraging. Yoon and Patel (1996) developed a k-ω model based on the study of Wilcox (1993) that avoids the use of wall functions. Their computed velocity and turbulence fields, as well as the pressure and friction distributions along the dune are in general agreement with existing detailed experimental data in a rectangular channel with two-dimensional dunes of typical but regular shape. Cheong and Xue (1997) developed a three-layer near-wall k- ε model to account for the separated and reattached region by treating the anisotropy between normal stresses with streamline curvature correction. They opined that standard wall function treatment appears to be inadequate in the separating-reattachment region but computations are fairly satisfactory in other regions.

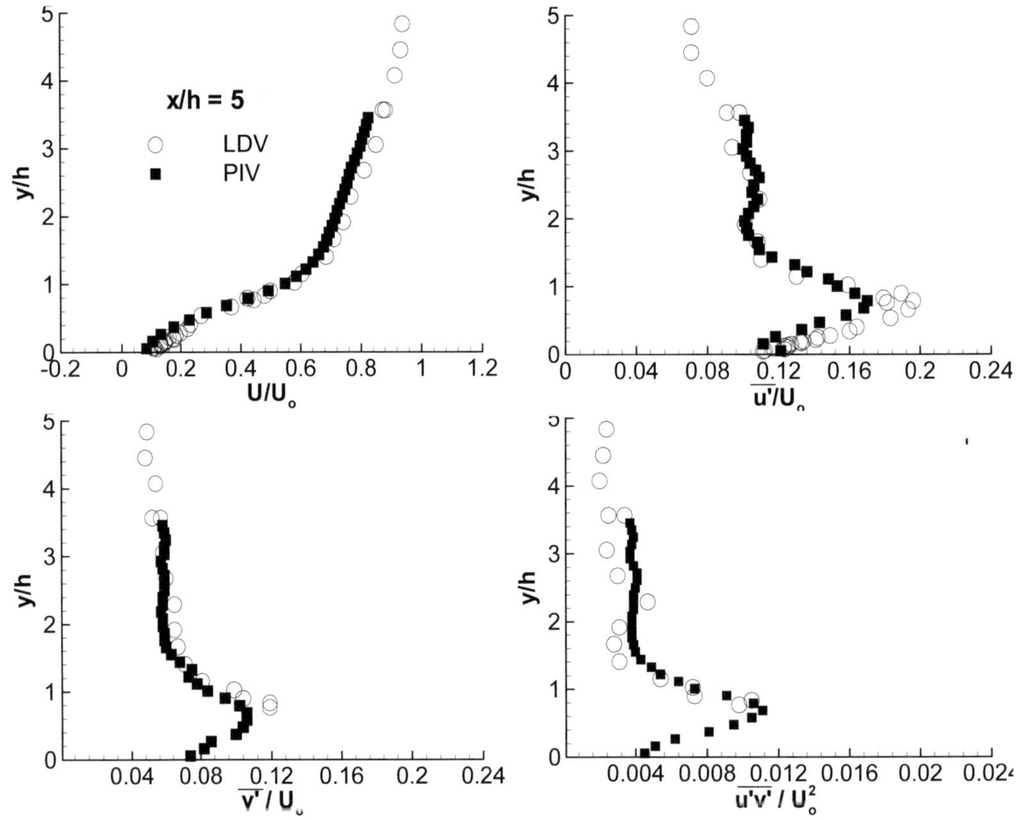

a) X/h = 5, X/λ = 0.25

Fig. 11. Comparison between PIV and LDV data (from Hyun et al. 2003, copyright permission of Springer)

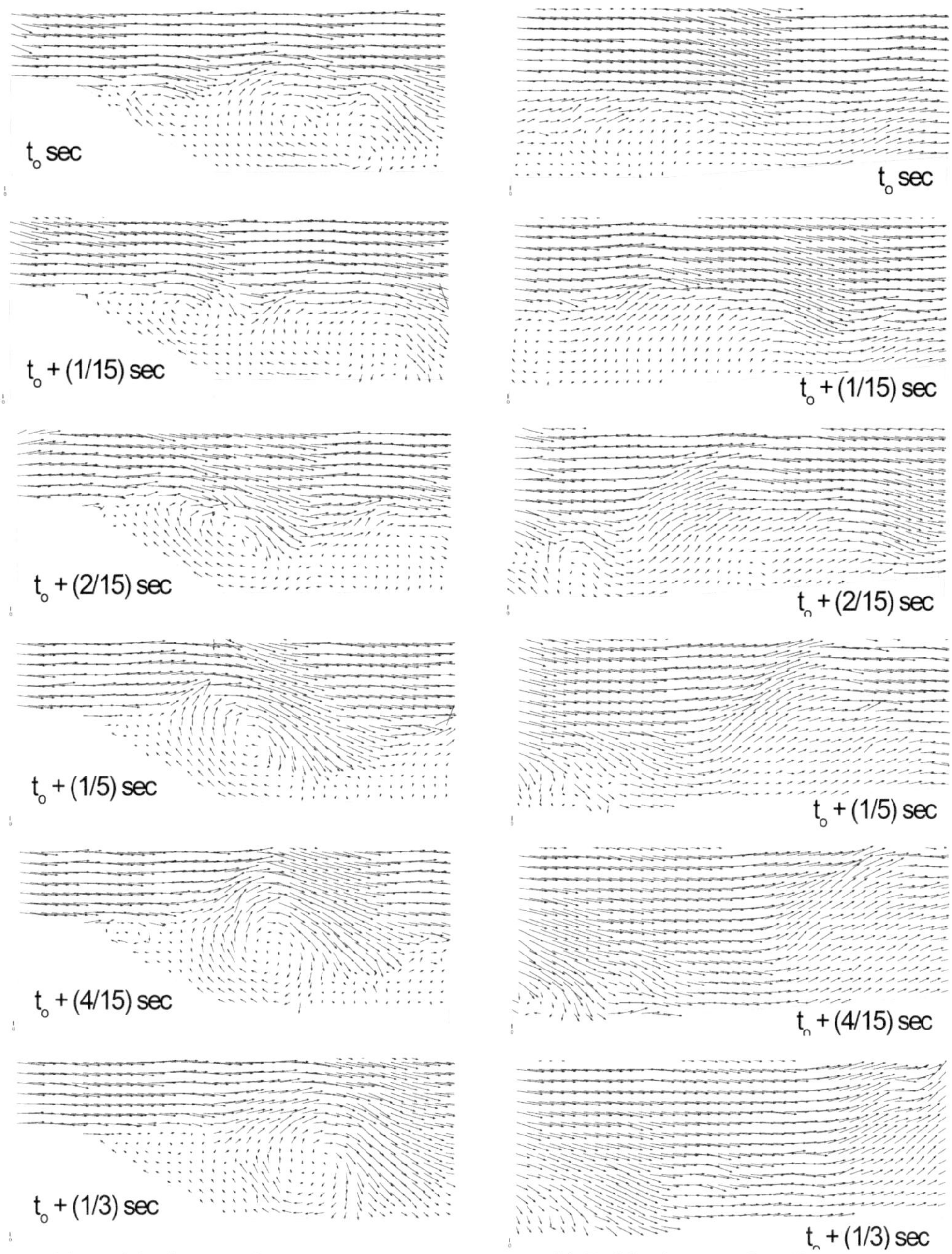

(a) Field-of-view #2 in Figure 11 (b) Field-of-view #3 in Figure 11

Fig. 12. Instantaneous velocity field obtained by PIV (from Hyun et al. 2003, copyright permission of Springer)

Cokljat and Kralj (1997) used a k- ε model and RSM to investigate the better turbulence model for prediction of flows over river bed forms. Cokljat and Kralj (1997) used Lyn's (1993) data for validation of their numerical model and found that k- ε model of turbulence is capable of modeling the main features of the flow to acceptable level of accuracy. Mendoza and Shen (1990) used a k-ε model with wall functions but with an algebraic stress model in place of the more commonly used eddy viscosity assumption. The model was validated using experimental data obtained by Raudkivi (1966) and concluded that model satisfactorily predicted U-velocity, $\overline{uv}$, and turbulent kinetic-energy profiles, pressure and shear stress distributions of the steady turbulent flow over two-dimensional dune-like artificial bed forms. The above-mentioned models used the rigid lid assumption. Bravo and Zheng (1999) studied the standard k- ε model, RNG (Renormalization Group) k- ε model and differential RSM (Reynolds stress model) model for their applicability to predict turbulent flow over artificial bed forms. They implemented both free surface and rigid lid models to study the effects of sand grain roughness, near wall modeling, free surface, body forces and grid resolution. They found that free surface flow models gave better flow predictions and take fewer iterations to convergence than the rigid lid models. Patel and Lyn (2004) illustrated the results of both RANS computations and large eddy simulations (LES), compared the mean velocities and turbulent statistics to those measured by Hyun et al. (2003). The above-mentioned numerical models with RANS based turbulence closures provide mean flow information, such as velocity profiles, Reynolds shear stress, and bed resistance, but do not give any information about the unsteady flow structures that are important features of such flows. Lyn (2002) mentioned coherent turbulent flow structures can only be predicted either by direct numerical simulation (DNS) or large-eddy simulation (LES).

Balachandar et al. (2002) carried out experimental investigations (PIV and LDV) and complemented them with numerical simulations (LES). The LES computations used the assumption of space periodicity and approximate treatment of the free surface by simulating the flow over a single dune in a closed channel formed by the mirror reflection of the open channel on the free surface. The finite-volume LES numerical code of Cui et al. (2000) was used in the study. No-slip conditions were applied at the top and bottom walls, periodic conditions were applied at the inlet and outlet, and also along the sides of the solution domain at each instant. Convergence was claimed at each time step when the sum of the pressure and viscous resistance was in balance with the imposed pressure difference between the duct inlet and outlet. After the flow reached "steady-state", the solution was continued for about five large-eddy turnover times ($2d/u_*$, where u_* is the friction velocity) to accumulate the statistics. Measurements were compared with simulations for time averaged flow and instantaneous flow. The normalized measurements mean velocity in the x-direction, the RMS velocity fluctuations in the (x, y) directions, and the Reynolds shear stress ($-\overline{u'v'}$) by LDV and PIV are in agreement but the LES results for the longitudinal fluctuations and the shear stress are lower than the experiments (shown in Figure 13). In the case of the instantaneous flow field, the PIV shows better-defined vortices just downstream of the crest below the shear layer.

A recent DNS on an artificial dune-type wavy bed was carried out by Hayashi et al. (2003) to examine coherent structures and "boil of the first kind" in a low Reynolds number flow

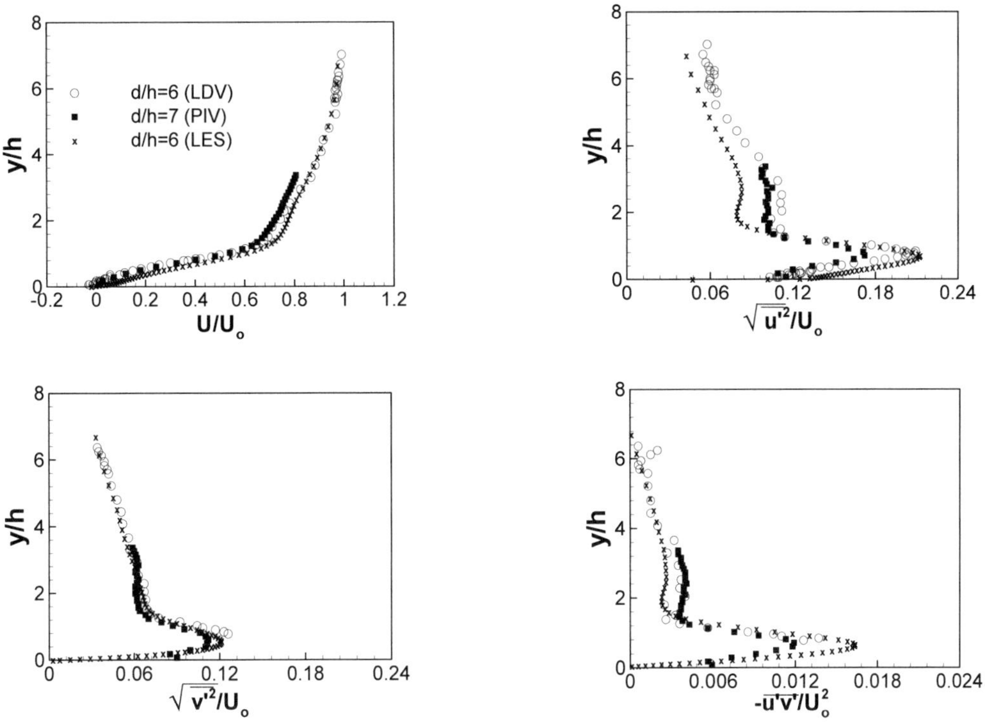

Fig. 13. Comparison of LDV, PIV and LES results at x/h=4 (from Balachandar et al., 2002)

(*Re* = 3300) and presented mean flow and turbulent structures of the open channel flow over bed forms. Hayashi et al. (2003) also found the presence of hairpin-shaped vortices, vortex patterns that showed both a normal and reverse hairpin shapes. Yue et al. (2006) used LES to study high-Reynolds number open-channel flow (*Re* = 5.7x10^4) over an idealized, two-dimensional laboratory-scale dune to investigate the unsteady flow structures as well as the mean-flow details and turbulence statistics. In a qualitative manner, they detected sweeps and ejections as well as formation of near-bed streaks and free surface structures and movement. They found that LES predicts the secondary peaks lying farther out from the bed, as observed in the experiments, their magnitudes are substantially under predicted. As the secondary peaks are remnants of the primary ones on the previous dunes, the discrepancy leads to the conclusion that the present LES does not adequately account for the upstream history. Williams (2007) carried out large eddy simulation of flow over a 2-D dune by using a surface adaptive curvilinear finite volume method to solve free-surface flows. The computed mean velocity profiles and normalized RMS plots of stream wise velocities were in good agreement with available experimental data of Balachandar et al. (2002) and PIV and LES offer a very graphic view of the instantaneous structures in the flow, as shown in Figure 14. Stoesser et al., (2008) investigated large eddy simulations of instantaneous flow over 2-D dunes (the geometry of the dunes is identical to that used previously by Balachandar et al. 2002) and found splats and evolution of hairpin vortices from kolk-boil vortexes around and beyond the point of reattachment.

6. Formation of coherent structures

Experimental and field investigations have documented the macro-turbulent characteristics of spatially varied flow over the bed forms and the formation of "kolks" and "boils" proposed by Matthes (1947). The "kolks" and "boils" are the upward tilting vortices of both fluid and sediment originating downstream of dune crests and at the point of reattachment. Bennett and Best (1995) suggested that the bursting events due to turbulent flow are associated with zone of Kelvin-Helmholtz instabilities developed at the zone of flow separation in front of ripple lee.

6.1 Turbulent structures by quadrant analysis

To evaluate the role of coherent structures, it is a common practise to divide the *u-v* sample space of the fluctuating velocities into four quadrants. Such representation has potential to reveal the relative contribution of the turbulent structures (events) from different quadrants to the total Reynolds shear stress. Flow visualization studies on rough wall bounded flows by Grass (1971) documented a streak that migrates slowly away from the wall and at some point detaches from it completely. This process of upward lifting of low-speed fluid away from the solid wall is commonly referred to as *ejection*. During the ejection process, the instantaneous local velocity (U_i) is lower than the time-averaged local velocity (U). Whenever ejection occurs, in order to satisfy continuity, high speed fluid moves towards the wall ($U_i > 0$) from outer flow regions and this event is called a *sweep*. It is only possible to detect ejection events when $u < 0$ and $v > 0$ are observed simultaneously. In the same way, sweep events can be detected by simultaneously observing a high-speed fluid parcel ($u > 0$), moving towards the bed ($v < 0$). There are two other events that also contribute to the total Reynolds shear stress. They are known as outward interactions ($u > 0$, $v > 0$) and inward interactions ($u < 0$, $v < 0$).

At every measurement location, the Reynolds shear stress is calculated and further decomposed as a sum of different events according to the procedure described by Lu and Willmarth (1973). By using the concept of a hyperbolic hole of size H, defined by $|uv| = Hu_{rms}v_{rms}$, the contribution from a particular quadrant can be written as

$$\overline{(uv)}_{Q_i,H} = \lim_{T\to\infty}\frac{1}{T}\int_0^T u(t)v(t)I(t)dt \quad (1)$$

Here, $I(t)$ is a detection function defined as:

$$I(t) = \begin{cases} 1 & when \;\; |uv|_Q \geq Hu_{rms}v_{rms} \\ 0 & otherwise \end{cases} \quad (2)$$

It is assumed that the velocity used to compute $\overline{(uv)}_{Q,H}$ is a function of time only. The parameter H defines a threshold value, which separates the extreme events from the random background turbulence. By increasing the value of H more extreme/strong events are identified. In the present study, quadrant decomposition yields Reynolds stress contribution to each quadrant for a given value of H denoted as $\overline{(uv)}_{Qi}(y)$ and is represented as Q_i (i = 1 to 4) for convenience.

Figure 15 shows the contributions by the four quadrants events (Balachandar et al., 2007) for the flow past the dune shown in Figure 4. At each station, the local maximum value of the mean shear stress is used as the normalizing scale. As expected, at X/h = -2, the contributions from Q_1 and Q_3 are small, and the fact that the contributions are negative indicates the presence of flow structures that transfer energy from the turbulence to the mean flow. All along the dune, the contributions by Q_1 events are more or less similar. With increasing distance from the dune crest, there is a very slight increase in the contributions from Q_1 events, especially along the separating shear layer, and immediately upstream and downstream of reattachment. As pointed out by Bennett and Best (1995), these events are probably associated with structures brought towards the bed along the separating shear layer. These events provide one possible mechanism by which sediment can be eroded.

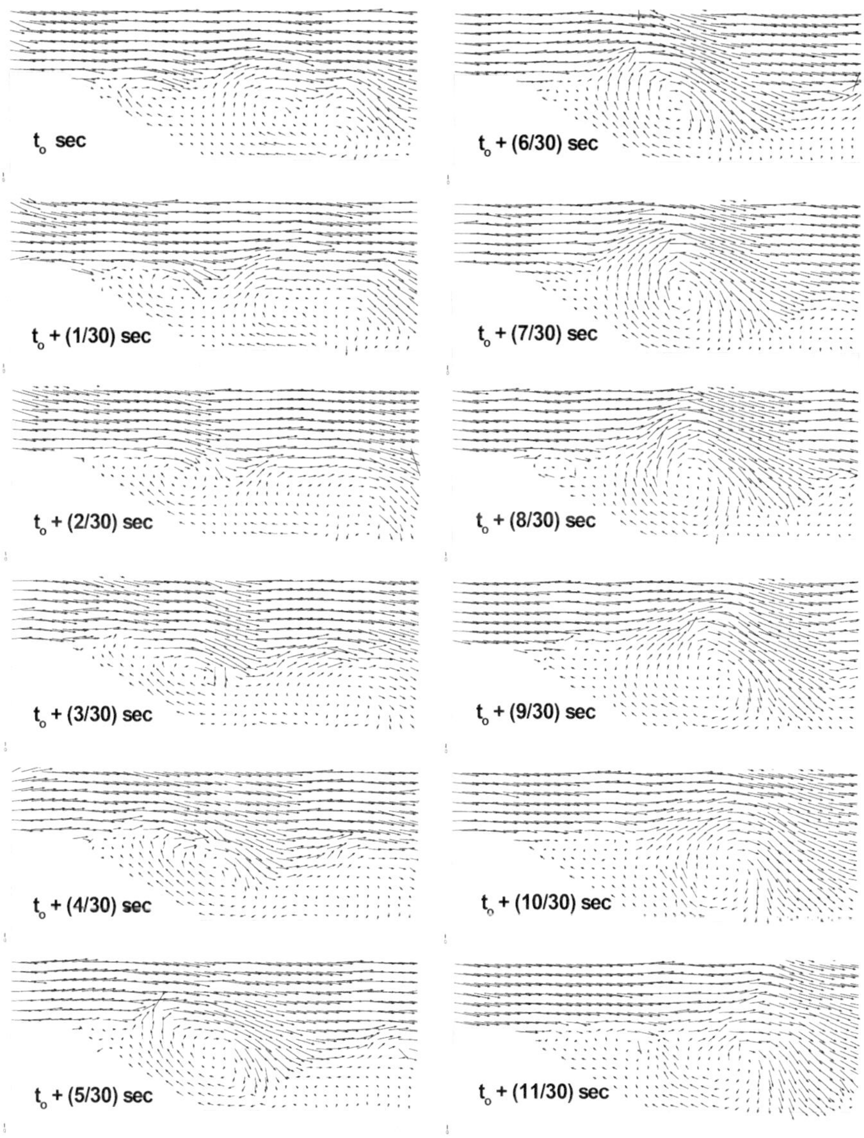

a) PIV

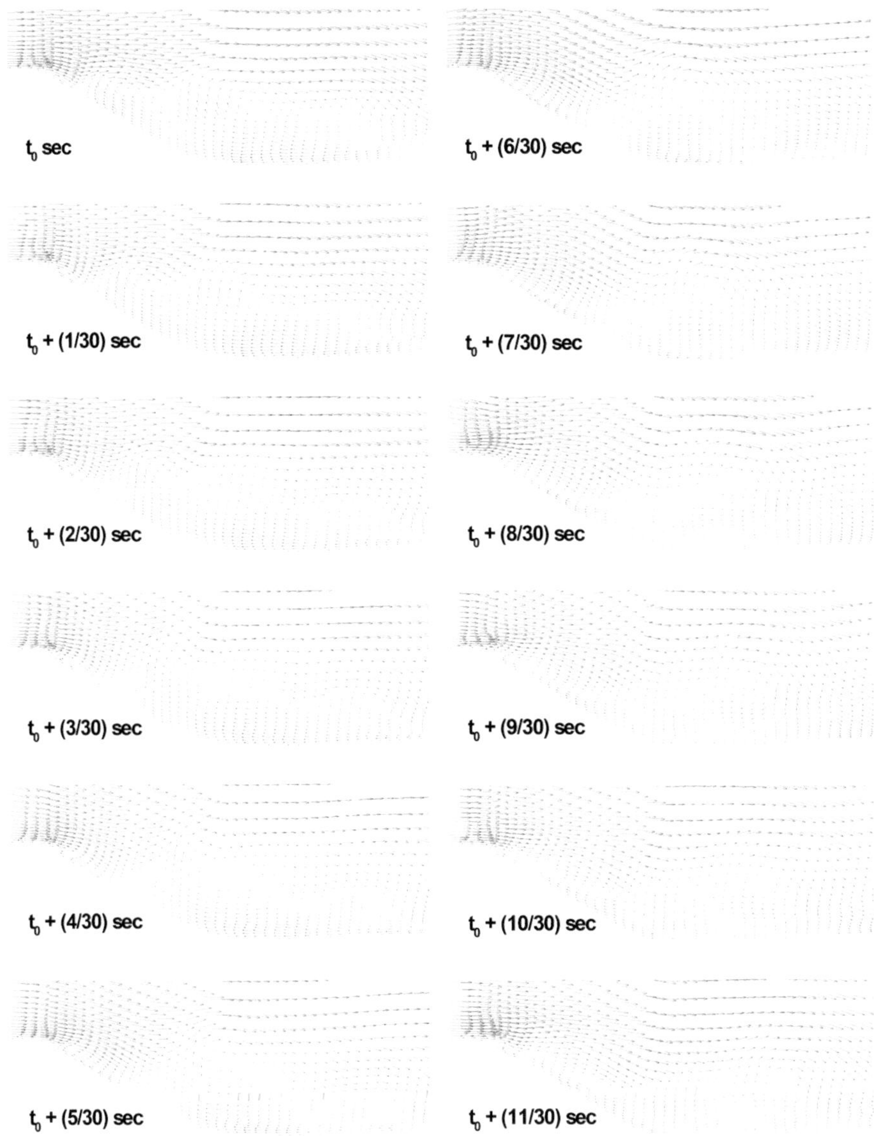

b) LES

Fig. 14. Time sequence of instantaneous velocity field (from Balachandar et al., 2002)

Detailed experimental observations (Balachandar et al., 2007; Best, 2005a, b; Kostachuck and Church, 1993; Rood and Hickin, 1989) revealed that Q_2 events are most likely associated with much of the entrainment and transport of sediment. The variation of Q_2 events over the wavelength of the dune was studied by Balachandar et al., (2007). It was found that the contributions of quadrant 2 events are dominant at all X/h locations as shown in Figure 15. In the figure, F represents the fractional contribution to the total shear stress by each quadrant. With increasing distance from the crest, the Q_2 profiles indicate large gradients, especially near the crest line. These high-magnitude events correspond to the source of the dune-related bursting phenomena reported previously (Jackson 1976). Similar to Q_2 events, the Q_4 events show large changes in the vicinity of the shear layer. In Figure 15a, for $y/d < 0.3$, the Q_4 profiles show a trend that is opposite to that of the Q_2 profiles. Ahead of reattachment (X/h = 2 and 4), and slightly above the crestline, the magnitude of the Q_4 events tends to be close to the Q_2 events.

Studying several results such as that in Figure 15, differences due to flow depth were noticeable in the contributions of all four quadrants. However, as noted with respect to Figure 15, the contributions of Q_2 events are very important and most likely associated with much of the entrainment and transport of sediment. Therefore, the effect of depth on Q_2 events is presented in Figure 16 over the wavelength of the dune. The results are shown in a two-column format with h as the normalizing scale in the left column and d as the normalizing scale in the right column. Ahead of separation (Figures 16a,c and e), the Q_2 profiles show a similar trend at all flow depths, each showing a near-wall peak in the vicinity of $y/h = 1$. This is identified with the remnants of the turbulence activity carried over from the previous dune. This trend continues well after reattachment, indicating the persistence of the Q_2 events on the stoss side of the dune. At stations ahead of the dune crest, a peak in the Q_2 profiles (Figures 16 d and f) around $y/d = 0.2$ indicates the sustenance of turbulence generated by the previous dune, carried over to the next dune. This value is not influenced by the bed roughness. Overall it is concluded that the characteristics of the shear layer, the length of the reattachment zone and the Q_2 events control the turbulence structure over the dunes.

Balachandar and Patel (2008) carried out LDV measurements to obtain more detailed information on the effect of bed roughness superimposed on a train of well-formed dunes. The fractional contribution to the mean Reynolds shear stress by quadrant 2 (Q_2: $u < 0$, $v > 0$) events are shown in Figure 17. The figure is presented in three-column format with one type of wall condition being represented in each column. In all the graphs, with increasing distance from the dune surface, the Q_2 values gradually increase towards the free surface. The profiles show large gradients, especially near the crest line. In all the graphs shown in Figure 17, the tendency is to have two local peaks (indicated as A and B, see, middle column of Figure 17). Except for the first row of graphs (X/h = -2), peak A, closer to the bed ($0.16 < y/d < 0.23$), corresponds to the location of the separating streamline; whereas peak B, away from the bed ($0.5 < y/d < 0.62$) is an indicator of a large-scale event in the outer region. The peaks at X/h = -2 (first row) correspond to the turbulence activities carried over from the previous dunes. Comparing the three columns of graphs, the peaks are located farther away from the bed with increasing bed roughness. As seen in the middle column of graphs, the high magnitude Q_2 events correspond to the source of the dune-related bursting phenomenon reported previously (Jackson 1976; Yalin 1972). It is also possible that the separating shear layer is associated with streamwise vortices generated along the braids between the Kelvin-Helmhotz instabilities (Lasheras and Choi 1988; Metcalfe et al. 1987). The Q_2 events influence the vertical convection of sediment and can be responsible for much of the vertical mixing in the rivers.

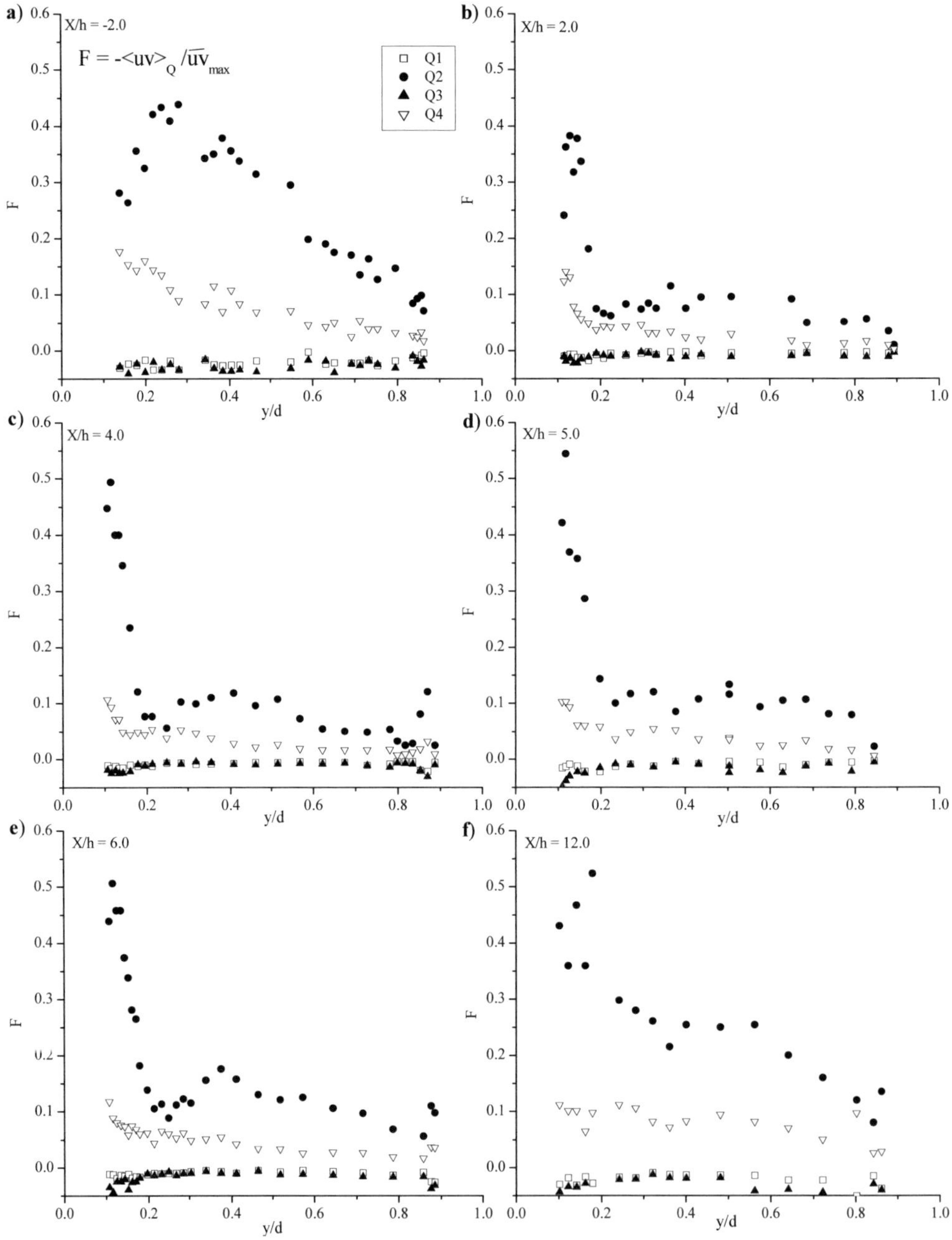

Fig. 15. Fractional contribution by the four quadrants ($d_1/h = 6$). (a) X/h = -2.0; (b) X/h = 4.0; (c) X/h = 4.0; (d) X/h = 5.0; (e) X/h = 6.0; (f) X/h = 12.0. (from Balachandar and Patel 2008, copyright permission of NRC research press)

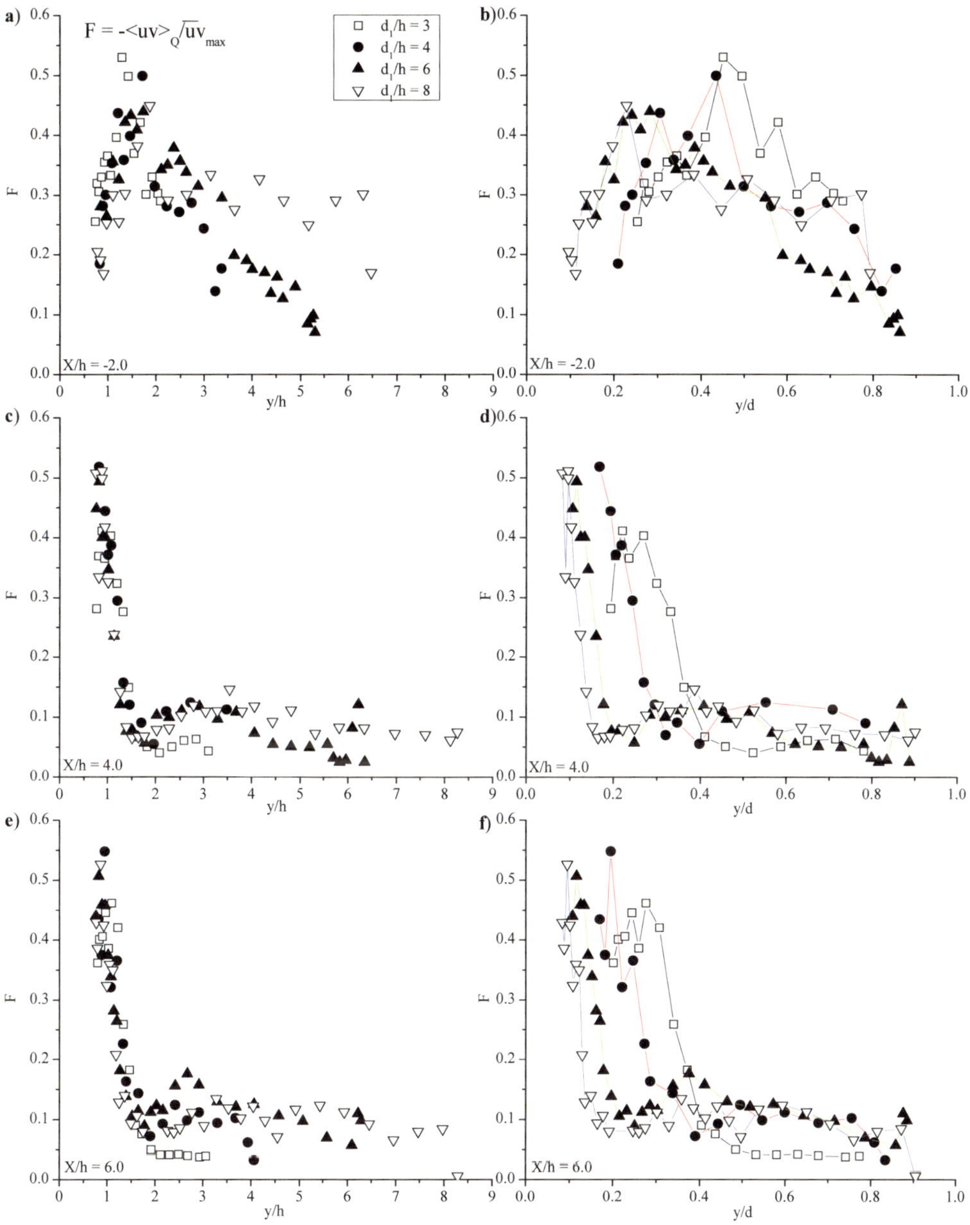

Fig. 16. Effect of depth on fractional contribution to quadrant 2: (a, c, e) using the dune height, h, as the normalizing length scale and (b, d, f) using the depth of flow d as the normalizing length scale. (from Balachandar and Patel 2008, copyright permission of NRC research press)

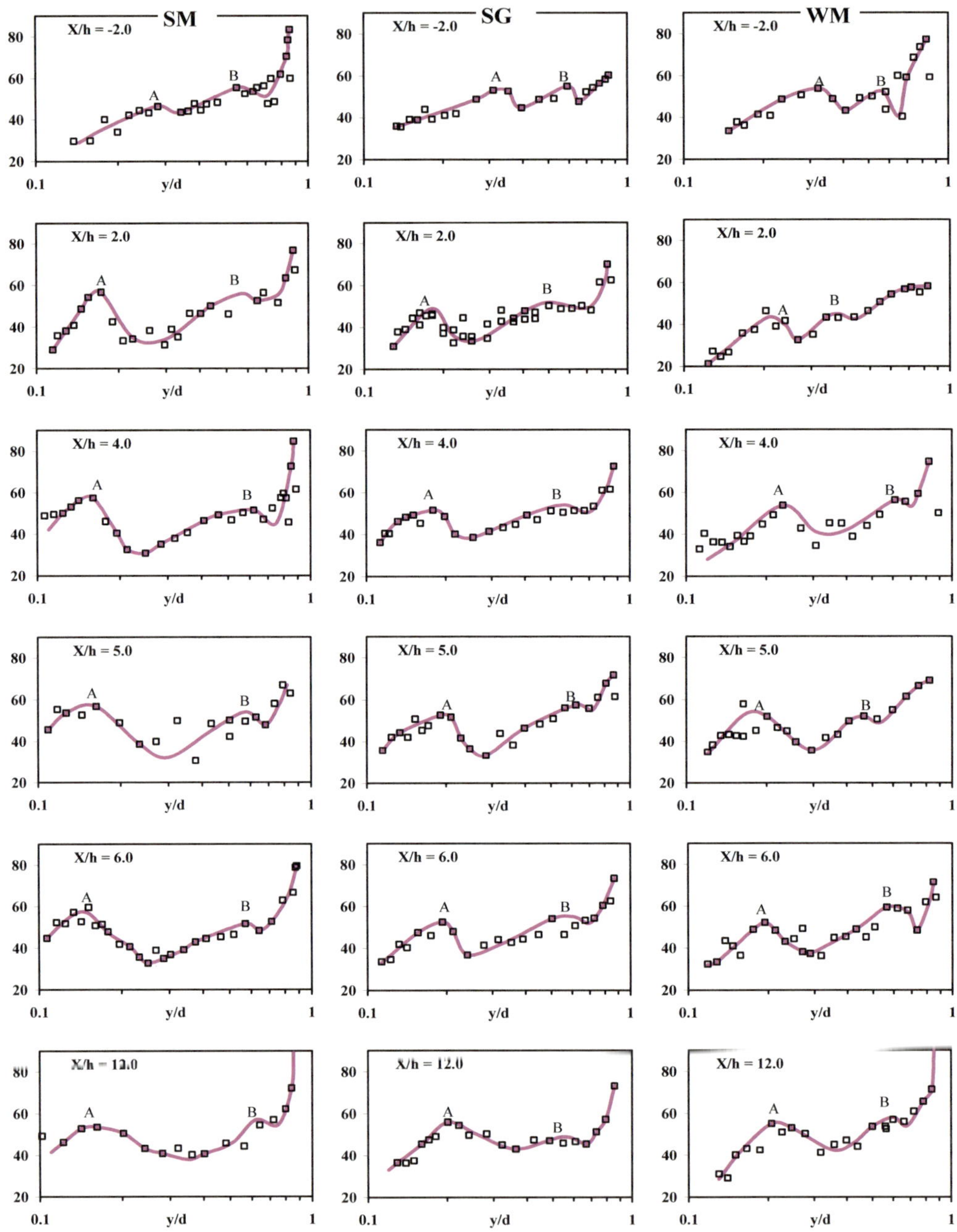

Fig. 17. Contribution to shear stress by quadrant 2. SM - train of smooth bed dunes manufactured from Plexiglas; SG - sand grain test; WM - wire mesh test. (from Balachandar and Patel 2008, copyright permission of NRC research press)

7. Future research topics

The ASCE taskforce subcommittee (2002) and Best (2005 a) have suggested several important research directions to further understand, quantify and model the dynamics of flow over dunes. The taskforce opined the need to synthesize the knowledge gained with the advances in measurement techniques and simulation technologies to quantify the development of dunes, amount of sediment transport and bed changes. Efficiency of the available theoretical models to simulate the formation of the bed forms with uniform and heterogeneous sediment needs to be ascertained for laboratory and field data. Presently, 2-D dunes are modeled in laboratory and simulations but changes in the transverse direction should be included in the simulations (Parsons et al. 2005). Most of the dune studies are concentrated on flow separation and flow field on lee side of dunes, however near-bed turbulence characteristics on the stoss side are not fully understood. Carling et al. (2000) found that evidence of large-scale, coherent low frequency flow structures and separated boundary layers above the stoss side of the large dunes. The knowledge gained in understanding of coherent structures should be utilized in future research (Stoesser et al. 2008 and Yue et al. 2005) on modeling of sediment transport and evolution of bed in open channels. Another important research direction is to study the scaling effects between laboratory and field studies of flow over dunes (Best and Kostaschuk 2002). Measurements and simulations of formation of bed forms should be studied during varying discharge regimes because most of the sediment transport and bed form changes takes place during the floods (Ten Brinke et al. 1999). Low angle dunes are common in big rivers such as Jamuna river in Bangladesh (Kostaschuk et al., 2004) therefore modeling and laboratory studies on low angle dunes are required. Venditti et al. (2005) concluded that movement of the entire bed form is controlled by superimposed bed forms therefore morphodynamics of super imposed bed forms over migrating dunes needs to be studied theoretically and experimentally. Yue et al. (2005) opined that more realistic simulations of the interaction of the free surface with the bed is needed to understand the coherent structures and sediment transport. Kostaschuk et al. (2008) suggested to carry out further research with respect to amount of suspended sediment on formation and as well as migration of dunes. The future research may also include simulation of 2-D and 3-D flow over dunes using DES (detached eddy simulation) which is computationally more efficient (Xu et al., 2007) than LES modeling.

8. References

Adrian, R. J., Meinhart, C. D. and Tomkins, C. D. (2000). "Vortex organization in the outer region of the turbulent boundary layer." J. Fluid. Mech., 422, 1-54.

ASCE Task Force (2002), Flow and transport over dunes, *J. Hydraul. Eng.*, 127, 726- 728.

Balachandar, R., Hyun, B. S., and Patel, V. C. (2007). Effect of depth on flow over a fixed dune, *Canadian Journal of Civil Engineering*, 34, pp. 1587-1599.

Balachandar, R. and Patel, V.C. (2002). "Rough Wall Boundary Layer on a Plate in an Open Channel". *J. Hydr. Engrg.*, ASCE 128(10), 947–951.

Balachandar, R. & Patel, V. C. (2005). "Velocity measurements in a developed open channel flow in the presence of an upstream perturbation", *J. Hyd.* Res., 43 (3), 258-266.

Balachandar, R., & Patel, V. C. (2008). Flow over a fixed rough dune, *Canadian Journal of Civil Engineering*, 35, pp. 511-520.

Balachandar, R., Polatel, C., Hyun, B. S., Yu, K., Lin, C. L., Yue, W., & Patel, V. C. (2002). "LDV, PIV, and LES investigation of flow over a fixed dune." *Proc., Symp. Held in Monte Verità: Sedimentation and Sediment Transport,* Monte Verità, Switzerland, Kluwer Academic, Dordrecht, The Netherlands, 171–178.

Best, J. L. 2003. The kinematics, topology and significance of dune-related macroturbulence: some observations from the laboratory and field. *In* Fluival Sedimentology VII, Proc. 7th International Symposium on Fluvial Sedimentology.

Bennett, S. J., & Best, J. L. (1995). ''Mean flow and turbulence structure over fixed, two-dimensional dunes: Implications for sediment transport and bed form stability.'' *Sedimentology,* 42, 491–513.

Bennett, S. J., & J. L. Best (1996), Mean flow and turbulence structure over fixed ripples and the ripple-dune transition, in Coherent Flow Structures in Open Channels, edited by P. J. Ashworth et al., pp. 281- 304, John Wiley, Hoboken, N. J.

Best, J. L. 2005a. The fluid dynamics of river dunes: A review and some future research directions. *J. Geophys. Res.,* 110, F04S02, doi:10.1029/2004JF000218.

Best, J. L. (2005b), The kinematics, topology and significance of dune related macroturbulence: Some observations from the laboratory and field, in Fluvial Sedimentology VII, edited by M. D. Blum, S. B. Marriott, and S. Leclair, Spec. Publ. Int. Assoc. Sedimentol., 35, 41- 60.

Best, J.L. & Kostaschuk, R.A. (2002) An experimental study of turbulent flow over a low-angle dune. *J. Geophys. Res.,* 107(C9), 3135–3154.

Best, J.L., Kostaschuk, R. A. & Villard, P. V (2001), "Quantitative Visualization of Flow Fields Associated with Alluvial Sand Dunes:Results from the Laboratory and Field Using Ultrasonic and Acoustic Doppler Anemometry," *Journal of Visualization,* Vol 4(4), 373-381.

Blackwelder, R. F. & Kovasznay, L. S. G. 1972 Large-scale motion of a turbulent boundary layer during relaminarization. J. Fluid Mech. 53, 61-83.

Blom, A., Kleinhans, M.G., 1999. Non-uniform sediment in morphological equilibrium situations. Data Report Sand Flume Experiments 97/98. University of Twente, Rijkswaterstaat RIZA, WL/Delft Hydraulics. University of Twente, Civil Engineering and Management, The Netherlands.

Blom, A., & Ribberink, J. S. (1999). "Non-uniform sediment in rivers: Vertical sediment exchange between bed layers." Proceedings, IAHR Symposium on River, Coastal and Estuarine Morphodynamics, G. Seminara, ed., Universiata degli studi di Genova, Italy, Springer, New York, 45-54.

Bravo H.R., and Zheng Y. H., (1999) "Free surface models of turbulent flow over sand dunes," *Intl. Journal of Applied Sc. and Computations,* 5(3), pp. 207 - 219.

Carling, P. A., J. J. Williams , E. Golz & A. D. Kelsey, (2000),"The morphodynamics of Fluvial sand dunes in the River Rhine, near Mainz, Germany. II. Hydrodynamics and sediment transport," *Sedimentology*, 47, 253-278

Cellino, M., & Graf, W. H. (2000). ''Experiments on suspension flows in open channels with bed forms.'' *J. Hydraul. Res.*, 38(4), 289–298.

Cheong, H.F and Xue, H. "Turbulence Model for Water Flow Over Two-Dimensional Bed Forms," J. of Hydr. Eng., 123(5), 1997, pages 402-409.

Cokljat, D. and Kralj, C. "On the Choice of Turbulence Model for Prediction of Flows Over River Bed forms,' J. of Hydraul. Res., Vol. 35, No.3, 1997, pages 355-361.

Coleman, S. E. & Eling, B. 2000 Sand wavelets in laminar open-channel flows. J. Hydraul. Res. 38, 331–338.

Coleman, S. E. & Fenton, J. D. 2000 Potential-flow instability theory and alluvial stream bed forms. J. Fluid Mech. 418, 101–117.

Coleman, S. E. & Melville, B. W. 1996 Initiation of bed forms on a flat sand bed. J. Hydraul. Engng 122, 301–310.

Coleman, S. E., Nikora, V. I., McLean, S. R., Clunie, T. M., Schlicke, T., & Melville, B. W. 2006. Equilibrium hydrodynamics concept for developing dunes. Physics of Fluids, 18 (10), 105104-1-12.

Cui, J., Patel, V.C. & Lin, C.L. 2000, Large-eddy simulation of turbulent flow over rough surfaces, Iowa Inst. Hydr. Res., The University of Iowa, IIHR Report 413.

Djenidi, L., Antonia, R. A., Amielh, M. & Anselmet, F. (2008). "A turbulent boundary layer over a two-dimensional rough wall", *Exp. Fluids*, 44(1), 37-43.

Engelund, F. & Fredsøe, J. 1982 Sediment ripples and dunes. Ann. Rev. Fluid Mech. 14, 13–37.

Gabel, S. L. (1993), Geometry and kinematics of dunes during steady and unsteady flows in the Calamus River, Nebraska, USA, Sedimentology, 40, 237- 269.

Guy, H.P., Simons, D.B. & Richardson, E.V. (1996). "Summary of alluvial channel data from flume experiments." Professional Paper 462-I, 1956-1961, U.S. Geological Survey.

Hayashi, S., T. Ohmoto, & K. Takikawa (2003), Direct numerical simulation of coherent vortex structures in an open-channel flow over dune type wavy bed, J. Hydrosci. Hydraul. Eng., 21(1), 1 -10.

Hyun, B-S., Balachandar, R., Yu, K., & Patel, V. C. 2003. Assessment of PIV to measure mean velocity and turbulence in open channel flow. Experiments in Fluids, 35: 262-267.

Jackson, 1976. R.G., Sedimentological and fluid-dynamic implications of the turbulent bursting phenomenon in geophysical flows. *J. Fluid Mech.* 77 (1976), pp. 531–560.

Jirka, G. H. & Uijttewaal, W. S. J. (2004) "Shallow flows", Taylor and Francis Group, London, ISBN 90 5809 700 5.

Johns, B., Soulsby, R. L., & Xing, J. (1993). "A comparison of numerical model experiments of free surface flow over topography with flume and field observations." *J. Hydr. Res.*, 31(2), 215-228.

Julien, P. Y., & G. J. Klaassen (1995), Sand-dune geometry of large rivers during flood, J. Hydraul. Eng., 121, 657- 663.

Kadota, A., & Nezu, I., Three-dimensional structure of space-time correlation on coherent vortices generated behind dune crest, *J. Hydraulic Research*, Vol. 37, pp. 59-80, 1999.

Klaassen, G. J., Experiments with graded sediments in a straight flume, *Rep. Q 778*, Toegenpast Onderz. Waterstaat Rivers, Delft Hydraulic., Delft, Netherlands, 1990.

Klaassen, G.J., 1991. Experiments on the effect of gradation and vertical sorting on sediment transport phenomena in the dune phase. Grain Sorting Seminar, 21–25 October, 1991, Ascona (Switzerland). Versuchsanstalt fu¨ r Wasserbau, Hydrologie und Glaziologie der Eidgeno¨ssischen Technischen Hochschule Zu¨ rich, pp. 127–146.

Klaassen, G. J., H. J. M. Ogink, & L. C. van Rijn (1986). DHL-research on bedforms, resistance to flow and sediment transport. In 3rd International Symposium on River Sedimentation, Jackson, Mississippi, U.S.A.

Klaassen, G. J., J. S. Ribberink, & J. C. de Ruiter, On the transport of mixtures in the dune phase, paper presented at 215 Euromech Colloquium Mechanics of Sediment Transport in Fluvial and Maritime Environment, Eur. Mech. Counc., Genova, Italy, 1987.

Kleinhans, M. G., 2002. Sorting out sand and gravel; sediment transport and deposition in sand-gravel bed rivers. Netherlands Geographical Studies Royal Dutch Geographical Society, Utrecht, The Netherlands, p. 293.

Kleinhans, M.G. (2004) Sorting in grain flows at the lee-side of dunes: review. Earth Sci. Rev., 65, 75–102. doi:10.1016/S0012-8252-(03)00081–3.

Kline SJ & Robinson SK (1989) Quasi-coherent structures in the turbulent boundary layer. Part I:status report on a community-wide summary of the data.In:Kline SJ, Afgan NH (ed) Near Wall Turbulence. Proceedings of Zaric Memorial Conference, pp. 200-217, New Yor; Hemisphere.

Kostaschuk, R. A., & M. A. Church (1993), Macroturbulence generated by dunes: Fraser River, Canada, Sediment. Geol., 85, 25-37.

Kostaschuk, R. A., & P. V. Villard (1996), Flow and sediment transport over large subaqueous dunes: Fraser River, Canada, Sedimentology, 43, 849–863.

Kostaschuk, R., P. Villard, & J. Best (2004), Measuring velocity and shear stress over dunes with acoustic Doppler profiler, J. Hydraul. Eng., 130, 932–936.

Kostaschuk, R.A., Shugar, D.H., J.L Best., D. R. Parsons, S. N. Lane, R. J. Hardy & O. Orfeo (2008) Suspended sediment transport over a dune, *MARID (Marine and River Dune Dynamics) 2008 Workshop*, 1-3 April 2008 - Leeds, United Kingdom pp. 197-201.

Krogstad, P.-A., Bakken, O. M. & Ashrafian, A. (2005). "An experimental and numerical study of channel flow with rough walls." *J. Fluid Mech.*, 530, 327–352.

Lanzoni, S., (2000), Experiments on bar formation in a straight flume 2. Graded sediment, Water Resources Research, 36(11), pp. 3351-3363.

Lanzoni, S., & M. Tubino, Grain sorting and bar instability, J. Fluid Mech., 393, 149–174, 1999.

Lanzoni, S., M. Tubino, & S. Bruno, Formazione di barre alternate in alvei incoerenti a granulometria non uniforme (in Italian), Paper presented at XXIII Convegno di Idraulica e Costruzioni Idrauliche, Idraul. e Costruz. Idraul., Napoles, Italy, settembre 20–25, 1994.

Lasheras JC; Choi H (1988) Three-dimensional instabilities in a plane free shear layer: an experimental study of the formation and evolution of the streamwise vortices. J Fluid Mech 189: 53-76.

Lin, C., Lai, W. L., & Chang, K. A., (2003) Simultaneous particle velocimetry and laser Doppler velocimetry measurements of periodical oscillatory horseshoe vortex near square cylinder-base plate juncture, *Journal of Engineering Mechanics* 129 (10), pp. 1173–1188.

Lisle, T. E., H. Ikeda, & F. Iseya, Formation of stationary alternate bars in a steep channel with mixed size sediment: A flume experiment, *Earth Surf. Processes Landforms, 16,* 463–469, 1991.

Lopez F, Fernandez R, Best J. 2000. Turbulence and coherent flow structure associated with bedform amalgamation: an experimental study of the ripple–dune transition. *Joint Conference on Water Resources Engineering and Water Resources Planning and Management.* American Society of Civil Engineers: Minneapolis, Minnesota.

Lyn, D. A. (1993). ''Turbulence measurements in open-channel flows over artificial bed forms.'' *J. Hydraul. Eng.,*119(3), 306–326.

Lyn, D. A. (2002). "Flow and transport over dunes." *J. Hydraul. Eng.,*128(8), 726–728.

Manes, C., Pokrajac, D. & McEwan, I. (2007). "Double-averaged open channel flows with small relative submergence", *J. Hydr. Engrg., ASCE* 133(8): 896-904.

Matthes, 1947. Macroturbulence in natural stream flow. *Trans. Am. Geophys. Union* 28 (1947), pp. 255–262.

McLean, S. R., Nelson, J. M., & Wolfe, S. R. (1994). ''Turbulence structure over two-dimensional bedforms: Implications for sediment transport.'' *J. Geophys. Res.,* 99(C6), 12729–12747.

McLean, S. R., S. R. Wolfe, & J. M. Nelson (1999), Spatially averaged flow over a wavy boundary revisited, Journal of Geophysical Research, 104 (C7), 15,743–15,753.

Mendoza, C., & Shen, H. W. (1990). ''Investigation of turbulent flow over dunes.'' *J. Hydraul. Eng.,* 116(4), 459–477.

Metcalfe, R.W., Orszag,S.A.,Brachet, M.E., Menon,S.,& Riley, J.J. 1987. Secondary instability of a temporally growing mixing layer. Journal of Fluid Mechanics, 184:207-243. doi:10.1017 /S0022112087002866.

Muller & Gyr, 1982. Visualization of the mixing layer behind dunes. In: M. Sumer & B.A. Müller, Editors, *Mechanics of Sediment Transport,* A.A. Balkema, Rotterdam (1982), pp. 41–45.

Muller, A., & A. Gyr (1987), On the vortex formation in the mixing layer behind dunes, J. Hydraul. Res., 24, 359–375.

Nelson, J. M., McLean, S. R., & Wolfe, S. R. (1993). ''Mean flow and turbulence fields over two-dimensional bed forms.'' *Water Resour.Res.,* 29, 3935–3953.

Nelson, J. M., & J. D. Smith (1989), Mechanics of flow over ripples and dunes, J. Geophys. Res., 94, 8146– 8162.

Nezu, I., Kadota, A. & Nakagawa, H. Experimental study on turbulent structures in unsteady open-channel flows, *Proc. Japan Soc. Civil Eng* 491/11-27 (1994) 81-88 (in Japanese)

Nezu, I. and Nakagawa, H. (1993). Turbulence in Open Channel Flows. IAHR Monograph, A. A. Balkema, Rotterdam, The Netherlands.

Nikora, V., Goring, D., McEwan, I. & Griffiths, G. (2001). "Spatially averaged openchannel flow over rough bed" *J. Hydr. Engrg. ASCE* 127:123–133.

Noguchi, K., Nezu, I., & Sanjou, M., (2009) "Large Eddy Simulation of Coherent Structures in Turbulent Flow over Fixed Two-Dimensional Dunes,"*33rd IAHR Congress: Water Engineering for a Sustainable Environment*, 9 - 14 August, 2009 Vancouver, British Columbia, Canada.

Nordin, C. F. (1971). Statistical properties of dune profiles. Sediment transport in alluvial channels. U.S. Geological Survey Professional paper, 562-F.

Patel, V. C., and Lyn, C. L. (2004). "Turbulence modeling in flow over a dune with special reference to free surface and bed roughness effects." *Proc., 6th Int. Conf. on Hydroscience and Engineering (ICHE-2004)*, Brisbane, Australia.

Parker, G., Dhamotharan, S. & Stefan, H. (1982): Model experiments on mobile, paved gravel-bed streams. Water Resources Research 18, 1395-408.

Parsons, D. R., J. L. Best, O. Orfeo, R. J. Hardy, R. Kostaschuk, and S. N. Lane (2005), Morphology and flow fields of three-dimensional dunes, Rio Paraná, Argentina: Results from simultaneous multibeam echo sounding and acoustic Doppler current profiling, J. Geophys. Res., 110, F04S03, doi:10.1029/2004JF000231.

Raudkivi, A. J. (1966), Bed forms in alluvial channels, J. Fluid Mech., 26, 507–514.

Raudkivi, A. J. (1997). "Ripples on streambed." *J. Hydraul. Eng.*, 123(1), 58–64.

Ribberink, J. S., Mathematical modelling of one-dimensional morphological changes in rivers with non-uniform sediment, *Rep. 87-2*, Commun. on Hydraul. and Geotech. Eng., Tech. Univ. Delft, 200 pp., Delft, Netherlands, 1987.

Robert, A. & Uhlman, W. 2001 An experimental study of the ripple-dune transition. Earth Surf. Process. Landforms 26, 615–629.

Rood, K. M., & E. J. Hickin (1989), Suspended sediment concentration in relation to surface-flow structure in Squamish River estuary, southwestern British Columbia, Can. J. Earth Sci., 26, 2172–2176.

Roussinova, V., & Balachandar, R. (2011). "Open channel flow past a train of rib roughness." *J. turbulence.*, (in review).

Roussinova, V., Shinneeb, A. M., & Balachandar, R. (2010). "Investigation of Fluid Structures in a Smooth Open-Channel Flow Using Proper Orthogonal Decomposition." *J. Hydr. Engrg.*, ASCE 136(3), 143-154.

Schindler, R. J., & A. Robert (2004), Suspended sediment concentration and the ripple-dune transition, Hydrol. Processes, 18, 3215-3227.

Schindler, R. J., & A. Robert (2005), Flow and turbulence structure across the ripple-dune transition: An experiment under mobile bed conditions, Sedimentology, 52, doi:10.1111/j.1365-3091.2005.00706x.

Schmeeckle, M.W., Shimizu, Y., Hoshi, K., Baba, H. and Ikezaki, S. (1999) Turbulent structures and suspended sediment over two-dimensional dunes. In: *River, Coastal and Estuarine Morphodynamics, Proceedings International Association for Hydraulic Research Symposium*, pp. 261–270. Genova, September.

Smith C R. 1996. Coherent flow structures in smooth-wall turbulent boundary layers: facts, mechanisms and speculation. In Coherent Flow Structures in Open Channels, Ashworth P, Bennett S, Best J, McLelland S (eds). Wiley: Chichester; 1–39.

Stoesser, T., Braun C., Villalba M., G., & Rodi, W., (2008), "Turbulence Structures in Flow over Two-Dimensional Dunes", *J. Hydraul. Eng.*, 134(1), 42-55.

Ten Brinke, W. B. M., A. W. E. Wilbers, & C. Wesseling (1999). "Dune growth, decay and migration rates during a large-magnitude flood at a sand and mixed sand-gravel bed in the Dutch Rhine river system," *Spec. Pub., Int. Assoc. Sediment.* 28, 15-32.

Tevez, A., Mattar, M., Toniolo, H., Fernandez, R., & Lopez,F. Mean flowand turbulence structure over dunes with superimposed smaller bedforms. In Proceedings of the 28th IAHR Congress, Graz, Austria, 1999.

Theodorsen, T. (1955). "The structure of turbulence." in 50 Jahre Grenzschictforschung, edited by H. Gortler & W. Tollmein (Friedrich Vieweg & Sohn, Braunschweig).

Vanoni, V.A. (ed.), Sedimentation Engineering, ASCE Manual No. 54, New York: ASCE, 1975.

Venditti, J. G., & Bennett, S. J. (2000). "Spectral analysis of turbulent flow and suspended sediment transport over fixed dunes." *J. Geophys.Res.*, 105(C9), 22035–22047.

Venditti, J. G., M. Church, & S. J. Bennett (2005), Morphodynamics of small-scale superimposed sand waves over migrating dune bed forms, Water Resour. Res., 41, W10423, doi:10.1029/2004WR003461.

Villard, P. V., & R. A. Kostaschuk (1998), The relation between shear velocity and suspended sediment concentration over dunes: Fraser Estuary, Canada, Mar. Geol., 148, 71-81.

Volino, R. J., Schultz, M. P. & Flack, A. K. (2009)."Turbulence structure in a boundary layer with two-dimensional roughness". *J. Fluid Mech.*, 635, 75-101.

Wiberg, P. L., & Nelson, J. M., (1992), Unidirectional Row Over Asymmetric and Symmetric Ripples, *J. Geophys. Res.*, 97(C8), 12,745–12,761, doi:10.1029/92JC01228.

Wilcock, P. R. (1992). Bed-load transport of mixed-size sediment. *Dynamics of gravel-bed rivers*, P. Billi, R. D. Hey, C. R. Thorne, & P. Tacconi (eds), John Wiley and Sons Ltd, Chichester, UK, pp. 109-131.

Wilcock, P.R. & Southard, J.B. 1988: Experimental study of incipient motion in mixed-sized sediment. Water Resources Research 24, 1137-51.

Wilcock, P.R. & Southard, J.B. 1989: Bed-load transport of mixed-sized sediment: fractional transport rates, bedforms and the development of a coarse bed-surface layer. Water Resources Research 25, 1629-41.

Wilcox, D. C. (1993). *Turbulence modeling of CFD,* DCW Industries, La Canada, Calif.

Williams, J.J.R., (2007), "Free-surface simulations using an interface-tracking finite-volume method with 3D mesh movement," Engineering Applications of Computational Fluid Mechanics, 1(1), pp 49-56.

Wu, Y., & Christensen K. T., (2006) "Population trends of spanwise vortices in wall turbulence." *J. Fluid Mech.*, 568, 55-76.

Xu, C. Y., Chen, L. W., & Lu, X.Y., (2007) Large-eddy and detached-eddy simulation of the separated flow around a circular cylinder[J], *Journal of Hydrodynamics, Ser. B* 19 (5), pp. 559–563.

Yalin M.S. 1964. Geometrical properties of sand waves, *ASCE* Vol 90, HY5 pp. 105-119.

Yalin, M. S. *Mechanics of Sediment Transport.* Pergamon Press, Braunschweig, Germany, 1972.

Yalin, M. S. (1977), Mechanics of Sediment Transport, 298 pp., Elsevier, New York.

Yoon, J. Y., & Patel, V. C. (1996). ''Numerical model of turbulent flow over a sand dune.'' *J. Hydraul. Eng.*, 122(1), 10-18.

Yue, W., Lin, C.L., & Patel, V.C. (2005). "Large Eddy Simulation of Turbulent Open-Channel Flows With Free-Surface Simulated By Level Set Method," Physics of Fluids, Vol. 17, pp. 025108:1-12.

Yue, W., Lin, C.L., & Patel, V.C. (2006). "Large-eddy simulation of turbulent flow over a fixed two-dimensional dune." *J. Hydraul. Eng.*, 132(7), 643–651.

3

Stochastic Nature of Flow Turbulence and Sediment Particle Entrainment over the Ripples at the Bed of Open Channel Using Image Processing Technique

Alireza Keshavarzi[1,2] and James Ball[2]
[1]*Water Department, Shiraz University, Shiraz*
[2]*School of Civil and Environmental Engineering, University of Technology Sydney, NSW, 2007,*
[1]*Iran*
[2]*Australia*

1. Introduction

The formation of bed topography is the result of a complicated interaction between flow and sediment particles along the bed. Bed form and the geometry of ripples are also a function of bed roughness, median diameter of sediment particles and flow characteristics for example shear stress, separation and Froude number (Mogridge et al. 1994). Ripple geometry and its interaction with flow structure has been studied by many investigators; for example, Bagnold (1946), Carstens et al. (1969), Mogridge et al. (1994), Yalin (1977), Khelifa and Ouellet (2000), Miller and Komar (1980), and Nielsen (1992). In these studies, the geometry of bed ripples was found to be a function of turbulent shear stress parameters. Raudkivi (1997) pointed out that the ripples and vortices within the shear layer are affected by the flow depth, velocity distribution and shear stress on the bed. Despite more than three decades of investigation, there is still insufficient information to characterize ripple-flow interaction in adequate detail and over a range of turbulent flow conditions.

One area where this lack of information exists is the application of image processing for particle entrainment and analyze of the turbulence characteristics and flow over ripples. To study the flow structure over the ripples, Sajjadi et al. (1996) found that the vortices that form in the lee of ripples are important for the entrainment of sediment particles. Keshavarzi and Ball (1999) used image processing technique to record entrained and deposited particles over flat bed and found that there is an intermittent nature for particles entrainment and deposition over the bed. Jeremy et al. (2005) used high resolution super-VHS video system to monitor the development of the sand bed over a flat bed and observed that initiation occurs at relatively low flow strengths, where sediment transport is patchy and sporadic. Lajeunesse et al. (2010) used high speed video imaging system to record the trajectories of the moving particles over flat bed and observed that entrained particles exhibit intermittent motion composed of the succession of periods of flight and rest. Bennett and Best (1996) conducted a series of experiment over fixed bed ripples and compared the spatial structure of flow over fixed symmetrical ripples to reveal the contrasts in the dynamics of the flow separation zone over ripples. Kostachuck and Church (1993), Julien and Klassen (1995), Kostaschuk and Villard (1996), Carling et al. (2000) and Kostachuck (2000) have conducted field studies and concluded that the shear related coherent structures were very important in bed form development and bed stability. Consideration of the results from these studies indicates that the turbulence characteristics have a direct influence on sediment entrainment and the ripple geometry.

2. Coherent structure of flow over the bed

Analysis of the turbulence characteristics is based on the concept of the bursting phenomenon which was initially introduced by Kline et al. (1967) as a means of describing the transfer of momentum between the turbulent and laminar regions near a boundary. They defined bursting process as; Sweep event ($u' > 0, v' < 0$); Ejection event ($u' < 0, v' > 0$); Outward interaction event ($u' > 0, v' > 0$); and Inward interaction event ($u' < 0, v' < 0$) (Figure 1). The four bursting events have been identified having different effects on the mode and rate of sediment transport (Bridge and Bennett 1992). Particle entrainment from the bed is closely correlated to the sweep event (Thorne et al. 1989; Nelson et al. 1995; Drake et al. 1988; Nakagwa and Nezu 1978; Grass 1971 and Keshavarzi and Ball 1997, 1999). Also the contribution of sweep and ejection events has been found to be more important than outward and inward interactions. Furthermore, sweep and ejection events occur more frequently than outward and inward interactions (Nakagwa and Nezu 1978; Thorne et al. 1989; Keshavarzi and Ball 1997). In addition to the above characteristics, the average magnitude of the shear stress during sweep event being much higher than the time averaged shear stress (Keshavarzi and Ball 1997). A number of studies, for example, Offen and Kline (1975) and Papanicolaou et al. (2002) investigated the characteristics of the bursting process and its effect on particle motion. Consideration of this research led Yen (2002) to point out the necessity to incorporate the bursting process into the modelling of turbulent flow and sediment transport. Jafari Mianaei and Keshavarzi (2008, 2009) found that at the stoss side of ripples, ejection and sweep events and at the lee side of the ripple outward interaction and inward interaction events were dominant. The studies by Ojha and Mazumder (2008) showed that the ratio of shear stress for sweep and ejection events along the dunes varied in an oscillatory pattern at the near bed region, whereas such patterns disappear towards the outer flow. Also they found that along the dune length, sweep events contribute to shear stress generation more than other events. Termini and Sammartano (2009) investigated the effects of the variation of bed roughness conditions on the vertical distributions frequency of the occurrence of ejection and sweep events and concluded that the occurrence of sweep events increases as the bed roughness increases.

The structure of turbulent flow over the ripples in the bottom of an open channel is important for understanding of sediment particle entrainment and its transport. Two important issues which need to be understood in sediment movement are the stochastic nature of instantaneous shear stresses over a ripple bed and how it influences on sediment entrainment and transport. Therefore remains the necessity for sophisticated laboratory equipment to understand stochastic nature of flow over the ripples together with the hydrodynamic properties. Using the Acoustic Doppler Velocity meter (Micro-ADV) to measure three dimensional velocities, CCD camera to record images of particle motion and image processing technique enables to provide an accurate measurement of flow structure and particle entrainment from the bed.

3. Quadrant decomposition of velocity fluctuations

The bursting process consists of four categories of event; these categories are defined by the quadrant of the event. As shown in Figure 1, the events are:

- Outward interaction event $u'>0, v'>0$
- Ejection event $u'<0, v'>0$,
- Inward interaction event $u'<0, v'<0$ and
- Sweep event $u'>0, v'<0$.

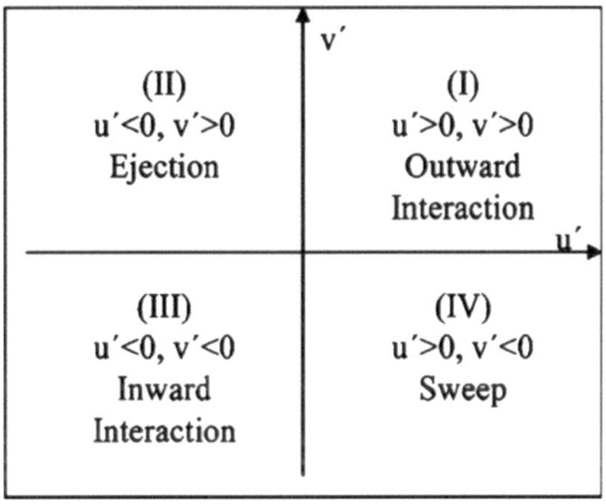

Fig. 1. Quadrant analysis of bursting process

The velocity fluctuations u' and v' are defined as variations from the time-averaged (mean) velocities in the longitudinal and vertical directions, ($\overline{u}$ and $\overline{v}$), respectively. Algebraically, they are defined by

$$u' = u - \overline{u} \text{ and } v' = v - \overline{v} \tag{1}$$

Where;

$$\overline{u} = \frac{1}{n}\sum_{i=1}^{n} u_i \text{ and } \overline{v} = \frac{1}{n}\sum_{i=1}^{n} v_i \tag{2}$$

n being is the number of instantaneous velocity samples.
The time-averaged instantaneous shear stress (Reynolds shear stress) at each point of flow is defined as:

$$\tau' = -\rho\overline{u'v'} \tag{3}$$

As shown by Jafari Mianaei and Keshavarzi (2008, 2009), the distribution of the instantaneous velocities is influenced significantly by presence of ripples.

4. Transformation of instantaneous shear stress

The magnitude of the time-averaged shear stress for each quadrant was found to be different, and to differ from the time-averaged instantaneous shear stress for the flow. The instantaneous shear stress for each event was normalized by the time-averaged shear stress at that point within the flow. Expressed algebraically, the non-dimensional instantaneous shear stress (C) was determined by

$$C = \frac{\tau'}{\overline{\tau}} \tag{4}$$

In order to transform the normalized data into a normally distributed parameter, Keshavarzi and Ball (1997) applied a Box–Cox transformation (Box and Cox 1964) to transform the instantaneous shear stress to a normal distribution. Therefore, a Box–Cox transformation was applied to determine the mean magnitude of the normalized instantaneous shear stress. For this study, a Box–Cox power transformation was applied to transform the magnitude of instantaneous shear stress into a normal distribution. A Box–Cox power transformation for the parameter (C) is defined for non-zero values of k by:

$$B(C) = \frac{(C+K)^{\lambda} - 1}{\lambda} \text{ for } \lambda \neq 0, \tag{5}$$

and

$$B(C) = \ln(C+K) \text{ for } \lambda = 0 \tag{6}$$

where K is a constant and λ is transformation power. If all of the values in the time series are greater than zero then the constant K usually is set to zero. Keshavarzi and Ball (1997) applied the Box-Cox transformation to a full data set and suggested that K is equal to 0.272 for sweep events and 0.28 for ejection events. The inverse transformation which is the transformation B^{-1} of B(C) is given by:

$$C = \left(\lambda B(C) + 1\right)^{1/\lambda} \tag{7}$$

Mean values of the transformed data B(C) for each quadrant were calculated using

$$\overline{B(C)}_k = \frac{1}{N}\sum_{i=1}^{n} B(C_i) \text{ k=1.....4} \tag{8}$$

The inverse Box–Cox transformation was applied to these mean values to enable determination of the time-averaged shear stress for each event and in a point of flow. Consideration of the distribution of time-averaged shear stress over the ripples indicates that there is an oscillatory pattern for the instantaneous shear stress along the bed.

5. Contribution probability of coherent flow and bursting events

Based on two dimensional velocity fluctuations, the occurrence probability of the bursting events for each quadrant is defined as;

$$P_k = \frac{n_k}{N} \tag{9}$$

$$N = \sum_{i=1}^{4} n_k \tag{10}$$

Where P_k is the occurrence probability of an event in a quadrant, n_k is the number of occurrences of each event, N is the total number of events and the subscript represents the individual quadrants (k=1…4). Using the above equations, the probability of each quadrant was computed at each point of flow within the depth. The contributions of coherent structures, such as the sweep (quadrant IV) and ejection (quadrant II) events, to momentum transfer have been extensively studied through quadrant analyses and probability analyses based on two-dimensional velocity information. Using similar techniques, the contributions of the four events to the entrainment and motion of sediment particles were determined from the experimental measurements.

The studies by Jafari Mianaei and Keshavarzi (2008, 2009) indicates that at the stoss side of ripples, the time-averaged instantaneous shear stress of ejection and sweep events were dominant to the outward interaction and inward interaction events and at the lee side of the ripple it was vice versa. In other studies for example by Ojha and Mazumder (2008) it is shown that the ratio of time-averaged shear stress for sweep and ejection events along the dunes vary in an oscillatory pattern at the near bed region, whereas such pattern seems to disappear towards the outer flow. Termini and Sammartano (2009) concluded that the occurrence of sweep events increases as the bed roughness increases.

6. Modelling of time series of bursting events using markov process

As mentioned in previous sections, bursting events can be categorised into four quadrant zones which are outward interaction, ejection, inward interaction and sweep. The probability of movement from one category or zone to another is important. This movement of the events was investigated using a time series model. Consideration of the data indicates that the bursting process occurred in zone 4 at the first time step. In the second time step, the bursting event occurred in quadrant 1 and then the next bursting event occurred in quadrant 4. The next 4 bursting events were in the same quadrant. As a result, the sequence of 20 bursting events was found to be IV, I, IV, IV, IV, IV, IV, I, II III, II, II, II, IV, IV, IV, IV, IV, I, I, I.

Hence, the turbulence of the flow results in a temporal sequence of bursting events. For the purposes of this study, the data were classified as discrete random variables and, hence, were analysed as a stochastic process or a Markov process. A discrete random variable {St} was defined as being a bursting event in a quadrant at time (t). Therefore, at time t, St can be in either quadrant 1, 2, 3 and 4, (outward interaction, ejection, inward interaction and sweep), respectively. In this study, the change in state was defined as a movement between quadrants and the transition probabilities for these changes were determined. It is worth noting that an event can be stable (the next bursting event occurs in the same quadrant) or can move to another quadrant.

A Markov analysis looks at a sequence of events and analyzes the tendency of one event to be followed by another event. In the study reported herein, zero, first and second order Markov processes were considered. Details of these are

- A zero order Markov process requires that the current situation does not depend on the immediately previous situation.
- A first order Markov process defines the current situation based only on the immediate past situation and can be expressed as

$$pr\{s_{t+1}|s_t, s_{t-1},...,s_1\} = pr\{s_{t+1}|s_t\} \tag{11}$$

- A second order Markov process defines the current situation, based on the two previous situations. It can be expressed mathematically as:

$$pr\{s_{t+1}|s_t, s_{t-1},...,s_1\} = pr\{s_{t+1}|s_t, s_{t-1}\} \tag{12}$$

According to the concept of a first order Markove process, the probability of the next situation depends on the current situation, but it does not depend on the particular way that the model system arrived at the current situation. The transition probabilities of first order Markov process can be computed as:

$$\hat{P}_{ij} = pr\{s_{t+1} = j|s_t = i\} = \frac{no.\ of\ j's\ following\ i's}{total\ no.\ of\ i's} = \frac{n_{ij}}{n_{i.}} \qquad i,j = 1,2,3,4 \tag{13}$$

where $n_{i,j}$ is the number of occurrences from situation i to situation j, $n_{i.}$ is the number of i's in the series followed by another situations, so that $n_{i.}= n_{i1}+n_{i2}+n_{i3}+n_{i4}$.The $p_{i,j}$ is estimated as the fraction of points for which $S_t = i$ is followed by points with $S_{t+1}=j$.
A second order Markov process defines the current situation, based on the two previous situations. The present situation S_{t+1} can be found based on the situations at S_t and S_{t-1}. It means that the situation at (t+1) depends on the situation at (t) and (t-1). The second order Markov process is defined mathematically as:

$$pr\{s_{t+1}|s_t, s_{t-1},...,s_1\} = pr\{s_{t+1}|s_t, s_{t-1}\} \tag{14}$$

The transition probabilities of second order Markov process were obtained from the conditional relative frequencies of transition counts. It can be computed as;

$$\hat{P}_{kij} = pr\{s_{t+1} = j|s_t = i, s_{t-1} = k\} = \frac{no.\ of\ j's\ following\ i's\ following\ k's}{total\ no.\ of\ i's\ following\ k's} = \frac{n_{kij}}{n_{ki.}} \qquad i,j,k - 1,2,3,4 \tag{15}$$

A zero order Markov process was examined also for comparison. If a zero-order Markov process exists then the current situation does not depend on the previous situation. The probabilities for a zero-order Markov process can be computed as:

$$\hat{P}_i = \frac{n_i}{n} \qquad i = 1,2,3,4 \tag{16}$$

where ni is number of situation i and n equal total number of sampling data.
To find the most appropriate model for prediction of the occurrence of bursting events, two criteria can be used for assessment of the best order of the Markov process; these criteria are:

- The Akaike information criterion (AIC) (Akaike 1974; Tong, 1975); and
- The Bayesian information criterion (BIC) (Schwartz, 1978; Katz, 1971).

Both of these criteria are based on the log-likelihood functions for the transition probabilities of fitted Markov process. These log-likelihood functions depend on the transition counts and the transition probabilities.
The log-likelihood functions for zero, first and second order Markov process are;

$$L_0 = \sum_{i=1}^{4} n_i \ln(\hat{P}_i) \tag{17}$$

$$L_1 = \sum_{i=1}^{4}\sum_{j=1}^{4} n_{ij} \ln(\hat{P}_{ij}) \tag{18}$$

$$L_2 = \sum_{k=1}^{4}\sum_{i=1}^{4}\sum_{j=1}^{4} n_{kij} \ln(\hat{P}_{kij}) \tag{19}$$

The AIC and BIC statistics are computed for each trial order m, using

$$AIC(m) = -2 * L_m + 2 * s^m * (s-1) \tag{20}$$

$$BIC(m) = -2 * L_m + s^m * (\ln(n)) \tag{21}$$

where s equals 4 and represents the four quadrant zones.
The order m which produces the minimum value in either Equation (20) or (21) is considered to be the most appropriate order of the Markov process.
Keshavarzi and Shirvani (2002) found that the minimum value for both the AIC and BIC criteria was found to occur for a first-order process. Therefore, Keshavarzi and Shirvani (2002) proposed that a first-order Markov process is an appropriate model for the occurrence of bursting events. Hence, the type of bursting event at time t^{n+1} depends only on the type of bursting event at time t^n and, conversely, the situation at time t^n depends only on the situation at time t^{n-1}.

7. Organization of the bursting events

The organization of bursting events and the associated coherent structures is still a major concern and the focus of many studies. To find the organization of coherent structures in the flow Keshavarzi and Shirvani (2002) applied a conditional probability analysis to the experimental data and therefore a conditional probabilities or empirical conditional probabilitiesy is proposed for organisation of the bursting process.
Base on the development of the conditional probabilities, Keshavarzi and Shirvani (2002) identified three movement categories. These categories of movement between bursting event states were:

- Stable transition - a stable transition or movement between sequential bursting events is defined as being events that remain in the same quadrant for two consecutive time steps, i.e. t_i and $_{ti+1}$.

- Marginal or lateral transition - a marginal or lateral movement between sequential bursting events occurs when a bursting event in the following time step occurs in one of the neighbor quadrants, for example between quadrants 1 and 4, or 1 and 2, or 2 and 3 (or vice versa).
- Cross transition – a cross transition or movement between sequential bursting events is defined as being events where sequential bursting events are diametrically opposed, for example from quadrant 1 to 3 or from quadrant 4 to 2.

These categories of movement between bursting events were applied in consideration of the occurrence of bursting events recorded in the experimental data. Points to note about the sequence of bursting events are:

- First, it was found that the most probable next bursting event in a sequence occurs with a stable transition. This arises from the stable transition being stronger (having a higher probability) than marginal and cross transitions. In other words, when an event occurs in one of the quadrant zones at time step t_i, the probability of the event to stay in the same zone at time step t_{i+1} is higher than other probabilities. Additionally it was apparent that, zone 4 is more stable than others. The stability of the events was sorted 4, 2, 1, and 3, in which the event in zone 4 (sweep event) is the most stable and zone 3 (inward interaction event) has the minimum stability.
- The second category of movement was the marginal or lateral transition. As previously noted, this movement was defined as bursting events occurring in neighbouring quadrants. Compared to other categories, this movement was not as strong as the stable category. Finally, the marginal transition occurred more frequently between quadrants 4 and 1 than between other quadrants.
- The third and final category was the cross transition. This category had the lowest probability compared to the stable and marginal categories.

8. Image processing and application in sediment transport

In spite of the importance of the initiation of individual sediment particles motion, no generally accepted definition exists due to the difficulty in defining of incipient of individual particle motion. One important reason for the lack of a general definition is the difficulty of observing sediment particles at the initiation of motion. Recently, the use of image processing techniques for observing sediment particles motion has been considered in some research studies. Examples of these studies which have considered image processing techniques include Keshavarzi and Ball (1999), Jafari Mianaei and Keshavarzi (2009), Nelson et al. (1995), Best (1992) and Drake et al. (1988) who used image processing techniques as a tool to investigate the intermittent nature of particle entrainment. Arising from these studies, it was considered that the application of image processing has potential to assist in understanding the processes influencing incipient particle motion. Motion picture photography is uniquely capable of detailed observation, quantitative tracking, and measurement of bed particle movement in clear water. This technique can show the entrainment of particles from the bed, settlement of particles, speed of particle movement, transport mode, and resting periods of a particle on the bed. With the capturing and collection of these data, it is possible to develop a statistical description of particle entrainment or sediment particle motion at the bed. In this study attention is paid to define the initiation of sediment particle motion over a movable bed with consideration of the bursting processes arising from flow turbulence. In order to understand the processes, image processing techniques were used to observe the particle movement over a designated area of the bed. The observed motion of sediment particles was characterised and a statistical description of the initiation of sediment particle motion and its correlation to the turbulence characteristics of the flow determined.

8.1 Image analysis of particles in motion

8.1.1 Basic concepts

The signal in image processing is a physical measure of the light intensity for the image. In a digital image, the signals are a series of digits or numbers, which are organised as an array in a file. Shown in Figure 2 is the concept of how an image can be converted to a digital signal. A signal may be continuous or discrete. In image processing the signals are in the form of a discrete signal. In order to analyse a series of sequential images spatially and temporally, it is necessary to apply statistical tools to these arrays to ascertain similarities or discrepancies.

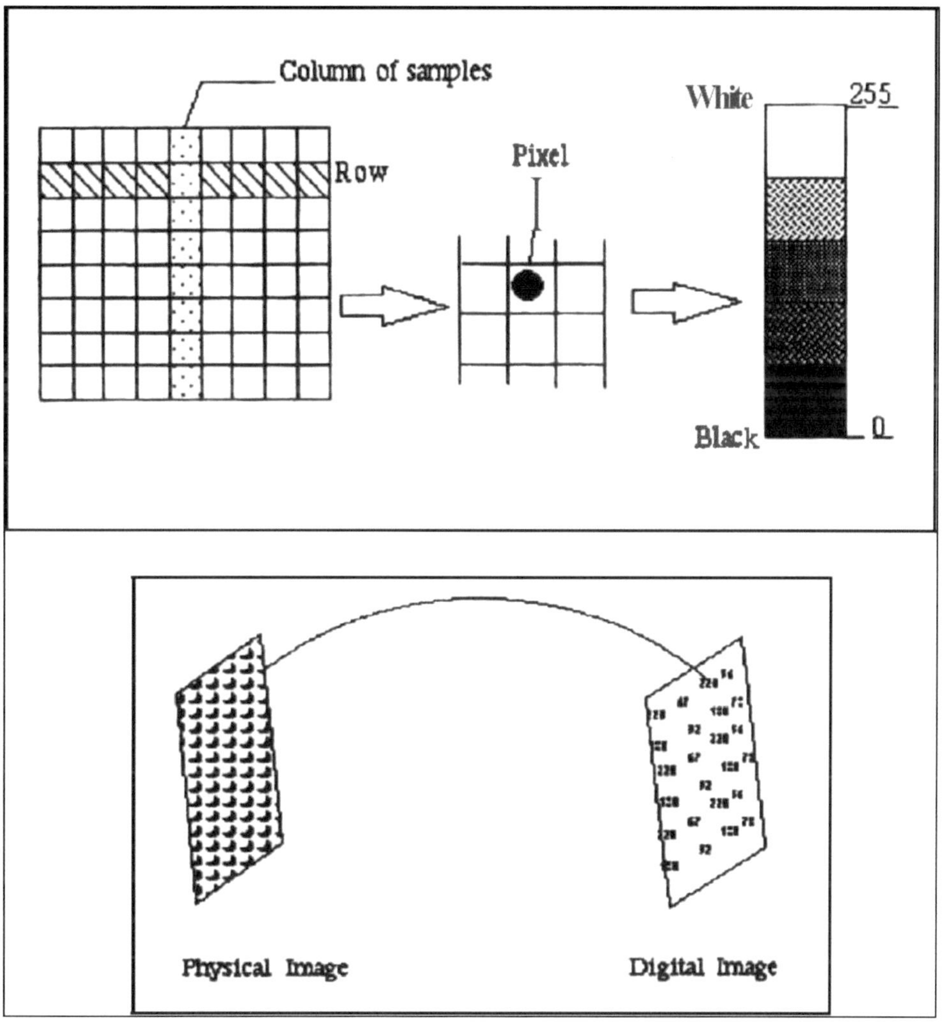

Fig. 2. The conversion process and digitising an image (after Keshavarzi and Ball, 1999)

8.1.2 Analysis techniques used

In order to analyse the captured images; two different techniques were used in this study. These techniques were:

- A probability analysis of the entrained particles determined by counting the number of particles in motion at an instant. This approach was useful to obtain an exceedance probability of particles in motion in time respective to the exceedance probability of shear stresses of the sweep events at the bed. The number of particles entrained into motion was obtained through determination of the difference between sequential images.
- Application of cross correlation and Fast Fourier transforms to determine the displacement of particle between images and hence the particle velocity.

8.1.3 The subtraction technique and particle counting

The difference between two images f(x,y) and g(x,y), can be expressed as;

$$h(x,y)= f(x,y)- g(x,y),$$

where h(x,y) is a new image. Two images can be compared by computing the difference between the light intensities at all pairs of corresponding pixels from image f(x,*y)* and image g(x,y) (Adrian 1991, Willert and Gharib 1991). Here, this technique was used and a sequence of images were compared to find the number of particles which were entrained and deposited over a specified area and in a given time increment. A computer program written in C++ was used to read a sequence of images in binary format and to produce an output image which was the difference between the images.

8.1.4 Viewing of particles in new image

In order to view the difference between two images as an image, the subtracted light intensities must be kept between 0 and 255, where the entrained and deposited particles are denoted by black and white spots respectively. To keep the light intensities in this range, a value of 255 must be added to the light intensities derived from subtraction of two images and then the sum divided by two. From this computation, the difference between two images can be obtained and the difference viewed as a new image. Thus result of this procedure is shown in Figure 3 for a hypothetical sediment particle.

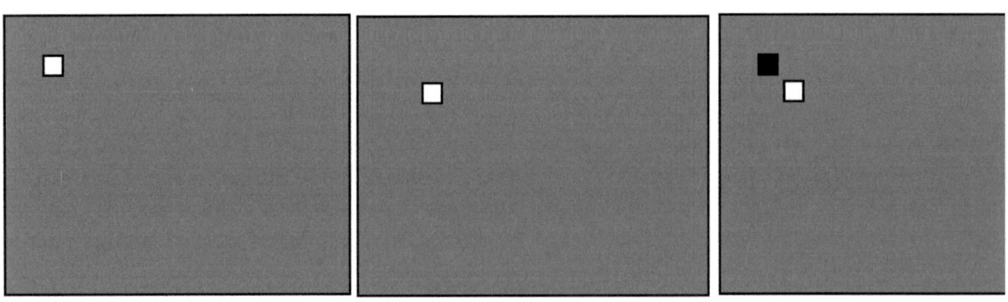

a) first image b) second image c) derived image

Fig. 3. A schematic illustration of image.

Shown in Figure 4 are two sequential images captured during experimental run with the computed difference between these two images. If the two images are compared, the differences between these images are apparent and can be observed in the produced image. As shown in derived image, it is clear how many particles were entrained or deposited in a time increment. In the derived image the entrained particles appear as black spots and deposited particles are displayed as white spots. For statistical analysis of the particle entrained, a sequence of 81 images was compared in each experimental run. A manual counting of entrained particles was preferred due to the need for interpretation to ensure accurate counting of the particles. This need for interpretation arises from the potential for particles to be agitated but not moved. In this case the movement appears as a curved shadow line in the produced image and, therefore, was not considered as a moved particle. An example of this effect is shown in Figure 4.

Keshavarzi and Ball (1999) were carried out an experimental study in a non-recirculating tilting rectangular flume of 0.61 m width, 0.60 m height and 35 m length. The sidewalls of the flume were made of glass, making it possible to observe and record the flow characteristics. It also made possible flow visualisation during the experimental tests. The bed of the flume was covered by sand particles of 2 mm nominal diameter. Construction of the flume enabled experimentation with different bed roughness and under different flow conditions. The movement of sand particles at the bed was observed and recorded with a high-resolution CCD camera. In order to get a clear picture; a slide projector was used to illuminate the specific area of the bed where the images were being recorded. Some precautions were made to ensure that a clear image was captured. For this analysis, a range of 5-25 frames per second were captured and digitized. The recorded video images were digitized into an array of 384 by 288 pixels with each pixel being quantized in 8 bits with light intensities ranging from *0* (for black) to 255 (for white). While the images were collected in colors, they were converted to a grey format for later analysis. The Grey format was selected due to its low requirements for storage and transfer to the computer. Using these derived images the numbers of the entrained and deposited particles in an instant of time were manually counted. In Figure 5 the time series of the number of entrained particles for a specified area are shown.

8.2 Relation between number of particle in motion and instantaneous shear stresses in sweep event

The entrainment of particles from a mobile bed in an open channel flow has been investigated in several studies; for example by Einstein and Li (1958), where it was pointed out that this process is stochastic in nature due to the effect of turbulence. The number of entrained particles over a specified area varies with time. The entrained particles at any time depend on the instantaneous turbulent shear stress arising from the velocity fluctuations and the instantaneous shear stresses at the bed. Shown in Figure 5 is the number of particles in motion, at an instant of time and consequently how the entrained number of particle varies with time. The entrainment process can be defined by considering the instantaneous shear stress in sweep events and also the instantaneous number of particles in motion. In a study of sediment entrainment from the bed, Williams et al. (1989) and Nelson et al. (1995) investigated and found a high correlation between the streamwise velocity component and the sediment flux. Additionally, Nelson et al. (1995) found that the transport rate tends to be higher when the vertical velocity and Reynolds momentum flux are angled towards the bed. They found the best correlation between the sediment flux and the streamwise velocity component to occur with a lag of 0.1 second and consequently they suggested a measuring frequency of 10 Hz would give the best results in terms of time scales. A frequency rate of 10 Hz therefore was selected for this investigation.

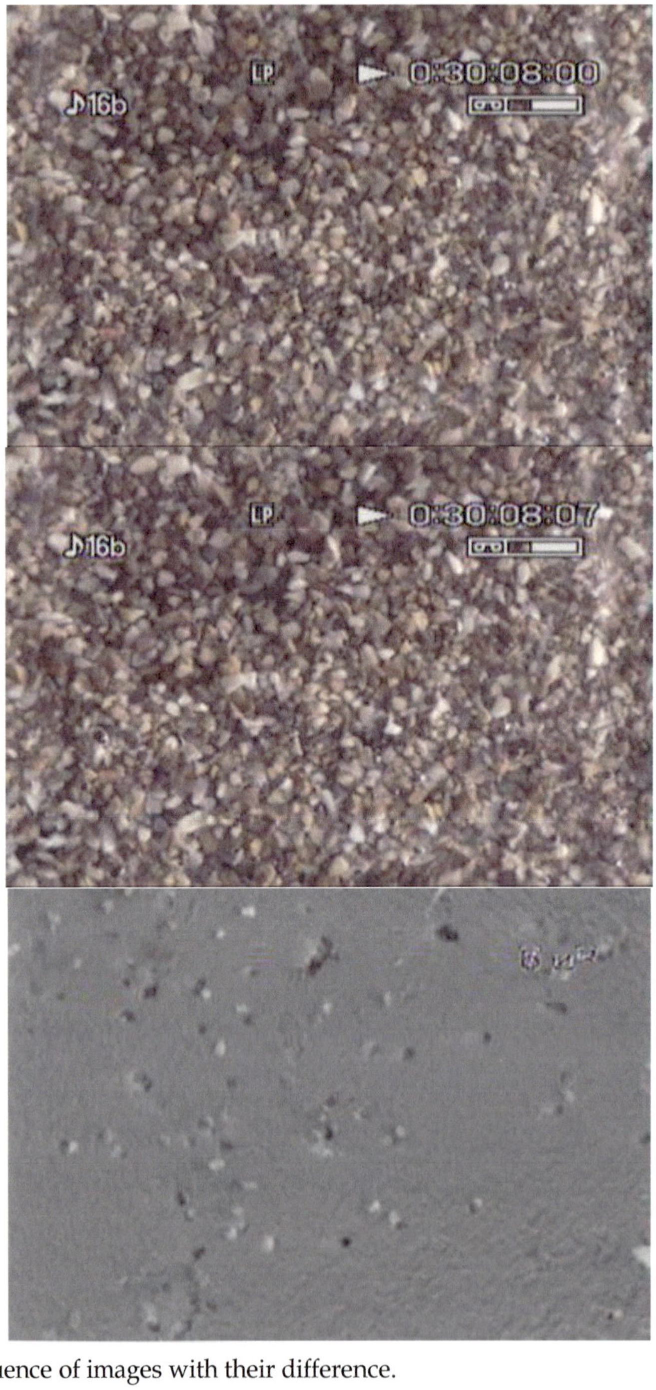

Fig. 4. Two sequence of images with their difference.

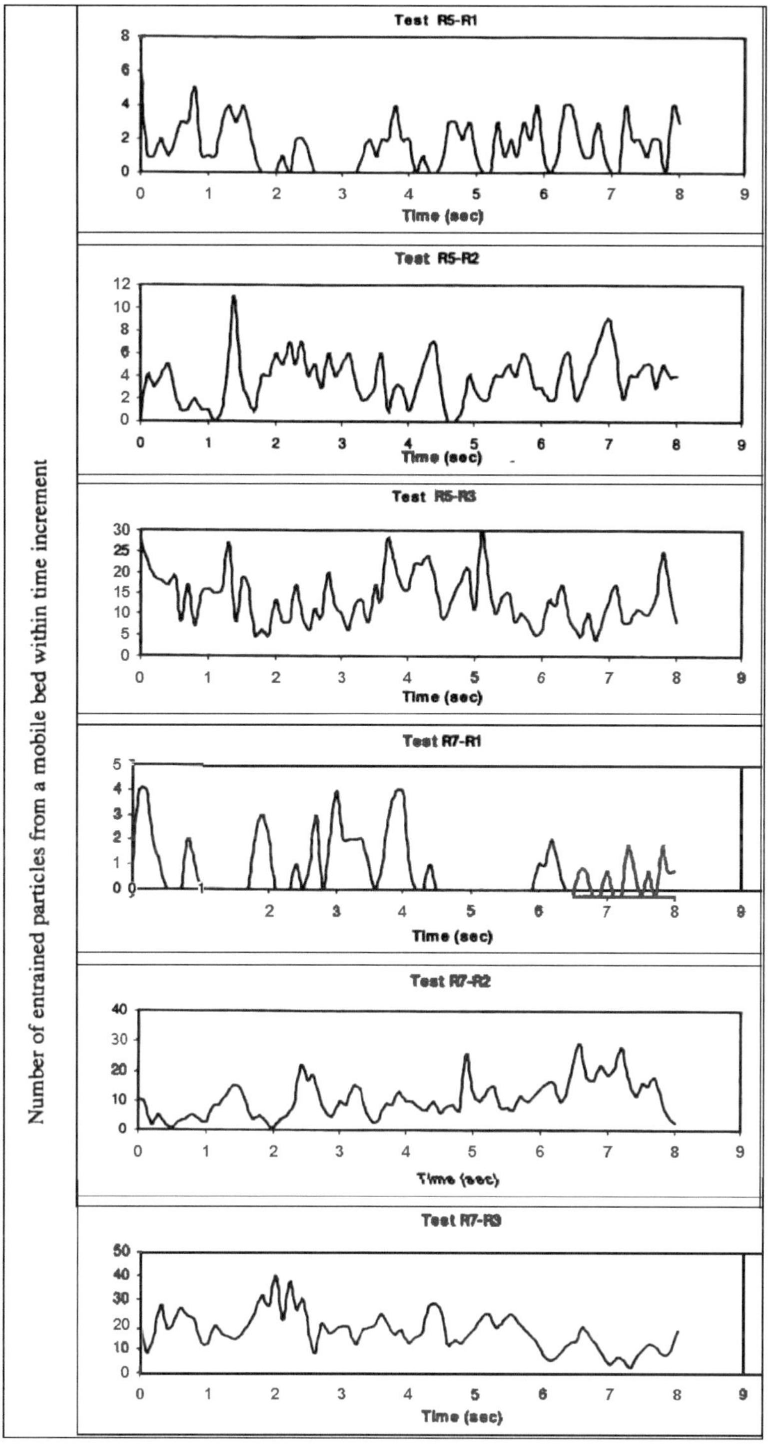

Fig. 5. Number of particles in motion in an increment of time (after Keshavarzi and Ball, 1999)

A cross-correlation analysis was undertaken between the number of entrained particles and the instantaneous shear stress in a sweep. Shown in Figure 6 are some examples the cross correlation between instantaneous shear stresses in sweep events and the instantaneous number of particles in motion. The above relationship was investigated using a cross correlation analysis between instantaneous shear stresses in sweep events and the number of particles entrained in a time increment. From this diagram, it is seen that no significant lag exists between the shear stresses in a sweep event and the number of particles in motion. In Figure 6, the horizontal axis depicts the lag while the vertical axis depicts the cross correlation coefficient obtained. The numbers of entrained particles were counted in a sequence of produced images derived from the subtraction of sequential recorded images. The dimensionless shear stress in sweep events was computed also from a time series of the velocity fluctuations, which was recorded simultaneously with the recording of the images. A good correlation was found between the number of particles in motion and the instantaneous shear stress in a sweep event. The percentage of the area, which was eroded from a defined area of the bed, was investigated with respect to the instantaneous turbulent shear stress of bursting events. It was assumed that the observation area occupied by particles of an arbitrary shape is proportional to the mean diameter of the size fraction. In order to compute the fraction of area entrained, the number of particles entrained was counted and compared with the total observed area. For this case the particles resting on the bed was 2 mm and the observation area was 25 square cm. Thus, each particle has an area fraction of where a is the area of the entrained particles and is defined as (a = and $A = L^2$ is the sample area of the bed.

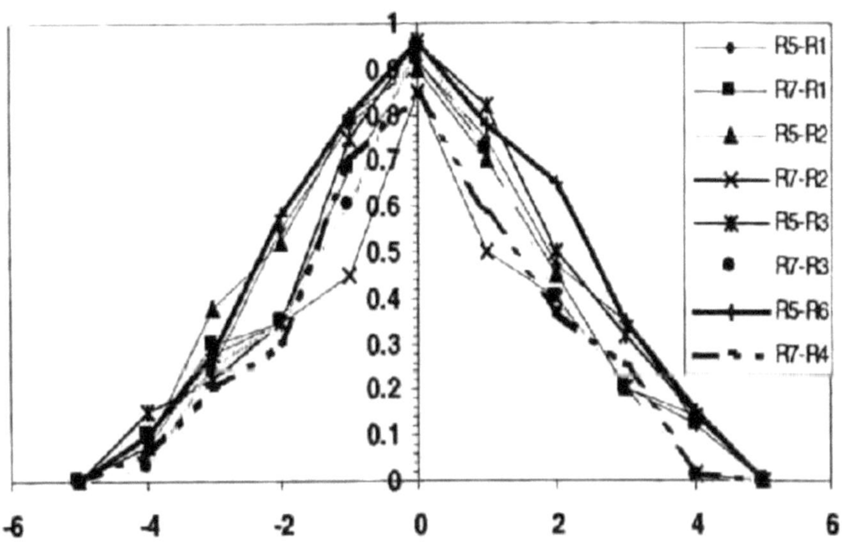

Fig. 6. Cross correlation of instantaneous shear stress in sweep event and entrained particles (after Keshavarzi and Ball, 1999)

Nabavi (2005) and Keshavarzi et al. (2010) carried out some experimental tests over symmetrical sinusoidal ripple bed form. A photo of symmetrical sinusoidal ripple bed form

is shown in Figure 7. The experimentall studies by Nabavi (2005) and Keshavarzi et al. (2010) were carried out in a non-recirculating glass flume located in the hydraulics laboratory at Shiraz University. The glass flume consisted of a rectangular cross-section with base width of 0.70m, height 0.60m and 15.50m length. The longitudinal slope of the flume was set to 0.0005. The flow rate at the end of the flume was measured using a pre-calibrated 90° V-notch weir. A downstream adjustable gate was installed at the downstream end of the flume to allow for adjustment of the flow depth and velocity within the flume. The experimental measurements were made for three different forms of ripples; the wave lengths of the ripples considered were 150, 200 and 250 mm while the height of the ripples were 30 mm for all wave lengths. The flume bottom was covered with an erodible bed material layer consisting of sand particles with mean average size (d50) of 0.62mm. A schematic of the ripple dimensions considered is shown in Figure 8.
The experimental tests of Nabavi (2005) and Keshavarzi et al. (2010) were performed with similar flow conditions. The flow rate for these tests was approximately 18.5 litres per second and the flow depth was constant at 145 mm. The flow velocity was measured in three dimensions using an Acoustic Doppler Velocity meter (Micro-ADV) at the points within the flow. The measurements were made at eight depths within the flow; these depths were 5, 10, 15, 20, 25, 30, 50 and 60 mm from the bed and at 16 longitudinal sections along the flume. The distribution of instantaneous shear stress were calculated from experimental data and they are shown in Figure 9 for 16 sections along the ripples. As it is shown in Figure 9 the pattern of the distribution of instantaneous shear stress varied in each section over the ripples.
Keshavarzi et al. (2010) determined the contributions of the four events to the entrainment and motion of sediment from the experimental measurements. Shown in Figures 10 to 13 are the frequencies of bursting events along the flume. From a quadrant analysis of velocity fluctuations, it was found that at a level of 5 mm from the bed downstream of the second ripple, Quadrants 1 and 3 were dominant when compared to Quadrants 2 and 4. This can be interpreted as an expectation that sedimentation should occur at this location. However, upstream of the ripple, Quadrants 2 and 4 were dominant to Quadrants 1 and 3. Therefore entrainment would be expected to occur at this location. These expectations are confirmed by measuring the bed profile over the ripples along the bed.

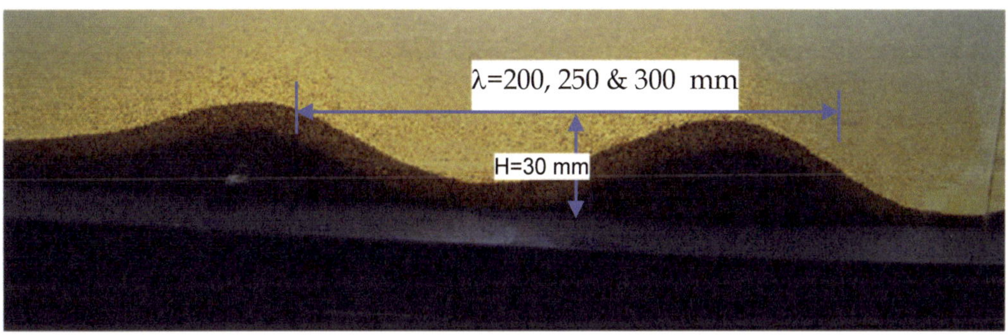

Fig. 7. A photo of the ripples

Jafari Mianaei and Keshavarzi (2009) used a ADV and CCD camera to record the entrained and deposited sediment particles with instantaneous velocity in 3 dimensions over the ripples at the bed of an experimental flume. They used similar technique used by Keshavarzi and Ball (1999) and measured the ratio of entrained particles to deposited particles along the channel and at the bed over the ripples.

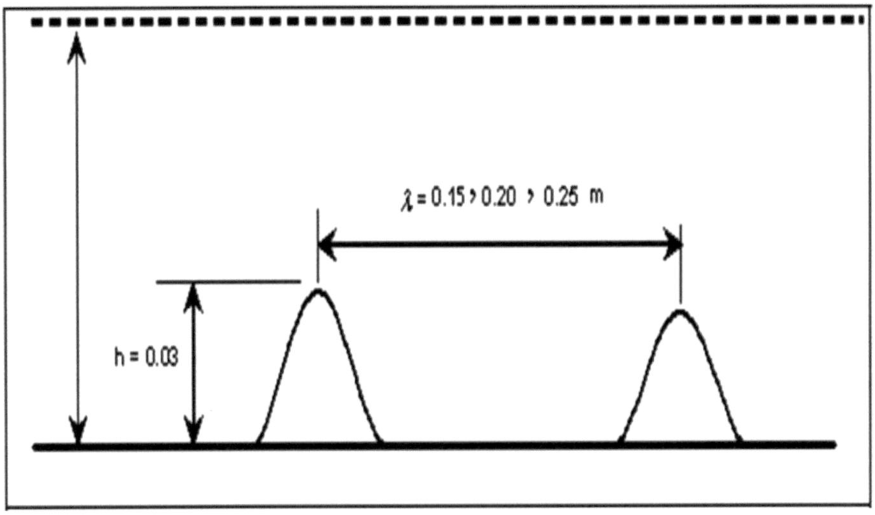

Fig. 8. A schematic of the ripple dimensions considered in the study by Keshavarzi et al. (2010)

The experimental test by Jafari Mianaei and Keshavarzi (2008, 2009) were carried out in a physical model in hydraulic laboratory of Shiraz University. The floor and lateral walls are made of glass, which allows better side viewing of the flow and sediment movement in the flume. The glass-wall flume consisted of a rectangular cross section with base width of 0.70 m, height 0.60 m and 15.50 m length. The flow rate at the end of flume was measured using pre-calibrated right angle V-notches. For the fixed bed experiment, to model the ripples in physical modeling two artificial ripples in each run were used. Artificial ripples were considered as fixed bed. These bed forms were constructed out of two very different materials; aluminum plate and sand. Bed forms made from bent aluminum plate produced a fixed bed, necessary for sampling the velocity structure of the flow field. The aluminum forms also served as the moulds for generating the movable sandy bed forms for mobile bed experiment, necessary for image capturing. According to dimension of ripples formed at nature, height and length of each artificial ripple was 0.03 and 0.20 m, respectively. The stoss side of each form was inclined at 11 degrees from the horizontal, while inclination of the lee side was 31 degrees (according to the angel of repose) for modeling ripples. A schematic diagram of the ripples used in this experimental study is shown in Figure 14.

Each bed form had no topographic variation in the cross-stream direction. The bottom of flume was covered with an erodible bed material layer consisting of sand particles with mean average size D50 of 0.62 mm for both fixed and mobile bed. These sand particles were used in order to fill the artificial ripples and shaping ripples at the bed of flume for making mobile bed. Three bed form configurations were used for both fixed and mobile experiments. In run (a), two bed forms were positioned back-to-back, the wavelength was 200 mm. In run (b), the bed forms were sufficiently separated and the wavelength was 250 mm. For the final configuration, run (c), the wavelength was 300 mm. Figure 15 shows three bed form configurations of ripples and measured points over the ripples for velocity measurement in fixed bed part and image capturing in mobile bed part. At the fixed bed part, the instantaneous velocities of flow in three dimensions were measured with an Acoustic Doppler Velocimeter (microADV), developed by SonTek. The ADV was operated on a pulse-to-pulse coherent Doppler shift to provide three-dimensional velocity components at a rate of 50 Hz. The acoustic sensor consisted of one transmitting transducer and three receiving transducers. The receiving transducers were mounted on short arms

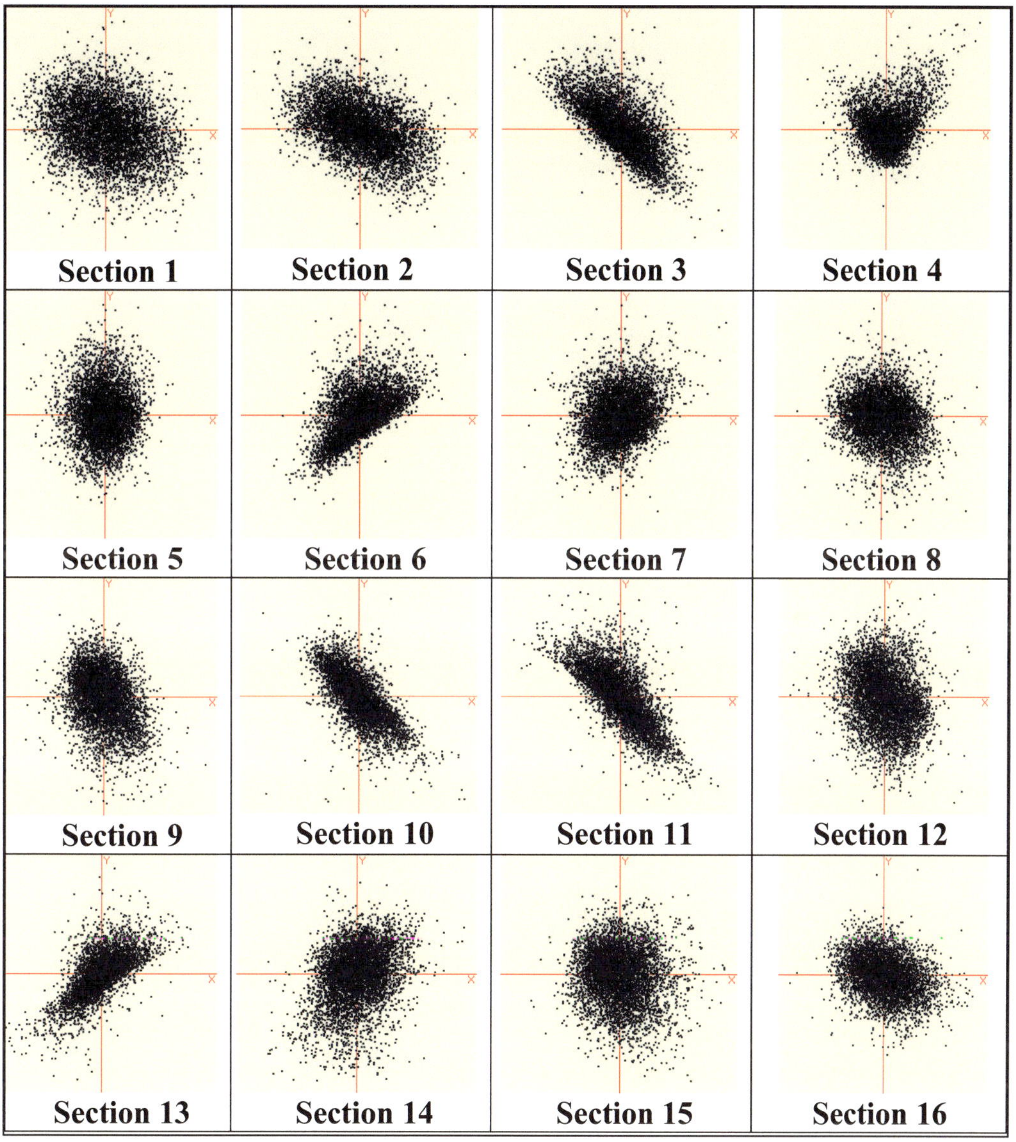

Fig. 9. Distribution of instantaneous shear stress over 16 sections along the ripples, Keshavarzi et al. (2010)

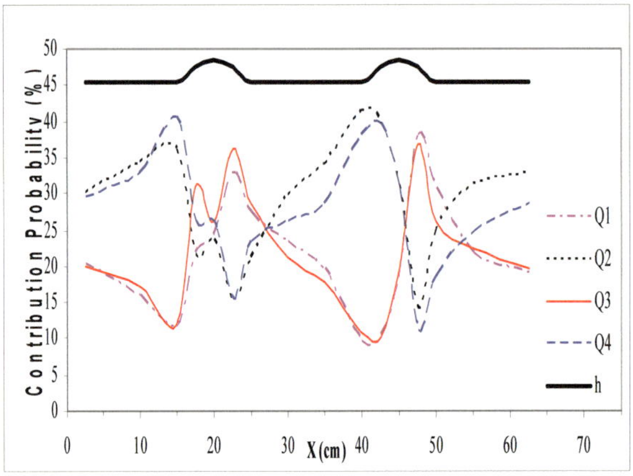

Fig. 10. Bursting events at 5 mm from the bed

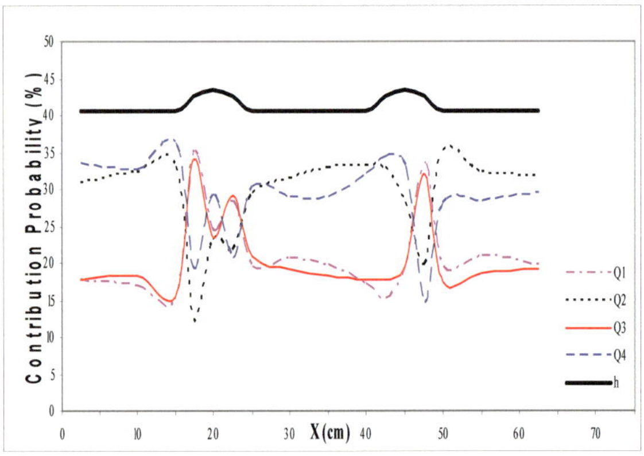

Fig. 11. Bursting events at 10 mm from the bed

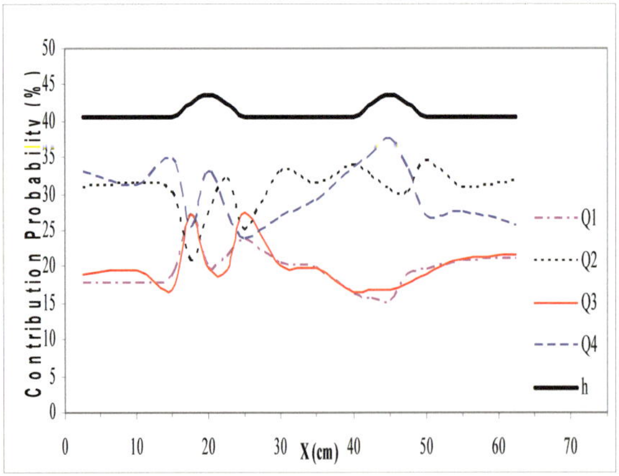

Fig. 12. Bursting events at 15 mm from the bed

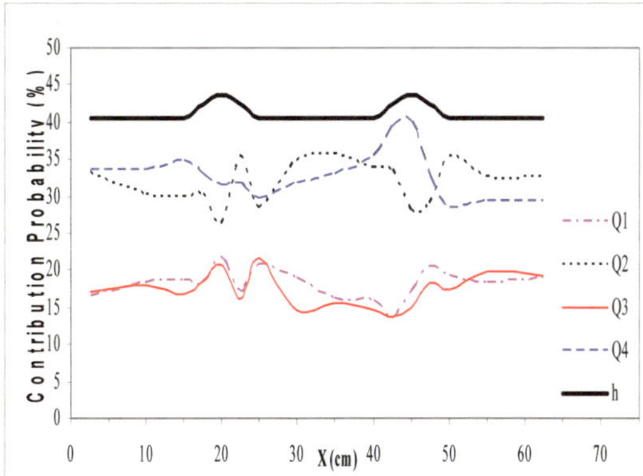

Fig. 13. Bursting events at 20 mm from the bed

around the transmitting transducer at 120 degree azimuth intervals. Acoustic beams were emitted from the transmitting transducers. The beams travelling through the water arrived at a measuring point, located 50 mm below the transducer. The sampling volume of the ADV is very small, providing high spatial resolution and allowing measurements to be taken close to within about 5 mm of the boundary. They were reflected by the ambient particles within the flow being received by the receiving transducers. The processing module performed the digital signal processing required to measure the Doppler shifts. The data acquisition software provided real-time display of the data in graphical and tabular forms. For increasing accuracy of velocity sampling, velocity range should be determined. The velocity range determines the maximum velocity that can be measured. Typically the lowest setting that will cover the expected range of velocities should be chosen to minimize inherent signal noise.

The results of analysis of the contribution of the probability of the bursting events are presented in Table 1, 2 and 3. According to analysis of bursting event it was found that at the stoss side of ripples, quadrants (II) and (IV) were dominant to the quadrants (I) and (III) and at the lee side of the ripple it was vice versa. Also the transition probabilities of the bursting events were determined. The results showed that stable organizations of each class of the events had highest transition probabilities whereas cross organizations had lowest transition probabilities. Additionally, an effort was made to find the average inclination angle of the bursting events in quadrants (II) and (IV). The results showed that the mean angle of events in quadrants (II) and (IV) increases at the downstream of stoss side to the crest in each experimental test. Also, at the lee side where the sediment particles were deposited, the inclination angles had the highest values.

At the mobile bed part, an image processing technique was used to determine amount of deposited and entrained sediment particles over the ripples. The bed form movement was recorded using a digital camera taking pictures through a clear Plexiglas sidewall. These photographs were taken 1500 picture in 60s. Captured photographs were digitized. Resulting digital images were then used to quantify both the deposited and entrained in selected points over ripples. The results of captured picture in image processing are shown in Tables 4 to 6. If the ratio of deposited to entrained particles be more than 1, it means that deposition is dominant to entrainment and for less than 1, it is vice versa. As could be seen in the Tables 4, 5 and 6, at the stoss side of ripples, ratio of deposited to entrained particles is less than 1 and at the lee side it is more than 1. It shows that at the stoss side of ripples, erosion is dominant to deposition and at the lee side it is vice versa. Variation of shear stress at the fixed bed part showed that at the stoss side of the ripple, domination of shear at quadrant (II) and quadrant (IV) causes sediment transport towards the crest of ripple and domination of shear at the quadrant (I) and quadrant (III) causes deposition at the lee side. Obtained results in this part, confirm the results of shear stress analysis at fixed bed part in the case of occurring erosion at stoss side and deposition at lee side of ripples. Therefore, the sediment bed load transport in the shape of ripples could be interpreted as entrainment of sediment particles at the stoss side toward the crest and deposition of the particles at the lee side. Frequently occurring such mechanism causes sediment transport in the shape of ripples at the bed of open channel and rivers.

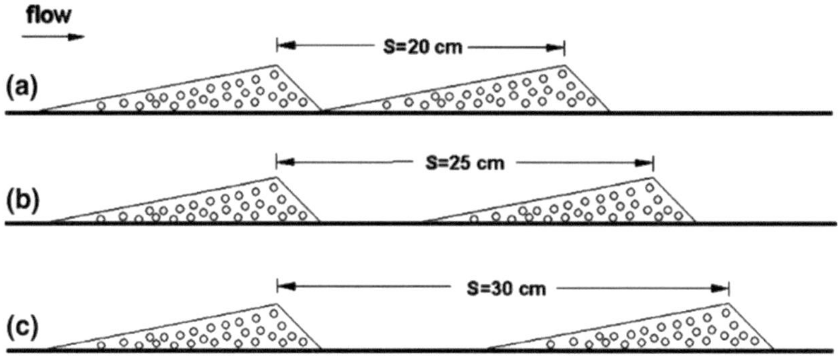

Fig. 14. A schematic diagram of the triangular ripples (after Jafari Mianaei. and Keshavarzi 2009)

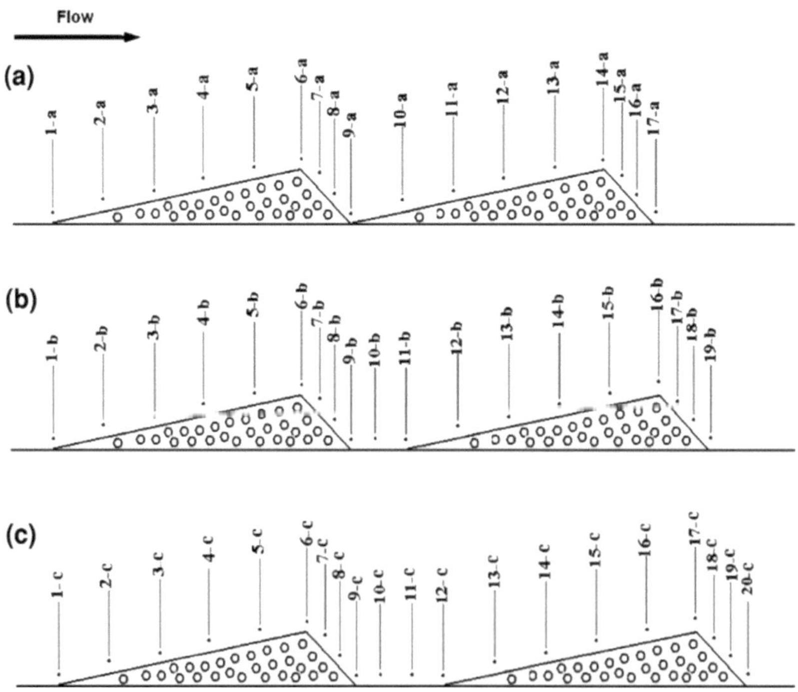

Fig. 15. Sections of velocity measurement over ripples (after Jafari Mianaei and Keshavarzi 2009)

Points	1	2	3	4	5	6	7	8	9	10	11	12	13	14	15	16	17
Quadrant (I)	19	18	19	22	24	24	26	25	26	18	16	16	18	19	27	27	31
Quadrant (II)	31	29	28	27	25	24	22	22	25	33	33	33	28	27	24	23	17
Quadrant (III)	19	20	21	22	25	23	27	29	27	16	17	15	20	20	26	26	34
Quadrant (IV)	32	33	32	28	27	28	26	24	22	32	33	35	33	34	24	24	17

Table 1. Contribution probability of the events in quadrant I, II, III and IV based on two-dimensional bursting processes over ripples for experiment (a)

Points	1	2	3	4	5	6	7	8	9	10	11	12	13	14	15	16	17	18	19
Quadrant (I)	16	19	21	24	19	23	25	36	34	16	17	18	17	19	19	27	30	31	32
Quadrant (II)	31	29	28	25	30	31	24	16	13	29	28	28	30	28	30	23	23	23	21
Quadrant (III)	19	21	22	25	20	20	26	32	39	16	18	18	19	21	19	25	27	28	27
Quadrant (IV)	33	31	29	27	31	26	24	16	15	39	36	36	34	33	31	25	21	18	20

Table 2. Contribution probability of the events in quadrant I, II, II and IV based on two-dimensional bursting processes over ripples for experiment (b)

Points	1	2	3	4	5	6	7	8	9	10	11	12	13	14	15	16	17	18	19	20
Quadrant (I)	22	21	23	21	20	21	27	27	27	20	17	16	16	16	20	18	21	26	28	36
Quadrant (II)	27	28	26	27	28	27	23	24	24	33	35	38	34	32	27	28	27	25	23	18
Quadrant (III)	23	23	22	23	22	24	28	26	26	19	17	15	17	19	21	21	21	26	25	28
Quadrant (IV)	28	28	28	30	30	28	23	24	24	28	31	31	32	33	32	33	31	23	24	18

Table 3. Contribution probability of the events in quadrant I, II, II and IV based on two-dimensional bursting processes over ripples for experiment (c)

Points	1	2	3	4	5	6	7	8	9	10	11	12	13	14	15	16	17
Deposited	316	288	286	269	365	307	285	289	283	366	234	230	263	290	279	284	297
Entrained	334	273	302	281	398	329	296	277	281	375	246	227	273	294	275	267	303
Ratio	0.94	1.05	0.94	0.95	0.91	0.93	0.96	1.04	1.01	0.98	0.95	1.01	0.96	0.99	1.01	1.06	0.98

Table 4. Ratio of deposited particles to entrained particles for run (a)

Points	1	2	3	4	5	6	7	8	9	10	11	12	13	14	15	16	17	18	19
Deposited	131	267	241	325	330	322	256	268	291	221	260	308	258	332	432	280	255	284	280
Entrained	122	277	247	332	353	350	244	248	263	242	281	358	275	356	460	339	261	249	251
Ratio	1.0	0.9	0.9	0.9	0.9	0.9	1.0	1.0	1.1	0.9	0.9	0.8	0.9	0.9	0.9	0.8	0.9	1.1	1.1

Table 5. Ratio of deposited particles to entrained particles for run (b)

Points	1	2	3	4	5	6	7	8	9	10	11	12	13	14	15	16	17	18	19	20
Deposited	239	248	309	371	305	329	379	453	499	554	450	409	433	328	348	404	299	242	263	379
Entrained	253	254	321	396	326	355	384	449	478	580	481	434	454	353	358	443	312	243	235	321
Ratio	0.9	0.9	0.9	0.9	0.9	0.9	0.9	1.0	1.0	0.9	0.9	0.9	0.9	0.9	0.9	0.9	0.9	1.0	1.1	1.1

Table 6. Ratio of deposited particles to entrained particles for run (c)

9. Conclusion

In this study, the flow structure over the ripples was investigated experimentally. The focus of this study was the measurement and analysis of the dominant bursting events and the flow structures over ripples in the bed of a channel. Two sets of ripples 1) Symmetrical bell shaped ripples with different sinuosity wave length and 2) Asymmetrical triangular shaped with different wavelength were tested in a comprehensive laboratory study. The velocities of flow over the ripples were measured at the threshold of motion in three dimensions using an Acoustic Doppler Velocity meter (Micro-ADV) with a sampling rate of 50 Hz. These velocities were measured at 5, 10 15, 20, 25, 30 50 and 60 mm from the bed at 16 longitudinal positions along the flume for a total of 128 points. At the same time the particle motion was recorded using CCD camera. An image processing technique was used to extract information from the recorded images. Consideration of these results showed that upstream of the first ripple, bursting events in quadrants 2 and 4 are dominant, however, downstream of the ripple, the bursting events in quadrants 1 and 3 are dominant and that sediment deposition occurred downstream of the ripple.

10. References

Adrian, R.J.. (1991). Particle-imaging techniques for experimental fluid mechanics. Annual Reviews Fluid Mechanics, 23: 261-304.

Akaike, H.,1974. A new look at the statistical model identification. IEEE Trans. Autom. Control, AC-19:716-723.

Bagnold, R. A. (1946) Motion of waves in shallow water: interaction between waves and sand bottoms, Proceedings of the Royal Society of London, Ser. A, 187, pp. 1-18.

Bennett S, Best J (1996) Mean flow and turbulence structure over fixed ripples and the ripple–dune transition. In: Ashworth PJ et al. (eds) Coherent flow in open channels. John Wiley, NJ, pp 281-203.

Best, J., (1992) On the entrainment of sediment and Initiation of bed defects: Insights from recent developments within turbulent boundary layer research. Sedimentology,39: 797-811.

Best, J., (1996). The fluid dynamics of small-scale alluvial Advances in dynamics and stratigraph Edited by P.A., and Dawson, M.R., John Wiley Sons, Chichester.

Box GEP, Cox DR (1964) An analysis of transformation. J Roy Stat Soc Ser B 26:211–252

Bridge JS, Bennett SJ. (1992). A model for entrainment and transport of sediment grains of mixed sizes, shapes and densities. Water Resour Res, 28(2):337–63.

Carling P, Golz E, Orr HG, Radecki-Pawlik A (2000). The morphodynamics of fluvial sand dunes in the Ripple–dune transition 647 Rhine River, near Mainz, Germany. II. Hydrodynamics and sediment transport. Sedimentology 47:253–278

Carstens, M. R., F. M. Neilson and H. D. Altinbilek (1969) Bed forms generated in the laboratory under an oscillatory flow: analytical and experimental study, U.S. Army Corps of Engineers, Coastal Engineering Research Center, Tech. Memo 28, p. 105.doi:10.1029/2009JF001628.

Drake TG, Shreve RL, Dietrich WE, Whiting PJ, Leopold LB (1988). Bed load transport of fine gravel observed by motion picture photography. J Fluid Mech 192:193–217

Grass AJ (1971) Structural features of turbulent flow over smooth and rough boundaries. J Fluid Mech 50(2):233–255.

Jafari Mianaei S. and Keshavarzi A. 2008. Spatio-temporal variation of transition probability of bursting events over the ripples at the bed of open channel. Stoch. Environ Res Risk Assess, 22:257-264.

Jafari Mianaei S. and Keshavarzi A. 2009. Study of near bed stochastic turbulence and sediment entrainment over the ripples at the bed of open channel using image processing technique. Stoch Environ Res Risk Assess, DOI 10.1007/s00477-009-0346-7.

Jeremy G. Venditti Michael A. Church Sean and J. Bennett. 2005. Bed form initiation from a flat sand bed. J. Geophysical Research, VOL. 110, F01009, doi: 10.1029/2004JF000149.

Julien P, Klassen G (1995). Sand-dune geometry of large rivers during floods. J Hydraul Eng ASCE 121:657–663

Katz, R.W. 1981. On some criteria for estimating the order of a Markov chain. Thechnometrics, 23:243-249.

Keshavarzi AR, Ball JE (1997). An analysis of the characteristics of rough bed turbulent shear stress in an open channel flow. J Stoch Hydro Hydraul 11(3):193–210

Keshavarzi AR, Ball JE (1999). An application of image processing in the study of sediment motion. J Hydraul Res 37(4):559–576.

Keshavarzi AR, Ball JE and Nabavi S. H. (2010). Study of Flow Structure over the Ripples in the Bed of Open Channel. 6th International Symposium on Management, Engineering and Informatics, MEI 2010, Orlando, Florida.

Keshavarzi AR, Shirvani A (2002) Probability analysis of instantaneous shear stress and entrained particles from the bed. In: Proceedings of the CSCE/EWRI of ASCE Environmental engineering conference, Niagara

Khelifa, A. and Y. Ouellet (2000) Prediction of sand ripple geometry under waves and currents, Journal of Waterway, Port, Coastal, and Ocean Engineering 126(1)14 - 22.

Kline SJ, Reynolds WC, Schraub FA, Runstadler PW (1967) The structure of turbulent boundary layers. J Fluid Mech 30(4):741–773.

Kostachuck R (2000) A field study of turbulence and sediment dynamics over subaqueous dunes with flow separation. Sedimentology, 47:519–531.

Kostachuck R, Church M (1993) Macroturbulence generated by dunes. Sed Geol 85:25–37

Kostaschuk R, Villard P (1996) Flow and sediment transport over large subaqueous dunes. Fraser River, Canada. Sedimentology 43:849–863

Lajeunesse, E., L. Malverti, and F. Charru (2010). Bed load transport in turbulent flow at the grain scale: Experiments and modeling. J. Geophysical Research, 115, F04001.

Miller, M. C. and P. D. Komar (1980) Oscillation sand ripples generated by laboratory apparatus, Journal of Sedimentary Petrology 50(1):173-182.

Mogridge, G. R., M. H. Davies, and D. H. Willis (1994) Geometry prediction for wave generated bedforms, Coastal Engineering 22, pp. 255-286.

Nabavi S.H. (2005). Study of flow structure over the ripples at the bed of open channel. MS thesis, Water Department, Shiraz University, Shiraz, Iran.

Nakagwa H, Nezu I (1978) Bursting phenomenon near the wall in open channel flow and its simple mathematical model. Fac Eng Kyoto University, Japan XL (4) 40:213–240

Nelson JM, Shreve RL, McLean SR, Drake TG (1995) Role of near bed turbulence structure in bed load transport and bed form

Nielsen, P. (1992) Coastal bottom boundary layers and sediment transport, World Scientific, pp. 324.

Offen GR, Kline SJ (1975) A proposed model of the bursting process in turbulent boundary layers. J Fluid Mech 70:209–228.

Ojha S.P. and Mazumder B.S., 2008. Turbulence characteristics of flow region over a series of 2-D dune shaped structures. Advances in Water Resources 31:561–576.

Papanicolaou AN, Diplas P, Evaggelopoulos N, Fotopoulos S (2002) Stochastic incipient motion criterion for spheres under various packing conditions. J Hydraul Eng ASCE 128(4):369–380.

Raudkivi, A. (1997) Ripples of stream bed, Journal of Hydraulic Engineering 123:1 pp. 58-64.

Sajjadi SG, Aldridge JN, Nicholas DJ (1996) Sediment transport mechanism over rippled sand beds. In: Spaulding ML, Cheng RT (eds) Estuarine and coastal modelling. ASCE, New York, pp 669–680.

Schwarz, G., 1978. Estimating the dimension of a model. Ann. Stat., 6, 461-464.

Sontek ADV Operation Manual (1997) Firmware version 4.0. Sontek, San Diego

Termini D. and Sammartano V. (2009). Sediment flow interactions and effects on bed forms mechanism. 7th ISE & 8th HIC Chile.

Thorne PD, Williams JJ, Heathershaw AD (1989) In situ acoustic measurements of marine gravel threshold and transport. Sedimentology, 36:61–74

Tong, H.,1975. Determination of the order of a Markov chain by Akaike's Information Criterion. J. Appl. Probability, 12:488-497.

Willert C.E. and Gharib, M., (1991). Digital particle image velocimetry. Experiments in Fluid; 10: 181-193.

Williams, J.J., 1990. Video observations of marine gravel transport. Mar. Lett.,10:157-164.

Williams, J.J., 1996. Turbulent flow in rivers. In Advances dynamics and stratigraphy. Edited by P.A., and Dawson, M.R., John Wiley &Sons. Chichester.

Yalin, M. S. (1977) On the determination of ripple length, Proceedings of the American Society of Civil Engineers 103:HY4, pp. 439-442.

Yen BC (2002) Stochastic inference to sediment and fluvial hydraulics. J Hydraul Eng ASCE 128(4):365–367.

4

Stochastic and Deterministic Methods of Computing Graded Bedload Transport

Faruk Bhuiyan
Bangladesh University of Engineering & Technology
Bangladesh

1. Introduction

Prediction of nonuniform sediment transport is an important issue for river engineers and morphologists. Presently available methods for sediment transport calculation involve empirical parameters and inexact ways of representing the graded materials and their threshold conditions in computations. The performances of the methods for predicting transport at various types of flows are uncertain. This reality entails verification and sometimes, adjustment of the parameters with reliable and correct data sets. On the other hand, majority of the equations have been developed from laboratory data of straight channels and, generally, verified either with simple straight flume data or field data from uncontrolled and, often poorly defined, natural environments. In this study, frequently used, as well as, recently developed bedload predictors are examined with data set from complex flows in strictly controlled laboratory experiments. The tests were carried out in the Flood Channel Facility, Hydraulic Research Wallingford, UK with graded materials. Local transport rates were sampled under complex flows in a meandering bend with bankfull and overbank conditions.

A meandering channel has straight reaches in the crossover region and curved reaches in the vicinity of the bend apex. Flow patterns in both reaches are complex due to formation of streamwise vortices and corresponding secondary flow cells. The velocity structure in the bend is fully three-dimensional. Consequently, sorting of grains and lateral transport occur in the vicinity of the bend which could persist up to the next crossover section (Dietrich and Smith, 1984; Parker and Andrews, 1985; Julien and Anthony, 2002; Bhuiyan and Hey, 2002). The present data of local graded sediment transport rate and corresponding hydraulic parameters are comprised of carefully carried out experiments under such complex flows. The performances of these methods are examined and issues related to prediction of graded sediment transport are presented in this paper.

2. Review of sediment transport predictors

Bed load transport predictors have a long history of development. Recent formulations are more oriented to incorporation of the probabilistic and stochastic concepts in representing flow turbulence and bed shear stress to compute the pickup rates of grains from channel beds composed of sediment mixtures. A brief review of the widely used classical as well as most recent methods and their limitations in applications are given here.

2.1 Meyer-Peter Muller (1948)

One of the old and widely used methods is the formula by Meyer-Peter and Muller (1948). This is based on extensive laboratory data with uniform and graded materials of various densities. The bed load transport is predicted from the effective bed shear stress taking into account of the form resistances. The equation is:

$$R(n'/n)^{3/2}S/\Delta d_m - 0.047 = 0.25\rho^{1/3}\gamma_s q_b^{'2/3}/\Delta d_m \quad (1)$$

where R and S are the hydraulic radius and slope, respectively; n and n' are the Mannings coefficient related to total roughness and that due to grain roughness, respectively; ρ, γ_s, Δ, d_m are the density of water, unit weight of sediment, submerged specific weight of sediment and representative sediment diameter, respectively. The nondimensional shield shear stress may assume other values than 0.047, e.g., 0.042 by Habersack and Laronne (2002).

2.2 Eisnstein (1950)

The Einstein's method (1950) of predicting bed load transport is a complex procedure which introduced the probability of particle movement, variable hiding effects of finer particles and fractional transport rates for first time. The flow (shear) intensity parameter (ψ_i^*) and transport intensity parameter (ϕ_i^*) are related as :

$$1-1/\sqrt{\pi}\int_{-0.143\psi_i^{*-2}}^{0.143\psi_i^{*-2}} e^{-t^2}dt = 43.5\phi_i^* / \left(1+43.5\phi_i^*\right) \quad (2)$$

The transport intensity parameter is a function of the specific bed load transport rate for a size fraction. Contribution from various sediment fractions are summed up to obtain the required bed load transport. Although this method is a detailed one, the performance for predicting total bed load transport of nonuniform sediment is not satisfactory. The method considers only the effect of sheltering of finer particles by the coarser ones but the effect of enlarged exposure of the coarser particles is not taken care of. This is the main reason for poor performance of the method for prediction of individual fractions in a mixture (Misri et al., 1984). Again, the method may underpredict transport rate due to overestimation of form roughness in gravel rivers. Gomez and Church (1989) proposed a modified iterative procedure to incorporate known hydraulic parameters in calculation.

2.3 White and Day (1982)

White and Day (1982) modified the Ackers and White (1973) formula for fractional bed load transport:

$$q_{bi} = G_{gri}\gamma_s U(U/u_*)^{n_w} d_{gr} \quad (3)$$

where G_{gri} is the transport parameter of the grain size class i; U and u_* are the mean velocity and shear velocity, respectively; d_{gr} is the grain size parameter on which other exponents of the equations depend.

2.4 van Rijn (1984, 1993, 2002)

The van Rijn's method (1984) is based on detailed physical process of grain transport. The transport for a particular size fraction is given by:

$$q_{bi} = 0.1\gamma_s \Delta^{0.5} g^{-0.5} d_i^{1.5} p_i T_i^{1.5} D_*^{-0.3} \tag{4}$$

where T_i is the transport parameter counting relative excess grain shear stress; D_* is the Bonnefille grain parameter; p_i is the percentage of size class i in the bed material. Later on van Rijn (1993) developed a stochastic method for predicting sediment transportation. Kleinhans and van Rijn (2002) used this method with the aid of laboratory experimental data and modified classical deterministic bed load equations (e.g., Mayer-Peter and Muller (1948) and van Rijn (1984) to stochastic predictors. The classical concept of turbulence is incorporated in this method and the shear stress responsible for grain movement is assumed to have a stochastic nature. The effective instantaneous bed shear stress (τ') is assumed to have a normal distribution with a standard deviation σ and a mean $\overline{\tau}'$ as:

$$F(\tau') = 1 / \sqrt{2\pi\sigma}\, exp\left(-1/2\left((\tau' - \overline{\tau}')/\sigma\right)^2\right) \tag{5a}$$

The probability that τ' will exceed critical shear stress for grain movement (τ_{ci}) in upstream or downstream direction can be obtained by integrating the probability density function for $|\tau'| > |\tau_{ci}|$ along these directions. Thus dimensionless bed load transport is expresses as:

$$q_{bi} / (\rho_s d\sqrt{\Delta dg}) = \int_{-\infty}^{-\tau_{ci}} P_{\tau'} \phi_{\tau' i} d\tau' + \int_{\tau_{ci}}^{\infty} P_{\tau'} \phi_{\tau' i} d\tau' \tag{5b}$$

where $P_{\tau'}$ is the time fraction at which instantaneous shear is active for grain movement; and $\phi_{\tau' i}$ is the dimensionless sediment transport of fraction i due to this instantaneous effective shear. The relative standard deviation of the instantaneous shear with respect to $\overline{\tau}'$ is normally taken as 0.4 in the above computation. A finite lower and upper bound of the above integration is chosen such that most of the distribution is covered (e.g., 6σ).

2.5 Patel and Raju (1996)

Patel and Raju (1996) modified the previous work by Misri et al. (1984) for predicting nonuniform bed load transport. Both extensive laboratory data as well as field data have been used to develop an empirical relationship between transport rate of individual size fraction and various parameters including hiding-exposure effects and effective grain shear stress. The final relationship between the product $\xi_B \theta_i'$ (i.e. the product of hiding-exposure parameter and dimensionless effective shear stress) and ϕ_B (dimensionless bed load transport) is presented in graphical format. In logarithmic scale this can be approximated with straight line for $\phi_B < 10^{-2}$ and for higher ϕ_B by cubic spline.

2.6 Sun and Donahue (2000)

Similar to Einstein (1950), the probability concept has been used by Sun and Donahue (2000) to develop bed load transport functions for nonuniform sediment. A theoretical attempt has been made by the authors to combine stochastic process with mechanics and measured data.

The near-bed exchange of graded materials is modeled by employing a Markov process with continuous time. The dimensionless average velocity and time of single-step motion are considered to be the functions of the flow intensity parameter which reflect the physical properties of the bed load transport rather than as constants expressed by Einstein. The probability of fractional incipient motion for nonuniform sediment α_i is derived considering random properties of forces and their moment arms as:

$$\alpha_i = 1 - \frac{1}{\sqrt{2\pi}} \int_{-2.7(\sqrt{0.0822\psi_i}-1)}^{2.7(\sqrt{0.0822\psi_i}-1)} e^{-0.5t^2}\, dt \tag{6a}$$

where flow intensity (ψ_i) is defined as the inverse of the non-dimensional grain shear stress corrected for the shelter-exposure effect of the nonuniform sediment mixture. The bed load formula for the *i*th class (percentage of this class on bed = P_{bi}) of the nonuniform sediment is:

$$\phi_i = 0.3 P_{bi} \alpha_i / \left(\psi_i^{3/4} (1 - \alpha_i)\right) \tag{6b}$$

The limitations of this method have been explained by Wu and Yang (2004) while developing their own model (see later). The main conceptual limitation is that although they used continuous-time Markov process of the two-state model (static and moving), the transitional probabilities used in this model are only suitable for discrete-time Markov process. Secondly, the rolling probability has been used for entrainment of particles but only saltation mode has been considered to calculate mean particle velocities. Although this method has been developed to apply for both fully mobilized and partial transport, mainly full transport data were used to calibrate various parameters.

2.7 Wu et al. (2000)

Wu et al. (2000) derived a semi-theoretical relation for calculating critical shear stress for incipient motion of size fractions in nonuniform sediment. The hiding-exposure probabilities of particles are related to the size and gradation of the bed material based on stochastic assumption. The exposure height of a particle on bed is taken as a random variable which follows a uniform probability distribution. An empirical relation between excess shear stress, T_i and non-dimensional fractional bed load transport is deduced as:

$$\phi_{bi} = 0.0053 \left((n'/n)^{1.5} (\tau_b / \tau_{ci}) - 1 \right)^{2.2} \tag{7a}$$

where τ_{ci} is a function of the hiding and exposure probabilities (p_{hi} and p_{ei}) of the particle d_i which are calculated summing over all fractions:

$$p_{hi} = \sum P_{bj} d_j / (d_i + d_j) \quad \text{and} \quad p_{ei} = \sum P_{bj} d_i / (d_i + d_j) \tag{7b}$$

2.8 Wu and Yang (2004)

Wu and Yang updated the concept of the stochastic method by Sun and Donahue (2000). The movements of bed sediment particles are represented by the pseudo four-state continuous-time Markov process. Both rolling and lifting modes have been considered to calculate probabilities of sediment entrainment. The final equation of this method is:

$$q_{bi} = \frac{\pi}{6}\rho_s P_{bi}\sqrt{(\rho_s-\rho)g}d_i^{1.5}Y_i P_{mi}V_{pi}(\Delta_i Y_i^{1.8}) \tag{8}$$

where Y_i and V_{pi} are the fractional mobility and dimensionless mean particle velocity, respectively; P_{mi} is the limiting probability (long-run moving probability) that a bed load particle of size class i is in the moving state. The last two terms in parenthesis of the above equation are the correction factors for the fully and the partially mobilized fractions. Laboratory and field data of both partial transport and fully mobilized transport have been used in developing this method. The fractional mobility Y_i is approximated by a cumulative lognormal distribution of effective dimensionless shear stress (θ_i') with its mean and standard deviation affected by the sand content of the sediment mixture.

3. Results and discussion

The transport rates are predicted with section-averaged parameters as well as local parameters (spaced at y/B = 0.0625; y and B are the lateral distance and width of channel, respectively). Figure 1a and Figure 1b show comparison of observed and predicted transport rates. Transport rates in five cross-sections in the experimental river bend are shown in the first figure. Observed transport rate by Halley-Smith sampler showed a low rate in the apex section which may be attributed to the complex sorting processes around the bend. Predictions are much scattered in that section, mainly underestimated. Local transport rates and velocities were sampled at various lateral locations spaced at y/B = 0.0625. For each cross-section, predictions were carried out using local data.

Figure 1(b) shows an example of results for the cross-over section. Predictions with both the original sediment mixture and the local sediment properties were examined. The predicted rates from the former shows large scatter indicating the importance of using local sediment characteristics for local transport prediction. In some cases, although predictions for total transport of the section appeared to be closer to the observed rates, local transport distribution across the section were not satisfactory. Again, the calculated local transport rates were much lower than observed values in the apex section.

To examine the relative performance of the methods, several parameters has been used. The discrepancy ratio, weighted discrepancy ratio and their geometric mean are used (Habersack and Laronne, 2002). The uniformity coefficient (η), root mean squire error (*rmse*) and relative error (ε_r) are also used :

$$\eta = rmse \Big/ \left(\left(1/n\sum q_{oi}^2\right)^{1/2} + \left(1/n\sum q_{pi}^2\right)^{1/?}\right) \tag{9a}$$

$$rmse = \left(\sum_{i=1}^{n} 1/n\left(q_{oi}-q_{pi}\right)^2\right)^{1/2} \text{; and } \varepsilon_r = 1/n\sum_{i=1}^{n}\left(q_{oi}-q_{pi}\right)/q_{oi} \tag{9b}$$

q_{oi} and q_{pi} are the observed and predicted transport rate of the fraction i and n is the total number of fractions for calculation. Based on such calculations, ranking or the performance matrix of selected predictors are shown in Table 1. Here the numbers indicate the position in the list (e.g., 1 = first, indicating the best performance) depending on certain parameter. The first column is based on the number of predicted data lying in the range 50% to 200% of the observed values. It is obvious that the ranking changes depending on the selected parameter.

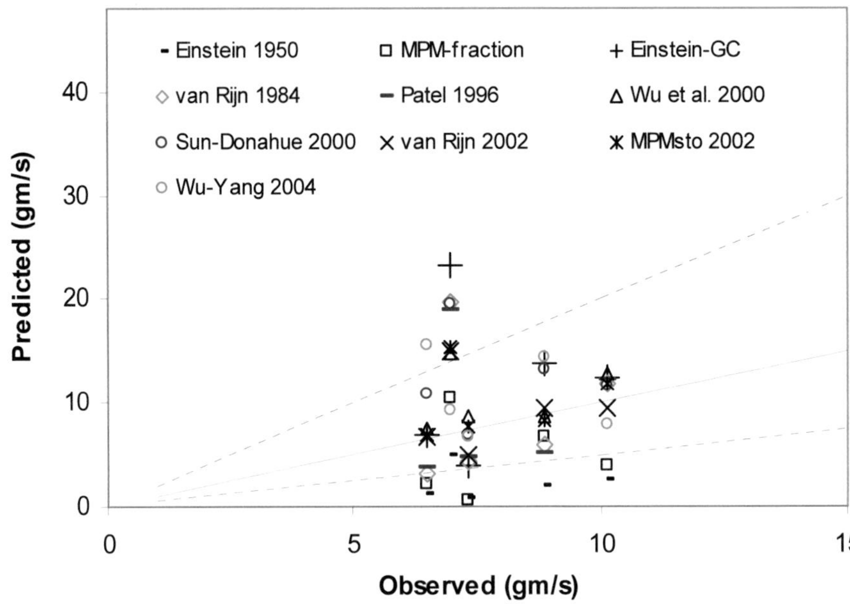

(a)

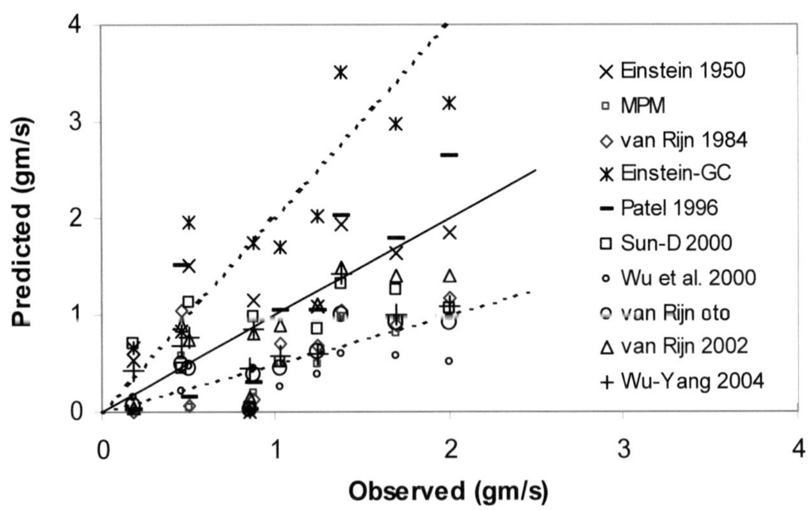

(b)

Fig. 1. Comparison of predicted and observed transport rates (a) Section-averaged rates (b) Local transport rates (solid line: perfect agreement; broken lines: 50% and 200% of the observed values)

	50 % to 200%	Uniformity coefficient	Mean discrepancy ratio, s	Weighted s	Root mean square error, rmse	Relative error
Van Rijn 2002	1	1	1	1	1	1
Wu-Yang	2	3	2	2	2	3
Sun-Donahue	3	4	3	3	3	2
Einstein-GC	4	6	9	8	10	10
Patel and Raju	5	2	4	4	4	6
van Rijn 1984	6	5	6	5	5	7
van Rijn (sto)	7	7	5	6	6	4
MPM	8	8	7	7	7	5
Wu et al.	9	9	8	9	8	8
White-Day	10	10	10	10	9	9

Table 1. Ranking of selected predictors

4. Conclusion

Nonuniform sediment transport prediction capabilities of classical as well as recently developed bed load transport predictors are evaluated. Data from controlled laboratory experiments of graded material transport in complex flows of meander bends have been used for this purpose. Predicting capabilities of majority of the methods are reduced in the vicinity of the bend apex. This indicates the complex flow pattern and sorting process around the bend. The methods based on probability concept appeared to be superior for predicting local transport of bedload. Although several deterministic methods show comparable performance for predicting total sectional transport rate, their performances are significantly reduced for predicting lateral variation of local transportation rates.

5. References

Ackers, P., and and White, W. R., Nov. 1973, “Sediment transport: new approach and analysis”, J. Hydraul.Div., ASCE, vol. 99, pp. 2041-2060.

Bhuiyan, F., Hey, R.D., “Sediment transport in a meandering channel with instream hydraulic structures”, 2004, Proc. International Conf. River Flow 2002, Sept. 2002, Ghent, Belgium, vol. 2, pp. 1091-1098.

Dietrich, W. E., and Smith, J. D., 1984, “Bed load transport in river meander”, Water Resour. Res., vol., 20, pp. 1355-1380.

Einstein, H. A., “The bed load function for sediment transportation in open channel flows”, 1950, Tech. Bulletin, No. 1026, United States Dept. of Agriculture.

Gomez, B., and Church, M., Jun. 1989, “An assessment of bed load sediment transport formulae for gravel bed rivers”, Water Resources Research, vol. 25, pp. 1161-1186.

Habersack, H. M., and and Laronne, J. B., May 2002, “Evaluation and improvement of bed load discharge formulas based on Halley-Smith sampling in an Alpine Gravel Bed River”, vol. 128, pp. 484-499.

Julien, P. Y., and Anthony, D.J., Feb. 2002, "Bed load motion and grain sorting in a meandering stream", J. Hydraul. Res., vol. 40, pp. 125-133.

Kleinhans, M. G., and van Rijn, L. C., Apr. 2002, "Stochastic prediction of sediment transport in sand-gravel bed rivers", J. of Hydraul. Engg., ASCE, vol. 124, pp. 412-425.

Lopes, V. L., Osterkamp, W. R., and Bravo-Espinoza, "Evaluation of selected bedload equations under transport and supply-limited conditions", March 2001, Proc. Seventh Fed. Int. Sed. Conf., pp. I - 192- I 198.

Meyer-Peter, E., and Mueller, R., "Formulas for bed load transport", 1948, Proc. 3rd Conf. Int. Association of Hydraulic Research, Stockholm, Sweden, 39-64.

Misri R.L., Ranga Raju, K. G., and Garde, R.J., March 1984, Bed load transport of coarse nonuniform sediment, J. Hydraul. Engg., ASCE, vol. 110, pp. 312-328.

Parker, G. and Andrews, E.D., 1985, "Sorting of bed sediment by flow in meander bends", Water Resour. Res., vol. 20, pp. 1361-1373.

Patel, P. L., and Ranga Raju, K. G., Mar. 1996, "Fractionwise calculation of bed load transport", J. Hydraul. Res., vol. 34, pp. 363-379.

Sun, Z., and Donahue, J., Feb. 2000, "Statistically derived bed load formula for any fraction of nonuniform sediment", J. Hydraul. Engg., ASCE, vol. 126, pp. 105-111.

Van Rijn, L. C., "Principles of Sediment transport in rivers, estuaries and coastal seas," 1993, Oldemarkt, Aqua, The Netherlands, pp. 7.41-7.43.

Van Rijn, L. C., Oct. 1984, "Sediment transport, part I: bed load transport", J. Hydraul. Engg., vol. 100, pp. 1431-1456.

White, W. R., and Day, T. J., 1982, "Transport of graded gravel bed materials", Gravel-bed Rivers, R. D. Hey et al., eds, Wiley, New York, pp. 181-213.

Wu, Fu-Chun, and Yang, Kuo-Hsin, 2004, "A stochastic partial transport model for mixed-size sediment: application to assessment of fractional mobility", Water Resource Res., vol. 40, paper no. W04501.

Wu, W., Wang, S. S. Y., and Jia, Y., Jun. 2000, "Nonuniform sediment transport in alluvial rivers", J. Hydr. Res., vol. 38, pp. 427-434.

5

Methods for Gully Characterization in Agricultural Croplands Using Ground-Based Light Detection and Ranging

Henrique Momm, Ronald Bingner, Robert Wells and Seth Dabney
USDA-ARS-National Sedimentation Laboratory
United States of America

1. Introduction

Soil erosion has long been recognized as the primary cause of soil degradation in agricultural fields (Wells et al., 2010). Overland flow in agricultural fields is the main process associated with soil erosion, which is often grouped into categories of: sheet erosion, rill erosion, and gully erosion (Smith, 1993). Traditionally, research has focussed on understanding and modelling sheet and rill erosion processes (Poesen et al., 1996). Recent research has begun to focus on addressing gully issues such as the understanding of the formation of gullies, their contribution to overall soil loss, development of tools to locate channel initiation, and appropriate measuring techniques (Poesen et al., 2003). The increased focus on gully research can be partially attributed to recent studies demonstrating that gully formation is very common on cropland, especially in conventional tillage systems (Gordon et al., 2008) and can be as significant as sheet and rill erosion in terms of sediment yield (Bingner et al., 2010). Without a good understanding of gully processes, technology cannot be developed that can provide information needed by watershed managers when evaluating and implementing effective conservation practice plans.

Gullies can be generally classified as ephemeral, classical, or edge-of-field. The Soil Society of America (2001) defines ephemeral gully as "small channels eroded by concentrated flow that can be easily filled by normal tillage, only to reform again in the same location by additional runoff events". As the headcut migrates upstream and the channel gets wider, faster than the interval between farming tilling operations, farming equipment is forced to operate around the gully and as result the gully becomes permanent (classical gully). Finally, as the name suggests, edge-of-field gullies are defined by channels where concentrated flow crosses earth bank (Poesen et al., 2003).

New methodologies are being researched to understand gully formation and estimate sediment yield (Souchere et al., 2003, Cerdan et al., 2002, and Woodward, 1999). Studies use Digital Elevation Models (DEMs) as the basis to formulate theories explaining the relationship between field topography and gully occurrence (Woodward, 1999, Parker et al., 2007, and Cerdan et al., 2002). These efforts greatly benefit from accurate and detailed topographic information which can aid in the understanding of where and when gullies form and how these features evolve over time (headcut migration).

Despite the availability of DEMs at regional and local scales (spatial resolution ranging from 1 to 30 meters), these datasets often do not offer the necessary spatial and/or temporal resolution for gully investigations at individual plot scale. In order to capture the micro-topography impacting gully formation and development over time it is necessary to work with DEMs with spatial resolution ranging between 5mm to 15cm (Schmid et al., 2004). Additionally, ephemeral and classical gullies are affected by precipitation events, management and conservation practices, crop phenology, and crop canopy development. In ephemeral gully investigations, timing becomes significant as farming practices can remodel the field's topography by filling the gully channel. Classic gullies are also dynamic, responding to condition changes through headcut migration, deepening, and channel widening processes. Standard DEM datasets produced either by government agencies on a national scale or by private companies on a regional scale can only capture a snap shot in time of gully development.

Recent developments in laser scanner technology provide new opportunities for scientific investigation of gullies located in croplands. Although, laser scanners have been previously used in similar investigations such as large-scale classical gullies in different locations such as mountain-side sites (Perroy et al., 2010), forest sites (James et al., 2007), desert sites (Park, 2008), and landslides (Hsiao et al., 2003 and Roering et al., 2009), its application to ephemeral and classical gully investigation is in the early stages of development. Ground-based laser scanners provide the tools for detailed multi-temporal analysis of micro-topography of gullies. With this detailed information it is possible to obtain digital terrain models (catchment area, slope, planform curvature, thalweg, and cross sections), estimate soil roughness, perform volume calculations, and others. This is especially important in research setups where the same location needs to be surveyed multiple times over extensive periods of time as conditions change due to rain events, field management changes, and/or different conservation practices investigations. Ground-based laser systems are capable of providing high levels of topographic and morphologic details.

Ground-based light detection and ranging (LiDAR) systems differ from airborne systems. During data collection, the scanner (LiDAR sensor) is located on a tripod at ground level (Figure 1). Proximity to the ground surface causes laser pulses to be sent with larger range of vertical incidence angles than the airborne system. Ground sampling density is influenced by operator controlled factors such as investigation area selection (vertical and horizontal incidence angles), instrument resolution (average point density at mid-range), and degree of overlap between scans. The large volume of points combined with irregular point distribution of ground-based LiDAR data sets makes production of digital elevation models computationally challenging. Additionally, the presence of vegetation, standing residues, and shadows (regions with no data) represents an extra level of complexity that often cannot be addressed by standard off-the-shelf commercial geospatial software packages.

In this chapter, we describe new and enhanced techniques to pre-process and evaluate detailed topographic information acquired with ground-based LiDAR, which were specifically devised to quantitatively and morphologically characterize gullies located in agricultural fields. In the following section (Section 2) background information about ground-based LiDAR systems is provided. In section 3, multiple techniques to pre-process LiDAR data in the original point cloud format are depicted, such as cross-validation between scans (geo-referencing accuracy investigation), point sampling pattern investigation in terms of density (quadrant methods) and spatial distribution (distance to nearest neighbours), removal of unwanted points (thinning and outlier removal), and surface reconstruction (surface smoothing). Morphological analysis through cross-section investigation follows in Section 4. An alternative method to overcome the high computational cost involved in the conversion of large point cloud into raster grid is addressed in Section 5. We provide a brief description of the implementation details as well as the computer programming language libraries utilized in Section 6. A study case of a gully located in a field of the Cheney Lake Reservoir Watershed (in South Central Kansas, USA) is included in Section 7 to illustrate some the techniques previously described. The chapter ends with discussion of the importance of this work and points to future research needs (Section 8). The material in this chapter can be used as a basis to develop guidelines on the use of ground-based LiDAR technology to characterize topographic conditions associated with the formation and evolution of agricultural gullies at field scale. This is especially important in research sites where the same geographical location needs to be surveyed multiple times over lengthy periods of time as conditions change due to precipitation and runoff events, field management, and implementation of conservation practices.

Fig. 1. Ground-based LiDAR utilized for gully formation and evolution studies in agricultural croplands.

2. Light detection and ranging technology

LiDAR technology, measures the laser pulse travel time from the transmitter to the target and back to the receiver allowing the distance to be calculated (Wehr and Lohr, 1999). In this process, a very accurate timing system is needed to guarantee an accurate resolution since laser pulses are sent at the rate of thousand times per second. Additionally, the transmitter and the receiver must be located at the same physical location, which is a single-ended system because the phase the incoming signal is compared to the phase of an internal clock at the same location (Measures R. M., 1984). This technology is often classified into air-borne and ground-based systems.

In airborne LiDAR systems, the aircraft position is recorded using accurate positional systems such as inertial navigation unit, GPS receivers located in the aircraft, and GPS base locations. With these combined equipment, it is possible to determine three-dimensional geo-referenced coordinates for each pulse and then correct for the aircraft positioning in terms of roll, pitch, and heading, thus improving the accuracy of the system.

Ground-based LiDAR systems, are simplified because the sender/receiver's location is known. They operate in a similar fashion to the reflectorless total stations. For each pulse, these systems can measure: (1) x, y, z coordinate values with respect to the position of the laser sensor; (2) the intensity of the returned signal as reflectance; and (3) RGB values from an integrated digital camera within the instrument.

These systems are capable of collecting information with a wide range of ground sampling densities as a result of operator controlled factors such as scan angle (area covered by individual scans), average point density, and degree of overlap between scans. Higher point density can be achieved by higher sensor resolution, smaller vertical field of view angles, and multiple scans of the same ground location. The instrument resolution is often controlled by an imaginary plane located at the middle range of the vertical field of view and is orthogonal to the sensor's normal sight. The vertical field of view angles also influence the point density as point sampling becomes sparser as vertical scan angles tends to the horizon. Multiple scans can be used to collect data over the same geographical location resulting in increased sampling density and overcomes problems such as shadowing and limited coverage due to vegetation.

3. Techniques to pre-process topographic information

Pre-processing is an important step introduced in a topographic data analysis scheme to minimize the influence of unwanted points and to maximize the influence of the desired ones, thus simplifying the computational work needed in the later stages of the analysis process in order to reach the investigation goals. Pre-processing is composed of operations such as filtering, enhancing, transformation, or combination of these processes. The pre-processing techniques described herein include: cross-validation to assess geo-referencing accuracy between multiple scans of the same geographical location, investigation of point sampling patterns through quadrant method, evaluating the internal distribution of points using the average distance to the nearest neighbour, unwanted point removal, and surface reconstruction through surface smoothing.

3.1 Cross-validation

During field surveys, such as the one conducted a field in the Cheney Lake watershed in Kansas, the same geographical location can be covered by different scans. This technique is often used to increase sampling density and to avoid problems such as shadowing and limited coverage due to vegetation. The overlap in LiDAR point sampling can also be used to evaluate survey final geo-referencing accuracy. Since each individual scan is collected using local coordinates, geo-referencing is the procedure to convert from local coordinate system into global coordinate system.

Given that the overlapping scans collected have sufficient point density, neighbouring points can be identified, and elevation differences between these points can be determined. Similar approach is used to evaluate the accuracy of airborne LiDAR sensors by comparing elevation values of nearby points collected by different flight paths in overlapping areas. For each combination of two scans, points were analysed to form a list of neighbouring points. For illustration, consider two scans s^i and s^j where: $i \neq j$. A pair of points P^i_{xyz} and P^j_{xyz} are recorded into the list when two conditions are met: a) the points are from different scans $i \neq j$ and b) the horizontal distance between the two laser points is smaller than the distance threshold $\left(d < d_{threshold}\right)$.

For example, in Figure 2, scans 4, and 12, 13, 14, and 15 overlap. Scans 12, 13, 14, and 15 were designed to increase point sampling at the main gully channel. Each of these scans is geo-referenced using a set of four pre-registered targets (the global coordinates of the targets are known prior to the geo-referencing procedure).

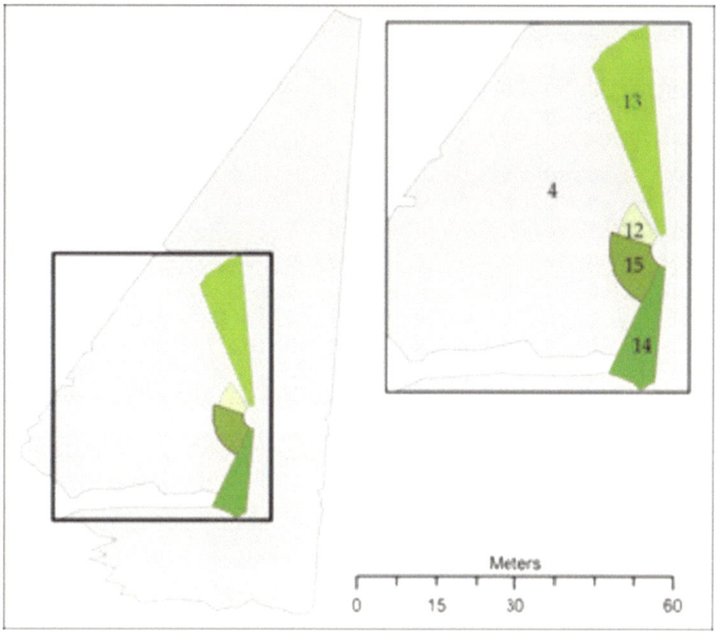

Fig. 2. Overlapping scans exemplifying the cross-validation procedure.

Using the cross-validation method for each pair of overlapping scans produces the tabular information presented in Figure 3. These values represent the upper quartile of the elevation difference between pairs of near points from different scans, multiplied by 1000. The values highlighted in Figure 3 exemplify geo-referencing problems between scans 4 and 12, 13, 14, and 15. Upon verification of the coordinates of the four targets used to geo-reference scans 12, 13, 14, and 15; the target with the largest RMS error was removed. A new cross-validation assessment between scan 4 and 12, 13, 14, and 15 was performed yielding new values of 17, 13, 16, and 17 respectively.

3.2 Quadrant method

A simplified and efficient way of exploring point pattern (spatial distribution and density) is to partition the location surveyed into sub-regions of equal area, or quadrants and then sum the number of laser points in each of the quadrants (Bailey and Gatrell, 1995). The division of the number of points by the area of the quadrant grid generates a measure of sampling intensity or laser points sampling density. The density map represents a two-dimensional frequency distribution (Bailey and Gatrell, 1995). This can be used to identify locations with high and low density, providing guidance to evaluate over and under-sampled areas. The density information can be correlated with topographic and land cover information for identification of areas with no laser point sampling due to shadowing and the influence of vegetation cover in the density information.

Scan	1	2	3	4	5	6	7	8	9	10	11	12	13	14	15	16	17	18	19	20	21	22
1	0																					
2	14	0																				
3	12	0	0																			
4	0	0	2	0																		
5	0	0	0	9	0																	
6	0	0	0	9	2	0																
7	0	0	0	9	5	0	0															
8	0	0	0	12	0	0	0	0														
9	0	0	0	12	0	0	4	0	0													
10	0	0	0	13	0	0	0	6	0	0												
11	0	0	0	16	0	0	0	6	0	3	0											
12	0	0	0	97	117	0	73	0	59	0	0	0										
13	0	0	0	69	0	0	0	60	0	40	0	0	0									
14	0	0	0	149	0	0	0	148	0	147	130	0	0	0								
15	0	0	0	88	0	0	0	92	0	84	0	0	3	30	0							
16	0	0	0	20	0	0	0	0	0	0	13	0	0	0	0	0						
17	0	0	0	21	0	0	0	18	0	17	17	0	0	161	0	3	0					
18	0	0	0	32	0	0	0	0	0	0	23	0	0	0	0	13	7	0				
19	0	0	0	0	0	0	0	0	0	0	0	0	0	0	0	0	0	0	0			
20	0	0	0	0	0	0	0	0	0	0	0	0	0	0	0	0	0	0	0	0		
21	0	0	0	26	0	0	0	0	0	0	28	0	0	0	0	18	0	0	0	0	0	
22	0	0	0	39	0	0	0	0	0	0	35	0	0	0	0	18	0	2	0	0	3	0

Fig. 3. Cross-validation values of the upper quartile obtained from the histogram of all the elevation difference between pairs of near points acquired from different scans (values in metres multiplied by 1000).

The quadrant method is influenced by the quadrant grid size. Larger quadrant grid sizes mask small density variations while small quadrant grid sizes create a larger range of density values with many quadrant grids containing no points. Using two different values of virtual quadrant grid, the gully represented by lighter colour in Figure 4A, is depicted by laser point sampling density associated with a virtual grid of 1.0 meter (Figure 4B) of 0.25 meter (Figure 4C). Both density maps reveal under-sampled locations.

3.3 Average distance to nearest neighbor

The use of the quadrant method provides the means to evaluate the spatial variation of the laser point sampling density. Density is often referred to as first order property of point pattern (Bailey and Gatrell, 1995). The use of the average distance to the nearest neighbours within each quadrant grid depicts an indication of the internal distribution of the laser points (Figure 5A). Based on this information binary maps can be generated of particular average distances (Figure 5B) to investigate sampling coverage. Although the average distance to the nearest neighbour provides an indication of how clustered or distributed laser points are in each quadrant grid, care should be used on analysis of large quadrant grids. In those analyses, two quadrants can produce the same average distance to nearest neighbour values, but one quadrant can contain a larger quantity of points uniformly distributed while the other quadrant contains fewer clustered points. To minimize this problem, a simple alternative is to combine the density map with the average distance to nearest neighbour map (Figure 5C).

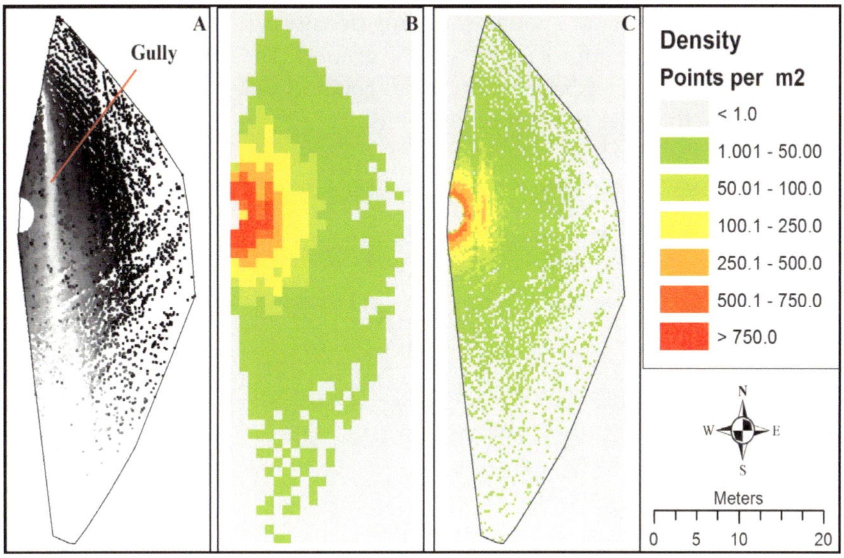

Fig. 4. Point pattern analysis using the quadrant method where the laser points (A) were sub-divided into quadrant grids of 1.0 and 0.25 meters (B and C respectively). The density map provides overall visualization of spatial variation of sampling density.

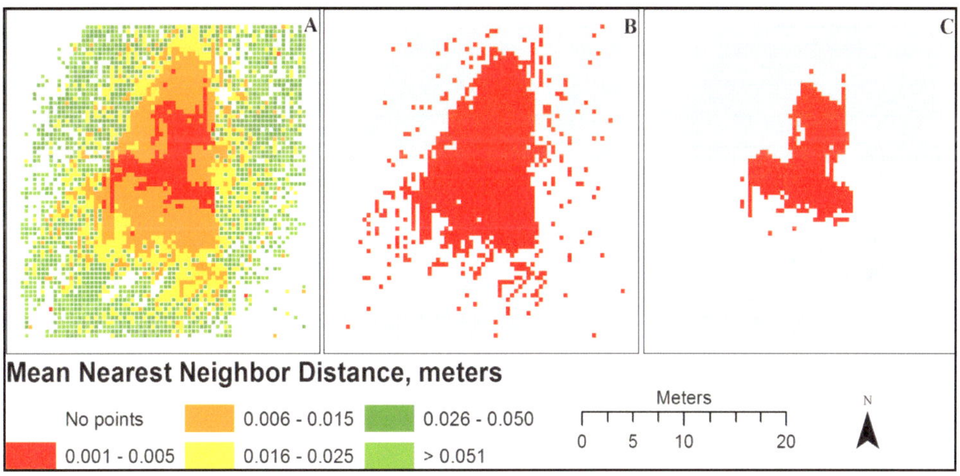

Fig. 5. (A) Average distance to the nearest neighbour of laser points within each quadrant grid. (B) Red symbology indicates the quadrant grids with average distance to nearest neighbour smaller than 1.5 cm. (C) Quadrant grids with average distance to nearest neighbour smaller than 1.5 cm containing greater than 500 laser points symbolized in red.

3.4 Thinning

Similarly to the quadrant method, thinning generates a two-dimensional virtual quadrant grid covering the entire extent of the point cloud, identifies the points within each quadrant, and keeps only the point with the lowest elevation in each quadrant. This technique, originally devised to filter trees out in order to obtain bare Earth data sets, is based on the assumption that in surveys with enough sampling density, vegetated areas will have points that can penetrate the tree canopies and reach the ground. The same concept can be adopted when working with terrestrial laser scanners. Vegetation foliage and residue stubs can be partially removed by thinning procedure (Figure 6).

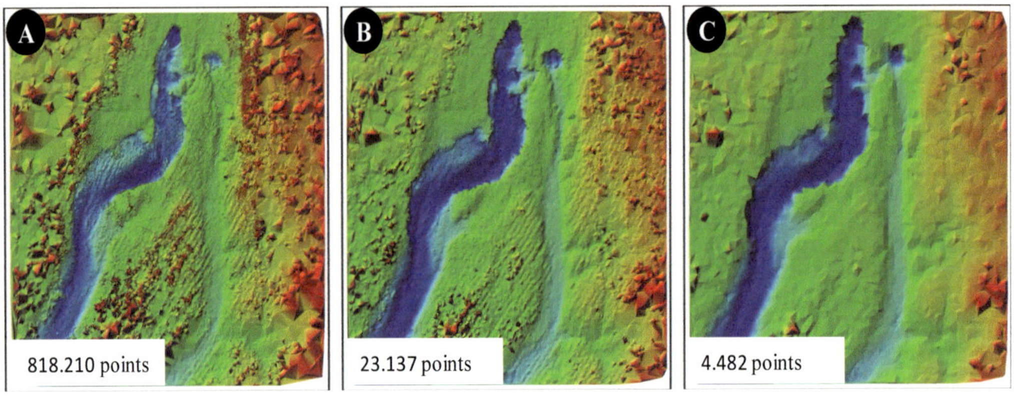

Fig. 6. Triangular Irregular Network (TIN) representation of point clouds before and after thinning procedure of a 18x18 meter section. The original surveyed section (A) is filtered through thinning using virtual grid cell sizes of 10 (B) and 25 cm (C), respectively.

3.5 Outlier removal

Gully investigations in cropland require artificial removal of outlier laser points to generate the true bare Earth representation. These points can constitute standing crop residue, vegetation canopy, or even individual incorrect registered scan lines. Two different approaches were considered here: square average distance (SAD) to the K-nearest neighbours (KNN) and distance to the fitted surface (DFS) to the KNN (Bäsken, 2010, Delage, 2010, Cazals and Pouget 2008). Unlike the previously described quadrant methods that compute two-dimensional distances, these two approaches classify a point as an outlier or not based on three-dimensional Euclidean distance.

In the SAD algorithm, for each point the K nearest neighbours are selected and the square average distance between these points is computed. If this value is larger than a user provided threshold the point is assigned to be removed. A subset of a gully channel with different outliers, standing crop residue (two sharp peaks) and small lines of elevated points were evaluated (Figure 7A). An initial attempt to remove these outliers used a square average distance threshold of 5 cm for the 50 nearest neighbours resulting in the removal of one point (Figure 7B). A second attempt using 10 nearest neighbours and a square average distance threshold of 1 cm removed an additional 14 laser points.

The DFS algorithm, identifies the K nearest neighbours for each laser point, fits a polynomial surface to these points, and computes the three-dimensional distance of the point being evaluated to the fitted surface. If the distance is greater than the user-specified threshold the point is then marked for removal. The polynomial order is also a parameter defined by the user. The same information shown in Figure 7A is filtered using a first order polynomial surface, planar surface, removing a total of 4312 points (Figure 7D). Visual comparison to the original dataset (Figure 7A) indicates the removal of most outlier points.

3.6 Smoothing

The presence of vegetation and/or standing crop residue can yield point clouds, and subsequently digital terrain representations, with large elevation variations (quantified by large standard deviations). In such noisy datasets, techniques can be devised for surface reconstruction (herein referred to as smoothing). The objective is to attenuate large local relief variations while maintaining the global relief of the terrain. Unlike the filtering procedures, no points are removed but rather three-dimensionally moved to a fitted surface to the KNNs of each point (Delage, 2010, Cazals and Pouget 2008). The output point cloud contains the same number of points than the original one, however, the coordinate values of each point is slightly modified. The parameters required are the k number of neighbours and the degree of the polynomial to generate the fitted surface. The lower the degree of the polynomial the smoother the reconstructed surface will be. The original surface (Figure 7A) can be reconstructed using a third order polynomial (Figure 7E) while a surface using a first order polynomial (plane) can also be constructed (Figure 7F). Visual inspection indicates a more realistic representation using the third order polynomial over than the first order.

3.7 Summary

Although the selection of the most appropriate algorithm along with their parameters varies according to the terrain (topography and land cover) and scale (flume, plot, or field), best results can be obtained by combining these procedures (filtering and smoothing). A list of the tools described in this section with the necessary inputs, the generated output, and the main purpose are listed in Table 1.

4. Morphological analysis

In scientific settings, different gullies are surveyed multiple times with the objective to capture a channel's dynamic changes. The objective is to quickly assess channel morphological changes such as headcut migration, widening, deepening, and deposition, as well as estimation of the volume of soil eroded/deposited. As the number of surveys increases and consequently the volume of data, an automated and standard procedure to extract key parameters is necessary.

4.1 Automated cross-section generation

Channel geometry can be quantified through a series of cross-sections. Although cross-section methods exist in commercial off-the-shelf Geographic Information Systems (GIS) software packages, the generation of a series of closely spaced cross-sections can become a time consuming process due to the necessary file format transformations (conversion of point cloud into either raster or TIN format) and the lack of automation (user has to define the individual location of each cross-section).

The procedure developed requires the user to define a centreline approximating the gully thalweg in the ESRI Shapefile file format and input parameters describing how the cross-sections will be generated (Figure 8A). The cross-section geometric parameters are: length of the cross-section, distance between cross-sections (*b* in Figure 8A), distance between points within the cross-section (*c* in Figure 8A), and search radius (*a* in Figure 8A). The final elevation for each point of the cross-section is defined by the Inverse Weighted Distance (IDW) algorithm (Equation 1).

$$z_{0.} = \frac{\sum_{i=1}^{s}\left(z_i \frac{1}{d_i^k}\right)}{\sum_{i=1}^{s}\left(\frac{1}{d_i^k}\right)} \tag{1}$$

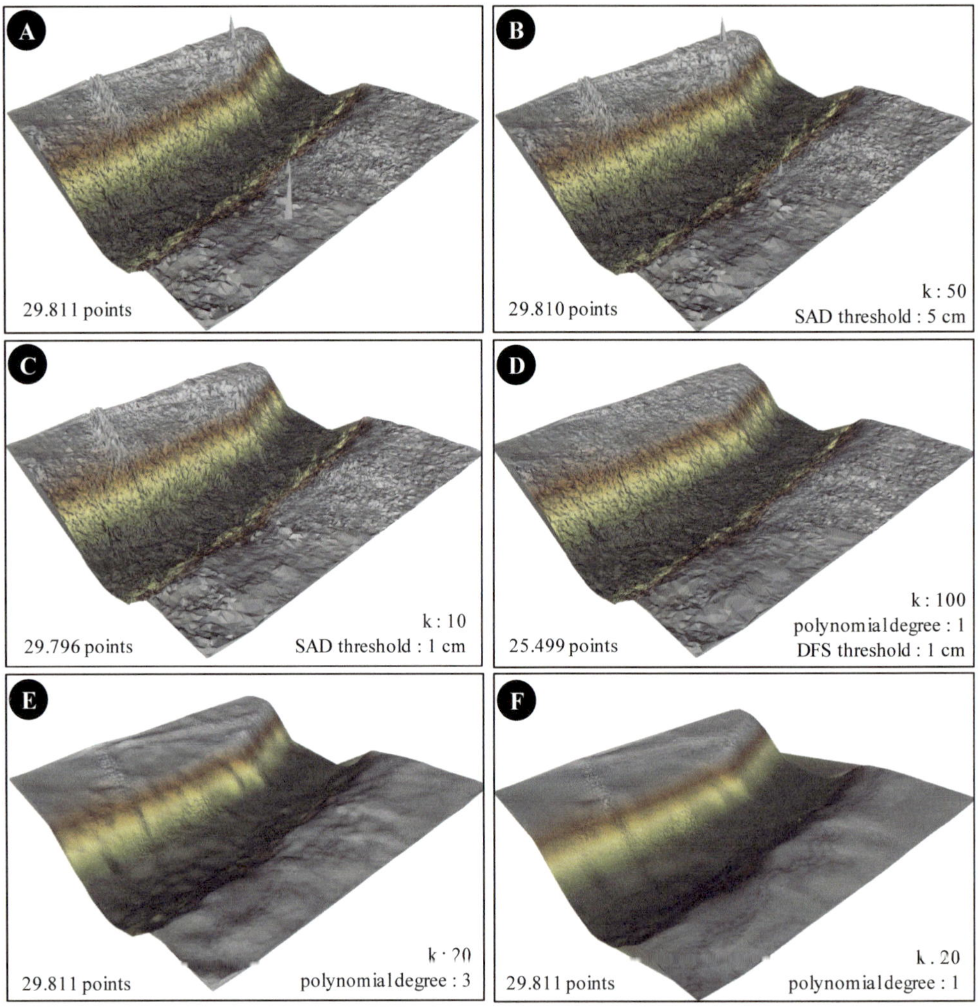

Fig. 7. Point cloud pre-processing algorithms: A – TIN representation of original point cloud; B – TIN representation of point cloud after outlier removal using a SAD of 5 cm and 50 KNN (k) points; C – TIN representation of point cloud after outlier removal using a SAD of 1 cm and 10 KNN points; D – TIN representation after outlier removal using distance to fitted surface (DFS) to KNN points (4312 points removed); E – TIN surfaces were reconstructed by three-dimensionally moving points to the surfaces fitted to the KNN points using a third order polynomial surface; and F – TIN surfaces were reconstructed by three-dimensionally moving points to the surfaces fitted to the KNN points using a first order polynomial surface.

The IDW algorithm is a local interpolation algorithm that assigns different weights to laser points based on their proximity to the unknown location (z_0). In Equation 1, the z_i is the elevation at the known point i, d_i is the distance between point i and the point being estimated (unkown location), and k is the power controling the degree of local influence (closer laser points are assigned higher weight) and is also an input parameter.

Module	Inputs	Outputs	Purpose
Cross-validation	• Two overlapping point clouds • Distance threshold	• List of 2D distances between selected pairs of points	Assessment of geo-referencing accuracy between overlapping scans
Quadrant method	• Point cloud • Virtual grid cell size	• Polygon feature class with laser point count for each polygon	Point sampling density investigation
Average distance to nearest neighbour	• Point cloud • Virtual grid cell size	• Polygon feature class with average distance to nearest neighbour of laser points within each polygon	Investigation of the spatial distribution of laser point within each virtual grid cell.
Thinning	• Point cloud • Virtual grid cell size	• Point cloud with reduced number of points	Filtering of vegetation and standing residue.
Square average distance to KNN	• Point cloud • Number of neighbours to consider • Degree of polynomial • Degree of Monge • Square average distance threshold	• Point cloud with selected laser points identified for removal	Filtering of unwanted laser points based on average distance to nearest neighbours.
Distance to fitted surface to KNN	• Point cloud • Number of neighbours to consider • Degree of polynomial • Degree of Monge • Distance threshold	• Point cloud with selected laser points identified for removal	Filtering of unwanted laser points based on 3D distance to fitted surface.
Smoothing	• Point cloud • Number of neighbours to consider • Degree of polynomial • Degree of Monge	• Modified point cloud where points were 3D moved.	Surface reconstruction.

Table 1. Summary of new and enhanced topographic information pre-processing tools developed for gully investigation in agricultural fields using ground-based LiDAR.

Based on using the same centerline and input parameters utilized in different surveys, gully channel geometrical changes between surveys can be quantified (Figure 9). The cross-sectional area difference between two cross-sections is computed by defining a reference elevation and computing the area of the closed polygon (red polygon in Figure 9). The change in cross-sectional area is determined by computing the difference between the individual cross-sectional areas . The thalweg is generated by simply connecting the point with the lowest elevation in each cross-section (Figure 10).

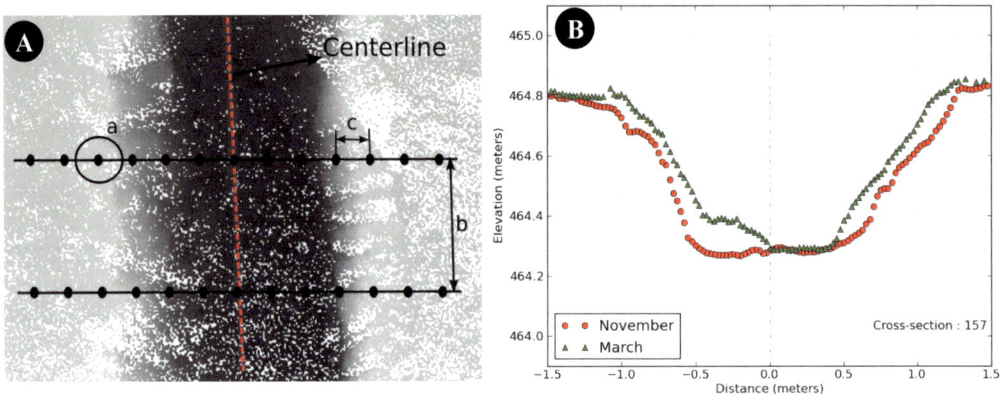

Fig. 8. (A) LiDAR point cloud with cross-sections generated. (B) Cross-section of a location at different periods of time.

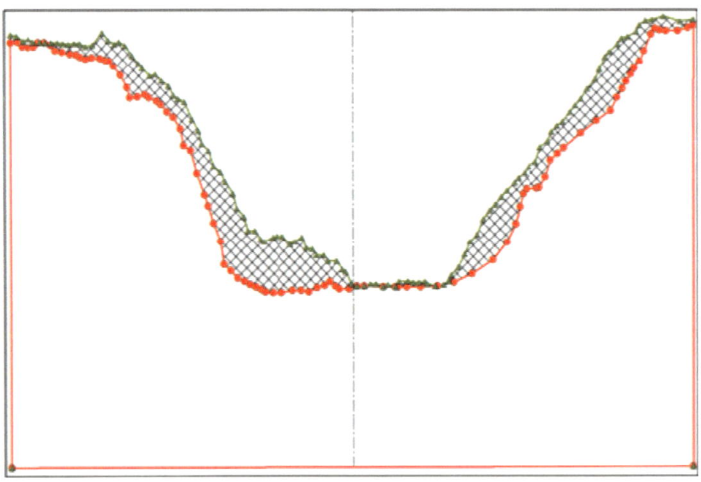

Fig. 9. Cross-sectional area generated with closed polygons by providing a base elevation.

4.2 Cross-section modeling

Quantification of gully channel geometrical properties is often performed by simplifying and fitting measured cross-sections through trapezoidal cross-section shapes. However, as the number of cross-sections grow as a result of gully channel extension, small cross-section interval, and high survey revisit time, identification of top-of-the-bank and channel toes manually can become tedious and time consuming. Alternatively, a simplified optimization algorithm can be used to fit a trapezoidal cross-section to the LiDAR derived cross-section in a semi-automated fashion.

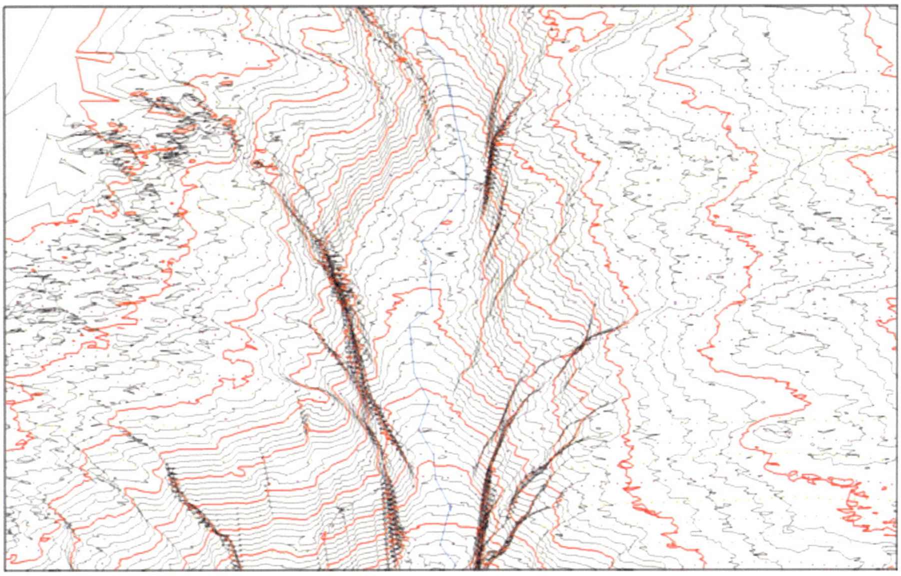

Fig. 10. Three-dimensional representation of the thalweg generated by connecting the cross-section point with the lowest elevation.

The algorithm iteratively fits two lines to each of the cross-section sides individually (left hand-side is shown in Figure 11). A series of scenarios are considered. Initially, the algorithm fits one line to the three points furthest from bank toe (blue points in Figure 11A) and another line to the remaining points (red points in Figure 11A). The root mean squared deviation (RMSD) of each of the fitted lines is computed and then the two combined (Equation 2). The process is repeated by increasing the number of points considered in the first line while reducing the points to the second line (Figure 11B), until only three points are left for the second line (Figure 11C). The optimum point is selected by identifying the minimum combined RMSD value trough minimization of Equation 2 (Figure 11D).

$$\text{Optimum Point} = \min imize\left\{\sqrt{\left(\sum_{i=1}^{i=j}\left[\left(Y_i - Y_i^a\right)^2\right] + \sum_{i-j+1}^{i=1}\left[\left(\left(Y_i - Y_i^b\right)^2\right)\right]\right)}\right\} \quad (2)$$

Where, Y_i^a is the elevation of the first fitted line *a* at point *i*, Y_i^o is the elevation of the second fitted line *b* at point *i*, *j* is the last point in the first set of points, and Y_i is the measured elevation at location *i*. A line is fitted to the gully bottom through least square error and the intercept of this line with the others generated with the optimization method (left and right banks) for a trapezoidal cross-section shape (Figure 12).

4.3 Summary

The main objective of the cross-section processes is to standardize and to expedite the cross-section generation and analysis. These procedures play an important role when conducting multi-temporal investigations devised to quantitatively assess gully changes over time (formation and evolution). The main components described herein are listed in Table 2, along with their required input, output, and the main purpose.

Module	Inputs	Outputs	Purpose
Cross-section generation	• Point cloud • Centreline (feature class) • Cross-section length • Distance between cross-sections • Distance between points in the cross-section • Search radius • IDW power • Minimum number of points to consider	• Point feature class with X, Y, and Z coordinates of each cross-section point plus cross-section number	Gully channel geometry investigations.
Gully thalweg generation	• Point feature class generated using the cross-section module	• Polyline feature class	Gully thalweg investigations.
Cross-section area computation	• Point feature class generated using the cross-section module • Datum elevation	• Table listing area for individual cross-section	Multi-temporal investigation of cross-sectional area change.
Cross-section modelling	• Point feature class generated using the cross-section module	• Table listing line equations and intersection	Systematic extraction of geometric parameters from cross-section.

Table 2. Summary of tools implemented for gully morphological analysis in agricultural fields using ground-based LiDAR.

5. Interpolation using local binning algorithms

Point clouds generated using Ground-based LiDAR systems at field-scale and even at flume-scale can yield large datasets (millions of points) with varying sampling densities and uneven point distributions posing a challenge to interpolation techniques based on global estimates such as a regularized spline with tension and ordinary Kriging algorithms. The algorithm proposed by Kim (2006) was adopted and expanded to generate grids of intensity, red, green, and blue values in addition to elevation values. This algorithm produces a tree data-structure (Quadtree) point cloud designed to recursively partition the two-dimensional space (X and Y coordinates) to expedite queries of laser points in the desired neighbourhood (Bäsken, 2010).

6. Implementation

The described algorithms were implemented in a combination of C++ and Python computer programming languages as a result of the need to have customizable tools that were not previously available in GIS software packages.

The point clouds were stored and manipulated using the American Society of Photogrammetry and Remote Sensing (ASPRS) LASer file format (LAS) through the libLAS library (http://liblas.org/). To overcome the large point cloud size (millions of laser points) the Computational Geometry Algorithms Library (CGAL) was selected to perform efficient nearest neighbour queries (Delage, 2010), compute distances, outlier removal (Allez et al., 2010), normal estimation, and fitting of three-dimensional polynomial surfaces (Cazals and Pouget, 2008). The cross-section modelling algorithm was developed using the Scientific Tools for Python (SciPy) due to the plotting capabilities offered. Additionally, the ESRI Shapefile file format was used when generating/manipulating cross-sections and thalweg files.

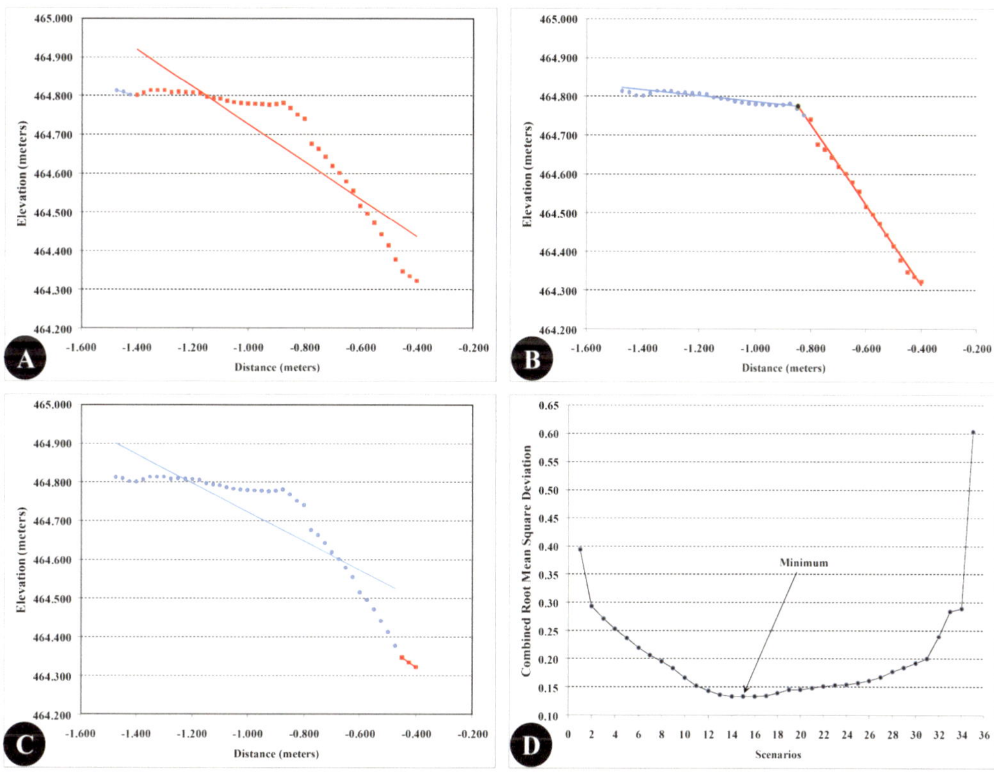

Fig. 11. Optimization procedures to fit a theoretical trapezoidal cross-section to the measured cross-section obtained with ground-based LiDAR points from Kansas field surveys. In the first scenario two lines are fitted, one to the first three points and the second to the remaining points (A). The procedure continues until the second line has only three points (C). The optimal fit is obtained by the minimum RMSD value (B) after all the scenarios are evaluated (D).

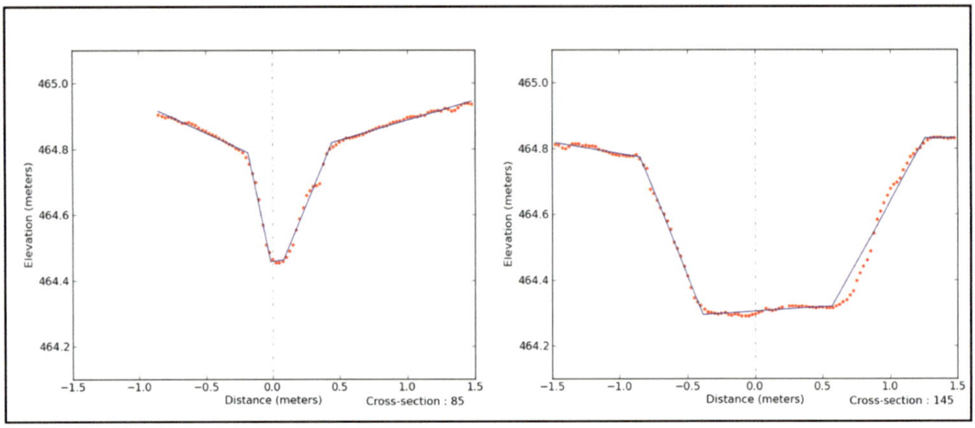

Fig. 12. Trapezoidal cross-sections fitted to measured data.

7. Study case

A gully located in South Central Kansas was selected for analysis. The gully is within the Cheney Lake Reservoir watershed and is approximately 96 metres long, 1.3 metres wide at the widest location, from 10 to 50 centimetres deep, and oriented North-South. The channel is free of vegetation and crop residues, while the surrounding field is covered by crop residues resulting from no-till management used in winter wheat followed by sorghum (milo) in the 2010 crop rotation. Historical cultivation practice records indicate that initially this gully did not disrupt farming operations; however, as no-tillage practices were adopted in 2005, the channel grew wider and deeper to the point that the farming equipment could not be used to travel across the gully and the ensuing cropping activity was performed around the main channel (Frees et al, 2010). The site was visited and surveyed in March and November of 2010. Measured precipitation records indicate two significant precipitation events occurred in June and July, 2010 (Figure 13).

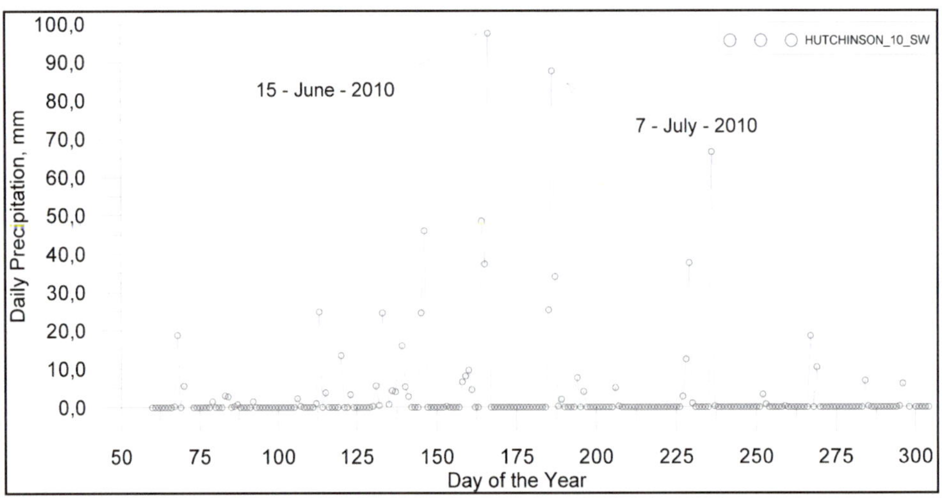

Fig. 13. Daily precipitation between March-2009 and November-2010 from a weather station located 4,5 Km from the field investigated.

The equipment used was a TOPCON GLS-1000 series that has a 4 mm single point accuracy distance (Table 3). The surveys are composed of 11 and 7 scans resulting in a total of 945,668 and 3,673,167 laser points for March and November, 2010, respectively. The surveys presented different laser point sampling densities (Figure 14). The survey conducted in March yielded higher variation in the sampling rate (density) in the gully channel with two clusters of high density. Conversely, the survey conducted in November generated a more uniform sampling rate specifically on the gully channel.

Parameter	Value	Unit	Condition
Single point accuracy distance	4.0	(σ) mm	1 to 150 meters
Single point accuracy vertical angle	6.0	seconds	1 to 150 meters
Single point accuracy horizontal angle	6.0	seconds	1 to 150 meters
Maximum scan rate	3000	Hz	
Scan density spot size	6.0	mm	1 to 40 meters
Scan maximum sample density	1.0	mm	up to 100 meters
Laser wavelength	1535	Nm	
Laser pulse duration	3.6	nano second	
Laser maximum pulse frequency	3400	Hz	

Table 3. General specifications of the TOPOCON GLS-1000 laser scanner

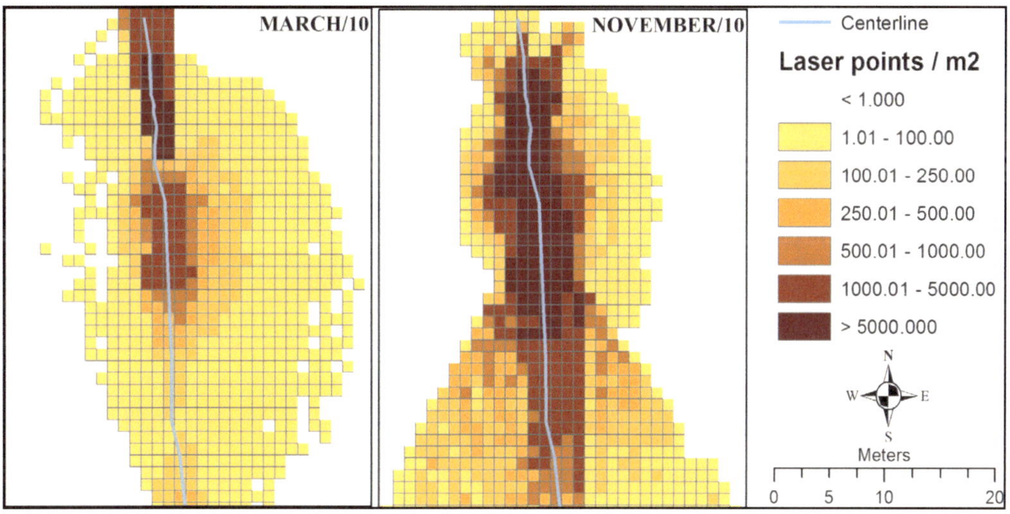

Fig. 14. Comparison of laser point sampling density between March and November, 2010 surveys.

Both surveys were subset to the same extent (gully only) to remove unwanted laser points regarding agricultural furrows and standing crop residues. Additional outliers were removed using the square average distance to KNN algorithm with threshold of 5 cm and k of 50 followed by a thinning algorithm using a virtual grid of 1.5 cm.

Using the local binning algorithm with a raster grid size of 2.5 cm, search radius of 5.0 cm and IDW power 2, two raster grids were generated (Figure 15). Areas were identified with elevation changes signifying different physical processes based on computing the difference between the two grids. The active headcut migration path is represented by the darker blue colour indicating the occurrence of incision processes between cross-section 50 and 90 (Figure 15). Further downstream, between cross-sections 99 and 145 (Figure 15), the predominant physical process is widening of the gully channel as result of encountering a less erodible layer.

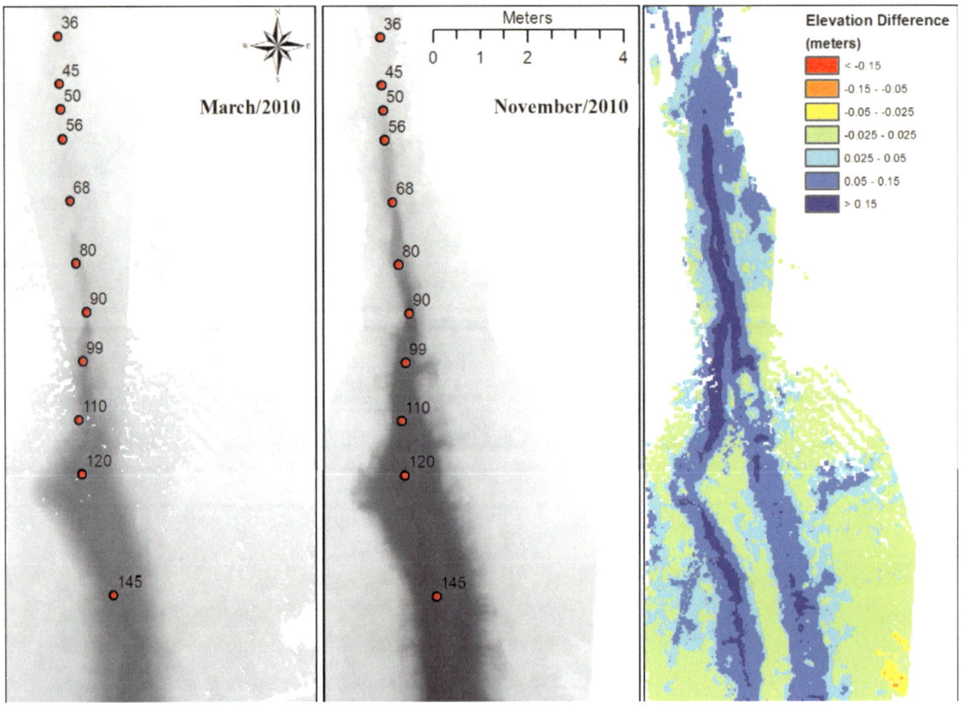

Fig. 15. Raster grid generated using local binning interpolation algorithm. Differencing of these grids illustrate elevation changes that occurred during the time period investigated. The red dots mark selected cross-sections.

Using the same user-provided centreline for both surveys a set of cross-sections were automatically generated and compared. Cross-sections at different locations of the gully were computed (Figure 16) based on various parameters (Table 4). A total of 448 cross-sections were generated for each survey varying in the extent and number of points as result of the non-uniform distribution of the laser points.

Parameter	Value
Cross-section size	3.0 meters
Distance between points in the cross-section	2.5 centimeters
Distance between cross-sections	10.0 centimeters
Power used in IDW	2.0
Search radius	5.0 centimeters

Table 4. Parameters considered in the generation of cross-sections for multi-temporal morphological analysis of gully evolution in croplands.

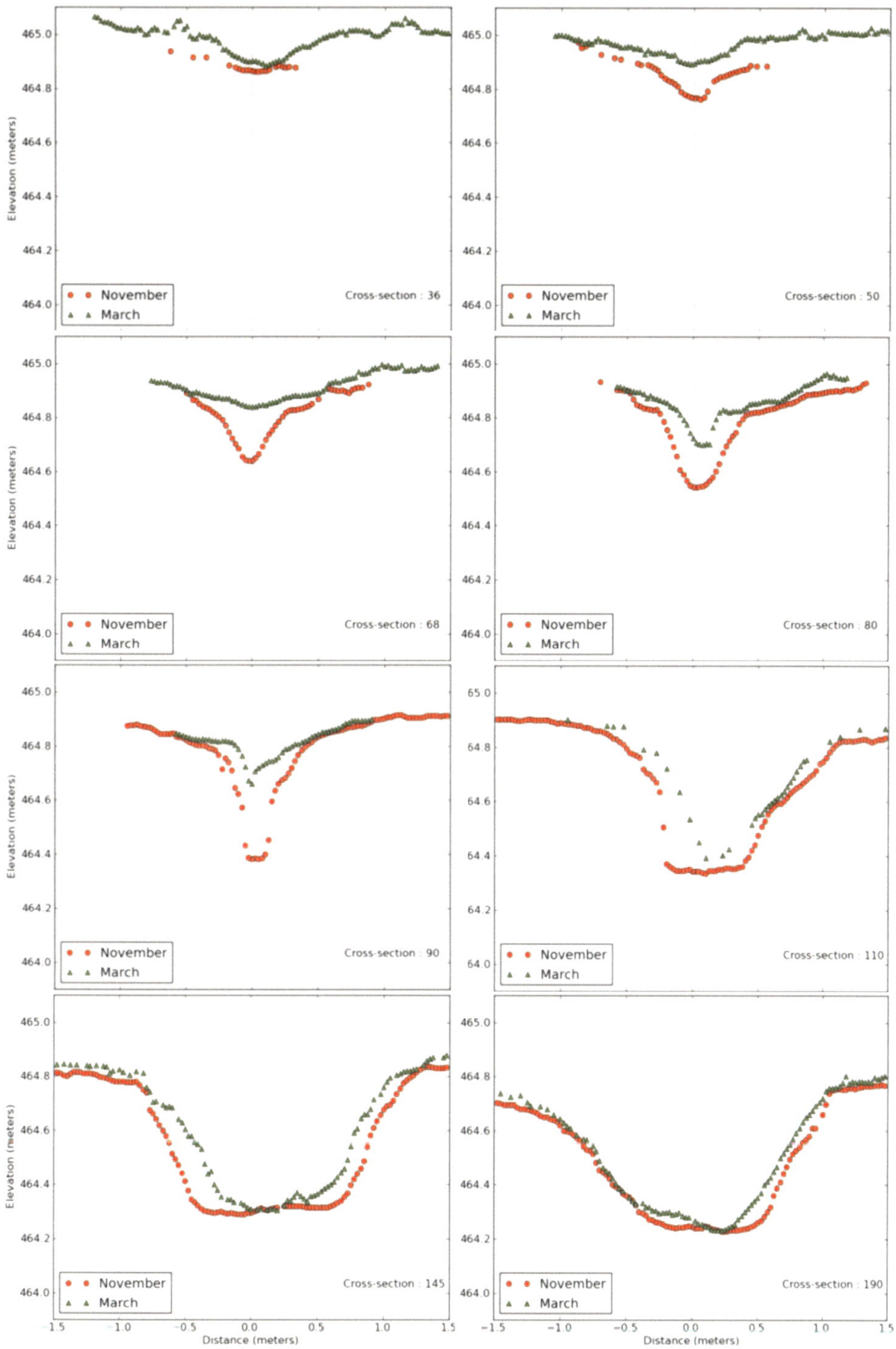

Fig. 16. Selected cross-sections measured in March and November of 2010.

The profiles, one for each of the surveys, were generated by deriving elevation information from the intersection of the user-defined centreline and each measured cross-section (Figure 17). Comparison between the two profiles indicates the rate of headcut migration (between distances 5 and 10) as well as the amount of incision in that section (Figure 15). Beyond 10 meters, the November profile (blue line in Figure 17) points to a situation of stable slope or a less erodible layer. The deposited material in March was likely removed as result of the large storm events.

Differences in cross-sectional area are plotted in Figure 18. Visual comparisons between the cross-section difference plot and the series of cross-sections identify erosional patterns. Between cross-section 70 and 90 the difference in cross-section area is attributed mainly by headcut migration and incision processes. From station 90 to 100 a combination of widening of the channel and incision are responsible for the differences in cross-section area. Beyond station 110 the change in cross-sectional area is mainly due to widening of the channel, 110-130 side walls only, 130-175 side walls and bed, and 175-250 widening of the channel bed.

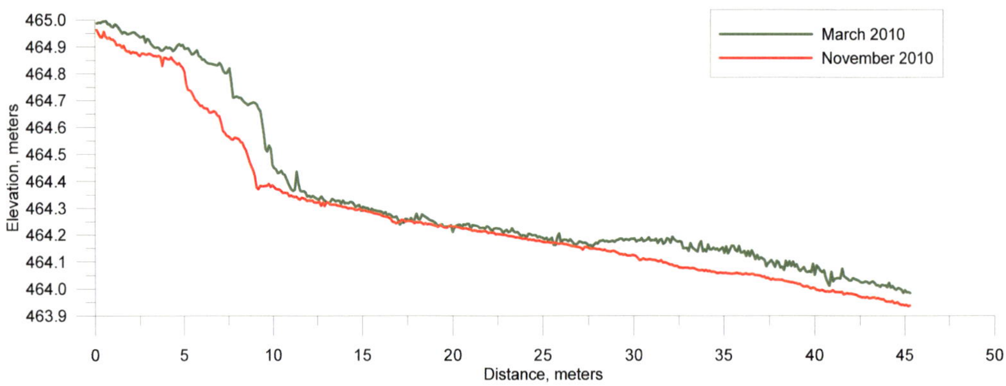

Fig. 17. User-defined centreline with Z elevation values obtained from the point of intersection with generated cross-sections for March (green) and November (red) of 2010.

8. Summary and closing remarks

The successful monitoring of gully evolution (ephemeral and classical) at individual gully scales in agricultural fields depends on accurate and detailed topographic information. The utilization of new technologies, such as ground-based LiDAR, provides new venues for scientific investigation of physical processes responsible for gully formation and evolution. Ground-based LiDAR can generate point clouds containing millions of points with point intervals varying from 1 mm to 200 mm depending on operator-defined parameters. These systems generate non-uniform spatial distributions of laser points (varying sampling density) because of the sensor's proximity to the target (equipment setup on tripods). The non-uniform distribution of points is influenced by relief, standing obstructions such as vegetation and crop residue, and vertical incidence angle. The format and massive amount of measured data represents a challenge for existing watershed modelling software applications that lack the necessary tools to pre-process the data and require the point clouds to be downgraded or interpolated in order to be used (Luccio, 2011). Watershed management planning will need these tools to implement economical and environmentally effective conservation practices.

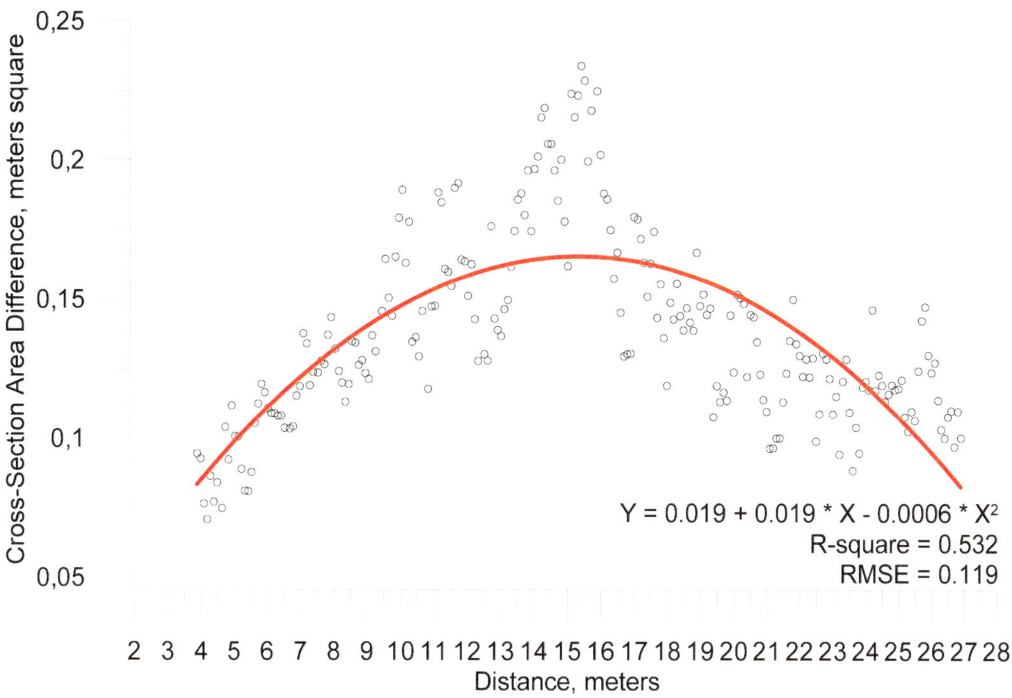

Fig. 18. Percentage change in cross-sectional area in square meters between March and November of 2010. Red curve represents the fitted second order polynomial function.

This chapter described a set of tools, devised primarily for gully investigation, to pre-process and extract morphological information from gully channels at different scales. These tools work with the point clouds in industry standard file format (LAS files) to minimize the need of file format conversions as well as to facilitate sharing between different systems. The tools also utilize advanced data structures and fitting algorithms to efficiently perform queries, compute distances between nearby laser points, and fit surfaces to KNNs, facilitating the removal of outlier points and the smoothing of the overall surface. Additionally, the extraction of morphological information in a semi-automated fashion was implemented through the analysis of cross-sections derived straight from the point clouds. This is an important feature providing a standard procedure to analyse multi-temporal data sets.

Future work includes the extension of the described system to include enhanced features such as classification of laser points based on elevation and intensity values, including combinations of each. In addition, correction of intensity values based on range and incidence angle, edge detection, and extraction of gully-specific geometric parameters are needed.

9. Acknowledgments

Acknowledgements are due to Lisa French and Howard Miller, at Cheney Lake Watershed, Inc., for contacting and coordinating with local land owners and for providing vital logistic support. Thanks are also due to Don Seale and Antonia Smith for their indispensible assistance during data collection. Finally, the authors would like to acknowledge the technical suggestions and discussions offered by Dr. Langendoen.

10. References

Bailey T. C. & Gatrell, A. C. (1995). *Interactive Spatial Data Analysis*, Prentice Hall, England.

Bäsken, Matthias. (2010). 2D Range and Neighbor Search. In CGAL User and Reference Manual. CGAL Editorial Board, 3.7 edition. Accessed on: February 2011. Available from: <http://www.cgal.org/Manual/latest/doc_html/cgal_manual/packages.html#Pkg:PointSet2>

Bingner, R. L., Wells, R. R., Momm, H. G., Theurer, F. D., & Frees, L. D. (2010). Development and Application of Gully Erosion Components within the USDA AnnAGNPS Watershed Model for Precision Conservation, *Proceedings of 10th International Conference on Precision Agriculture*, July 18-21, Denver, Colorado, USA.

Cerdan, O., Souchère, V., Lecomte, V., Couturier, A. & Le Bissonnais, Y. (2002). Incorporating soil surface crusting processes in an expert-based runoff model: Sealing and Transfer by Runoff and Erosion related to Agricultural Management *Catena*, Elsevier, Vol. 46, pp. 189-205.

Delage, C. (2010). Spatial Sorting. In CGAL User and Reference Manual. CGAL Editorial Board, 3.7 edition, 2010. Date of access: February 2011, Available from: <http://www.cgal.org/Manual/latest/doc_html/cgal_manual/packages.html#Pkg:SpatialSorting>

Cazals, F. & Pouget, M. (2008). Algorithm 889: Jet_fitting_3:---A Generic C++ Package for Estimating the Differential Properties on Sampled Surfaces via Polynomial Fitting. *ACM Transactions on Mathematical Software*, Vol. 35 No. 3, pp. 1-20.

Frees, L. D., Theurer, F. D., Bingner, R. L., & French, L. (2010). Cheney Lake CEAP Project: Validation/Calibration, Nutrients. *Proceedings of 9th Federal Interagency Sedimentation Conference*, Las Vegas, NV, June 27 - July 1.

Gordon, L. M., Bennett, S. J., Alonso, C. V., & Bingner, R. L. (2008) Modelling long-term soil losses on agricultural fields due to ephemeral gully erosion. *Journal of Soil and Water Conservation* Vol. 63, No. 4, pp. 173-181.

Hsiao, K.; Liu, J.; Yu, M. & Tseng, Y. (2004). Change detection of landslide terrains using ground-based lidar data. *Proceedings of XXth ISPRS Congress*, Istanbul, Turkey, Commission VII, WG VII/5.

James, L. A.; Watson, D. G., & Hansen, W. F. (2007). Using LiDAR data to map gullies and headwater streams under forest canopy: South Carolina, USA. *Catena*, Elsevier Science, Vol. 71, pp. 132-144.

Kim, H., Arrowsmith, J. R., Crosby, C. J., Jaeger-Frank, E., Nandigam, V., Memon, A., Conner, J., Badden, S.B., Baru, C. (2006). An Efficient Implementation of a Local Binning Algorithm for Digital Elevation Model Generation of LiDAR/ALSM Dataset, Eos Trans. AGU, 87(52), Fall Meeting, San Francisco, CA, USA.

Luccio, M. (2011). LiDAR Update - Terrestrial Explodes, Software Still Lags, Imaging Notes, Winter, Vol. 26, No. 1, pp. 24-31.

Measures, R. M. (1984). *Laser Remote Sensing Fundamentals and Applications*. John Wiley & Sons, Inc.

Park, A. (2008) Evaluation of Terrestrial LIDAR for Monitoring Geomorphic Change at Archaeological Sites in Grand Canyon National, Technical Report of the United States Geological Survey. USGS, no. 1384.

Parker, C., Thorne, C., Bingner, R., Wells, R., & Wilcox, D. (2007). Automated Mapping of Potential for Ephemeral Gully Formation in Agricultural Watersheds Laboratory Publication, National Sedimentation Laboratory, No. 56.

Perroy, R., Bookhagen, B., Asner, G., & Chadwick, O. (2010). Comparison of gully erosion estimates using airborne and ground-based LiDAR on Santa Cruz Island, California *Geomorphology*, Vol. 118, pp. 288-300.

Alliez, P., Laurent, S., & Salman, N. (2010). Point Set Processing. In CGAL User and Reference Manual. CGAL Editorial Board, 3.7 edition. Accessed on November: 2010. Available from: <http://www.cgal.org/Manual/latest/doc_html/cgal_manual/packages.html#Pkg:PointSetProcessing>

Poesen, J. W., Nachtergaele, J., Verstraeten, G., & Valentin, C. (2003). Gully erosion and environmental change: importance and research needs. *Catena* Vol. 50, pp. 91-133.

Poesen, J. W., Vandaele, K., & Wesemael, V. (1996) Contribution of gully erosion to sediment production on cultivated lands and rangelands. *Erosion and Sediment Yield: Global and Regional Perspectives*, IAHS Publ. Vol. 236, pp. 251-266.

Roering, J. J., Stimely, L. L., Mackey, B. H., & Schmidt, D. A. (2009). Using DInSAR, airborne LiDAR, and archival air photos to quantify landsliding and sediment transport. *Geophysical Research Letters*, Amer Geophysical Union, Vol. 36, No. 19, pp. L19402.

Schmid, T., Schack-Kirchner, H., & Hildebrand, E. (2004). A case study of terrestrial laser-scanning in erosion research: calculation of roughness indices and volume balance at a logged forest site. *International Archives of Photogrammetry, Remote Sensing and Spatial Information Sciences*, Vol. 36, No. 8, pp. 114-118.

Smith, L. (1993) Investigation of ephemeral gullies in loessial soils in Mississippi. (GL-93-11), U.S. Army Corps of Engineers, Storming Media, Vicksburg, MS.

Soil Science Society of America. (2001). Glossary of Soil Science Terms. Soil Science of Americal, Madison, WI. Date of access: November 2010, Available from: <http:://www.soils.org/sssagloss/>.

Souchere, V., Cerdan, O., Ludwig, B., Le Bissonnais, Y., Couturier, A., & Papy, F. (2003). Modelling ephemeral gully erosion in small cultivated catchments. *Catena*, Elsevier, Vol. 50, pp. 489-505.

Wehr, A. & Lohr, U. (1999). Airborne laser scanning - An introduction and overview, ISPRS *Journal of Photogrammetry and Remote Sensing*, Vol. 54 No. 2-3, pp. 68-82.

Wells, R., Bennett, S., & Alonso, C. (2010). Effect of Upstream Sediment Inflow on the Morphodynamics of Headcuts. *2nd Joint Federal Interagency Conference*, Las Vegas, Nevada, USA, June 27 - July 1.

Woodward, D. (1999). Method to predict cropland ephemeral gully erosion. *Catena*, Elsevier, Vol. 37, pp. 393-399.

6

Modeling Channel Response to Instream Gravel Mining

Dong Chen
Desert Research Institute, Las Vegas
USA

1. Introduction

Sand and gravel on riverbeds have been considered as an attractive (high quality and low cost) source of building material for centuries (Kondolf, 1994; Gill, 1994; Rinaldi et al., 2005). Detrimental effects of in-stream mining have been documented in literature including riverbed degradation or incision (Rinaldi et al., 2005), stream-bank instability (Chang, 1987; Kondolf, 1997), destruction of bridges and channelization structures (Kondolf, 1997; Rovira et al., 2005), etc. On the other hand, removal of in-stream sediment can be beneficial, for instance, it can serve as a maintaining way of the navigation water depths (Fredsøe, 1978). To minimize the detrimental effects and maximize the beneficial impacts, channel response due to gravel mining or dredging has been studied by experiments (Fredsøe, 1978; Kornis and Laczay, 1988; Lee et al., 1993; Lee and Chen, 1996; Neyshabouri et al., 2002), field observations (Kondolf, 1997; James, 2004; Neyshabouri et al., 2002; Rinaldi et al., 2005), and simplified analytical models (Cotton and Ottozawa-Chatupron, 1990). With the rapid development of computational fluid dynamics (CFD) since late 1980s, sophisticated numerical modeling has become a practicable tool for a quantitative understanding of the channel response due to sand and gravel mining (Van Rijn, 1986; Chang, 1987; Yue and Anderson, 1990; Gill, 1994; Cao and Pender, 2004; Chen and Liu, 2009). However, the inherent complexity of sediment transport and channel changes makes the firm, specific prediction of mining effects on rivers impossible at present. For instance, sediment transport around the mining pit area behaviors a distinct non-equilibrium state due to the sharp inflection of streamlines around the upstream and downstream ends of mining pits. However, existing numerical models either choose equilibrium sediment transport formulas or bear much uncertainty due to the various parameters selection which must be introduced to close the non-equilibrium sediment transport formulas. Therefore, specific laws and regulations regarding the safe in-stream mining have not been provided for users and officials despite extensive investigation made in the past (Kondolf, 1997; Neyshabouri et al., 2002).

This paper aims to simulate the channel response to instream gravel mining. First, the feasibility of the two-dimensional depth-averaged model (CCHE2D) in modeling mining-induced bed change was examined by comparing the calculated results to the data measured in two sets of published laboratory experiments. Thereafter, the two dimensional model (CCHE2D), with appropriate non-equilibrium adaptation parameters, was applied to examine the impact of deep sand and gravel mining pits (with a depth of 10 meters) on inundated area and riverbed change of the Rio Salado, Salt River, Phoenix, Arizona. Due to the lack of field measurement, the calculated results of CCHE2D were compared to modeling results of HEC-RAS, a one dimensional hydrodynamic and sediment transport model.

2. Mining-pit evolution

The evolution of a mining pit is a complex morphodynamic process resulting from the interactions between streamflow, sediment, and movable boundaries. As water flows over a mining pit, the dividing streamlines separate and converge at the upstream and downstream ends of the pit, respectively. Streamline separation causes eddy rollers and headcut erosion at the upstream end, while streamline convergence causes bed degradation at the downstream end of the pit. Concurrently, incoming sediment from upstream is trapped in the upstream portion of the pit. (Figure 1). The overall effect is downstream migration of the gravel pit as deposition occurs at the upstream front while the tail end degrades from local scour. Such patterns have been observed in both natural streams (Chang, 1987; Kondolf, 1997; Neyshabouri et al., 2002; Rinaldi et al., 2005) and laboratory experiments (Fredsøe, 1978; Kornis and Laczay, 1988; Lee et al., 1993; Lee and Chen, 1996; Neyshabouri et al., 2002). A sketch of the flow structures and initial pit migration is shown in Figure 1. Pit geometry includes the width (B_p), length (L_p) and depth (H_p). L_u and L_d present the lengths of up- and downstream impacted reaches, respectively. Because of the asymptotic nature of the phenomenon it is difficult to fix the exact location of the endpoint of impacted reaches. A common definition is that , the bed deformation should be less than 1% of the flow depth at the end of impacted reaches at both up- and downstream of the pit.

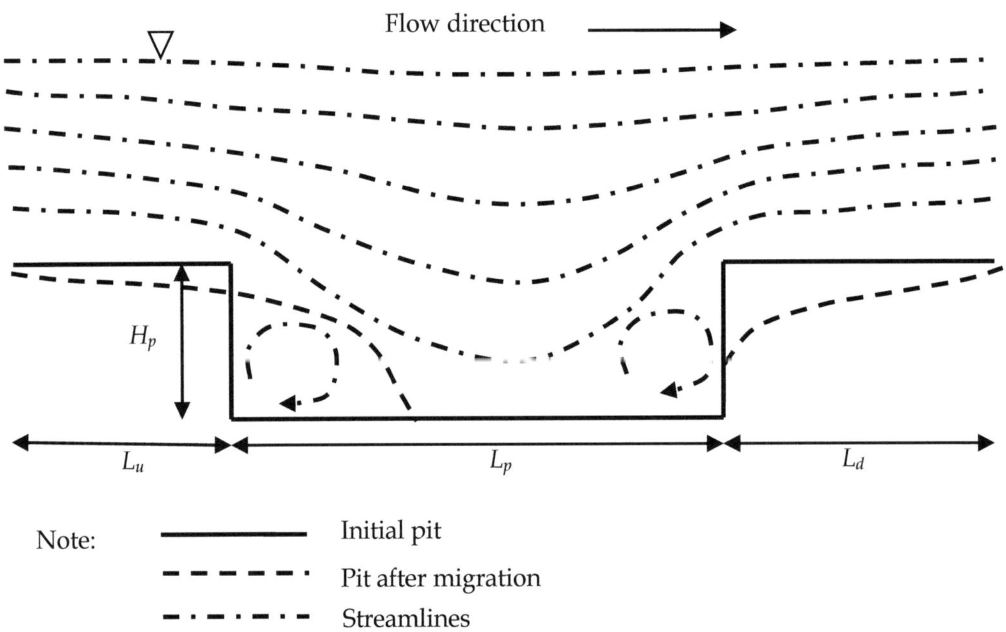

Fig. 1. Sketch of flow structure and initial pit evolution

Lee and Chen (1996) and Lee et al. (1993) conducted a series of experiments and found that the migration process of mining pits could be divided into two periods, namely convection period and diffusion period. Convection period is from beginning of deformation to the moment when the upstream boundary of the mining pit moves to the original downstream end of the pit. Diffusion period starts from that moment onward. The channel bed tends to attenuate sharp inflection on its longitudinal profile during both periods. Unlike the decreasing of scour depth during the "diffusion period", the depth of pit remains more or less constant during the "convection period" until the frontal surface of the filling sediment reach the downstream end. In this paper we continue to use names of the two periods, however, the same governing equations and non-equilibrium parameters were applied in both periods, since either convection or diffusion process of sediment transport takes import role in pit migration during both periods.

3. Models and approach

3.1 Flow and sediment transport models

Flow hydraulics, sediment transport, and channel morphological changes were simulated using CCHE2D and HEC-RAS. CCHE2D is an integrated software package for two-dimensional simulation for analysis of river flows, non-uniform sediment transport, morphologic processes, coastal processes, pollutant transport and water quality developed at the National Center for Computational Hydro-Science and Engineering at the University of Mississippi. These processes in the model are solved using the depth integrated Reynolds equations, transport equations, sediment sorting equation, bed load and bed deformation equations. The model is based on Efficient Element Method, a collocation approach of the Weighted Residual Method. Internal hydraulic structures, such as dams, gates and weirs, can be formulated and simulated synchronously with the flow. A dry and wetting capability enables one to simulate flows with complex topography. There are three turbulence closure schemes in the model, depth-averaged parabolic, mixing length eddy viscosity models and k-ε model. The numerical scheme can handle subcritical, supercritical, and transitional flows (NCCHE, 2011). Sediment transport formulas by Wu et al. (2000) were selected for all the calculations. A detailed description of the model was not included in this paper but can be found in the manual of CCHE2D (Jia and Wang, 2001; Wu, 2001). The feasibility of CCHE2D in simulating mining-induced bed change was discussed in Section 4.

HEC-RAS program was developed by Hydraulic Engineering Center (HEC) of U. S. Army Corps of Engineers. The latest HEC-RAS model provides a module for sediment transport analysis. This model was designed for modeling one-dimensional sediment transport, and can simulate trends of scour and deposition typically over periods of years or alternatively, for single flow events. For unsteady flow events, it segments the hydrograph into small time periods and simulates the channel flow for each time interval assuming a steady state flow in the whole channel. The non-equilibrium sediment transport approach included in the module makes the sediment transport process more realistic. The sediment transport potential is computed by grain size fraction so that the non-uniform sediment can be represented more accurately. The model can be used for evaluating sedimentation in fixed channels and estimating maximum scour during large flood events among other purposes (USACE, 2002). The HEC-RAS sediment transport module provides the option of several different sediment transport functions, thus users can select the most appropriate function according to the site conditions. The one-dimensional model offered a quick simulation although it could not provide accurate information other than the longitudinal direction.

The applicability and limitations of one- and two-dimensional models in modeling channel response to mining were discussed in Section 7.

3.2 Non-equilibrium sediment transport mode

Streambed deformation was calculated using various forms of the sediment continuity equation. The channel bed tends to be adjusted to re-meet the flow capacity, for instances, incision may propagate up- and downstream of the mine and deposition may occurs inside the mine. However, those processes could not be done instantaneously, i.e., the flow requires a finite length of bed to erode or deposit sufficient bed material to satisfy its equilibrium transport capacity. Wu (2008) stated that the assumption of local equilibrium transport is usually unrealistic and may have significant errors in the case of strong erosion and deposition. Bell and Sutherland (1983) conducted a series of experiments and concluded that the predictions of mathematical models are poor in the local scour region if an equilibrium transport formulation is used. Therefore, it is necessary to introduce non-equilibrium sediment transport schemes when modeling mining-pit migration (Yue and Anderson, 1990; Guo and Jin, 1999; Wu, 2008).

CCHE2D implements a non-equilibrium transport model for bed-material load including suspended-load and bed-load (Wu, 2001):

$$(1-P)\frac{\partial Z_c}{\partial t} = \alpha\omega_s\left(C - C_*\right) \tag{1}$$

$$(1-P)\frac{\partial Z_c}{\partial t} = \frac{1}{L}\left(Q_s - Q_{s*}\right) \tag{2}$$

where Z_c = calculated bed elevation (m); P = porosity of bed material; ω_s = settling velocity of suspended sediment (ms^{-1}); C = depth-averaged volumetric suspended-load concentration; C* = equilibrium depth-averaged volumetric suspended-load concentration; Q_s = volumetric bed-load transport flux per unit width (m^{2}s^{-1}); Q_{s*} = equilibrium volumetric bed-load transport flux per unit width (m^{2}s^{-1}); α = adaptation coefficient for suspended load; and L = adaptation length for bed-load (m).

The non-equilibrium adaptation length L characterizes the distance for bed-load to adjust from a non-equilibrium state to an equilibrium state; while α represents, theoretically, the ratio between the near-bed and depth-averaged suspended sediment concentrations. Coefficient α can also be represented by defining an equivalent adaptation length, L_s, as (Wu, 2001).

$$L_s = \frac{q}{\alpha\omega_s} \tag{3}$$

where q is the flow rate per unit width (m^{2}s^{-1}).

Both L and α are related to not only flow strength, sediment size and non-uniformity but also to the "extent of non-equilibrium", i.e., the difference between sediment load and the sediment transport capacity of flow. Researchers have reported a wide range of values for L and α. Bell and Sutherland (1983) investigated non-equilibrium sediment transport by discontinuing sediment supply at the upstream end of their flume. They found that the length for bed-load sediment to adjust from a non-equilibrium state to an equilibrium state was about the length of the first occurrence of a sand dune or scour hole, although the sand dune or scour hole extended and migrated progressively downstream throughout the experiment. Soni (1981) also found that L was related to flow condition and it changed with time in an experimental case of bed aggradation. Galappatti and Vreugdenhil (1985) found that the adaptation length for which the mean concentration approaches the mean equilibrium concentration is depended on sediment size and Chézy coefficient. Armanini and di Silvio (1988) also indicated that L should vary with sediment size and flow characteristics (flow depth, Chézy coefficient, etc.). In the experiments conducted by Wang (1999), the adaptation length was determined by the so-called "bed inertia", which represents the difference between sediment load and the sediment transport capacity of flow. In modeling practice, Phillips and Sutherland (1989) and Wu et al. (2000) adopted the non-equilibrium adaptation length as the averaged saltation step length of bed material particles approximated as a hundred times d_{50} for bed-load. Rahuel et al. (1989) gave much larger values by estimating L as two times the numerical grid length when dealing with natural channels. As for the parameter α, Han et al. (1980) and Wu and Li (1992) suggested α is 1 for strong scour, 0.25 for strong deposition; and 0.5 for weak scour and deposition.

4. Model feasibility

The feasibility of the two-dimensional depth-averaged two-dimensional hydrodynamic and sediment transport model, CCHE2D, in simulating mining-induced bed change was examined by comparing the calculated results to the data measured in two sets of published laboratory experiments: 1) a set of experiments by Lee et al. (1993); and 2) a set of experiments by Delft Hydraulics Laboratory (Galappatti and Vreugdenhil, 1985; van Rijn, 1986; Guo and Jin, 1999). Both experiments were conducted with steady flow and uniform rectangular cross sections except near the artificial mining areas. The two sets of experiments were chosen as representative of bed-load-dominated and suspended-load-dominated cases.

4.1 Experiments by Lee et al. (1993)

Lee et al.'s (1993) experiments were conducted using a 17 m long by 0.6 m wide recirculation flume. The rectangular pit was 54 cm long and 4 cm deep, with the upstream end located about 9.5 m from the flume entrance. The width of the pit was equal to the width of the flume i.e., 0.6 m. No sediment was supplied from upstream. Most of the sediment movement was in the bed load transport mode and no significant bed forms occurred. The flow conditions were subcritical flow. Flow parameters used in the flume study (Lee et al., 1993) and in the present modeling study are summarized in Table 1. Q is flow rate; h is flow depth; U is velocity; F_r is Froude Number; H_p and L_p are the depth and length of the mining pit, respectively. Since no sediment was supplied from upstream, the experiments were conducted in a non-equilibrium condition. The present model simulated bed-load transport only and the adaptation length L = the length of a numerical grid, i.e., 2 cm. Comparison of computed and measured bed change due to gravel mining is shown in Fig. 2. As can be seen from this chart, the numerical results agree with the experimental results, and the R^2 value for predicting bed elevation after 2 and 5 hours are 0.74 and 0.77, respectively.

4.2 Experiments by DHL

The present model is also applied to a flume experiment carried out by Delft Hydraulics Laboratory (Galappatti and Vreugdenhil, 1985; van Rijn, 1986; Guo and Jin, 1999). The experiment produces a uniform flow over a gentle-sided (1:10) trench in a 30 m-long, 0.5 m-wide, and 0.7 m-deep flume. The trench was 0.16 m deep initially. The mean flow velocity and the flow depth were 0.51ms^{-1} and 0.39m, respectively. The bed consisted of fine sand (d_{50} = 0.16mm). Only suspended sediment transport was simulated and the non-equilibrium adaptation coefficient, α, is calculated as 4.5 by Arminini and de Silvio's (1988) method:

$$\frac{1}{\alpha}=\frac{a}{h}+(1-\frac{a}{h})\exp\left[-1.5\left(\frac{a}{h}\right)^{-1/6}\frac{\omega_s}{u_*}\right] \tag{4}$$

where h is the flow depth (m); a is the thickness of bed-load layer. Table 2 shows the flow parameters used in the DHL flume study (Galappatti and Vreugdenhil, 1986) and the present model. The width of the pit was set the same as the width of the flume, i.e., 0.5 m. Since the pit has a side-slope 1:10, both upper and bottom lengths of the pit are shown (6.2 m and 3.0 m, respectively) in Table 2. Comparison of computed and measured bed change due to gravel mining is shown in Fig. 3. The agreements are quite satisfactory - the R^2 value for predicting bed elevation after 7.5 and 15 hours are 0.92 and 0.94, respectively.

Based on sections 4.1 and 4.2, the two-dimensional model, CCHE2D, is capable in simulating mining-induced bed change as long as the non-equilibrium parameters, i.e., L and α, being appropriately selected. Applicability of CCHE2D for our study reach, the Oeste reach of Rio Salado, has been proved by Chen and Liu (2009) and Chen et al. (2008; 2007). Sensitivity analysis of L and α in modeling pit migration was performed by Chen et al. (2010).

	Q (m3/s)	h (cm)	U (m/s)	Fr	H_p (cm)	L_p (cm)	d_{50} (mm)	L (cm)
Experiment	0.031	10.6	0.501	0.500	4	54	1.4	-
Model	0.031	10.0	0.519	0.518	4	54	1.4	2

Table 1. Flow condition in the experiments (Lee et al., 1993) and the present model

	Q (m3/s)	h (cm)	U (m/s)	Fr	H_p (m)	L_p (m)	d_{50} (mm)	α
Experiment	0.09945	39.0	0.51	0.26	0.16	6.2/3	0.16	-
Model	0.09945	38.4	0.62	0.32	0.16	6.2/3	0.16	4.5

Table 2. Flow condition in the DHL experiments (Galappatti and Vreugdenhil, 1986) and the present model

5. Application to the Rio Salado

5.1 Study reach

The Salt River (Rio Salado in Spanish) drains 14,500 square miles of mountainous desert terrain in central and eastern Arizona and is the largest tributary to the Gila River in Arizona. The river originates in eastern Arizona and flows westward to its confluence with

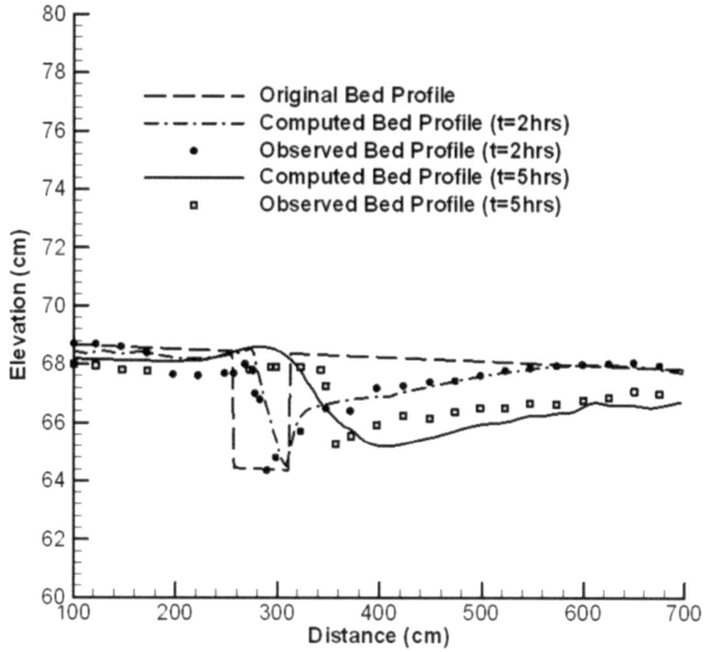

Fig. 2. Simulation results of bed change due to gravel mining (bedload transport only, observed data from Lee et al., 2003)

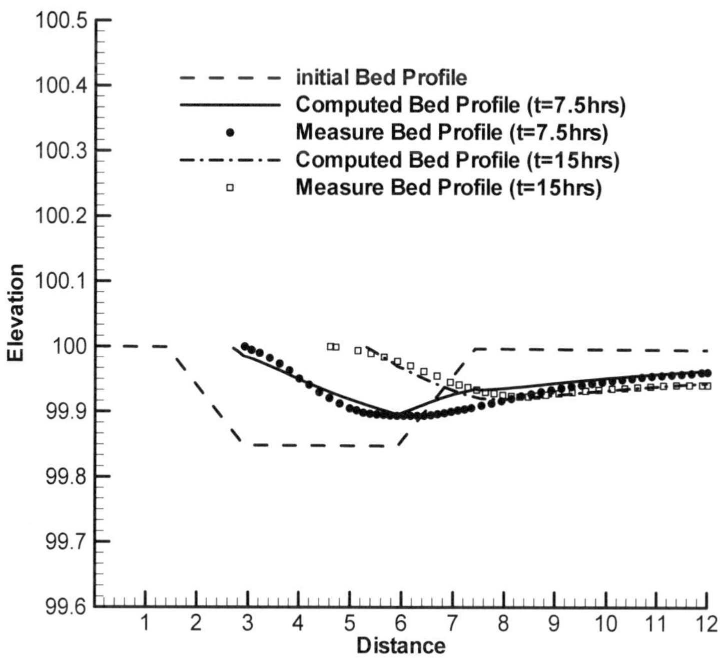

Fig. 3. Simulation results of bed change due to sand mining (suspended load only, observed data from Galappatti and Vreugdenhil, 1986)

the Gila River west of downtown Phoenix (Figure 4). Prior to agricultural development and urbanization of the Phoenix metropolitan area, the Rio Salado was a perennial stream fed by snowmelt from mountains in eastern Arizona. Flow in the river had a distinct seasonal pattern, with highest flows occurring in December and January and lowest flows in October. The yearly-averaged discharge was about 570 cfs before 1938. However, perennial flows on the Salt River have ceased due to dam and diversion construction in the early 1900s. In the early part of the 20th century, major modifications to the river system occurred as part of the Salt River Project, which placed several dams along the Salt River to allow diversions of water for agricultural and urban uses. Sand and gravel mining operations and development along the river induced additional changes to the river channel and hydrology. The materials extracted from the river have been used extensively throughout the development of the Phoenix Metropolitan area. Since 1965, the channel has carried a yearly-averaged discharge of only 400 cfs, with less than 14 cfs in almost three-fifths of the years (USACE, 2005). Now the water in the channel is dominated by the releases from upstream dams. The highest recent discharge occurred on February 13, 2005, of approximately 35,000 cfs as recorded by the USGS gauge at 51st Avenue (http://waterdata.usgs.gov/az/nwis/rt).

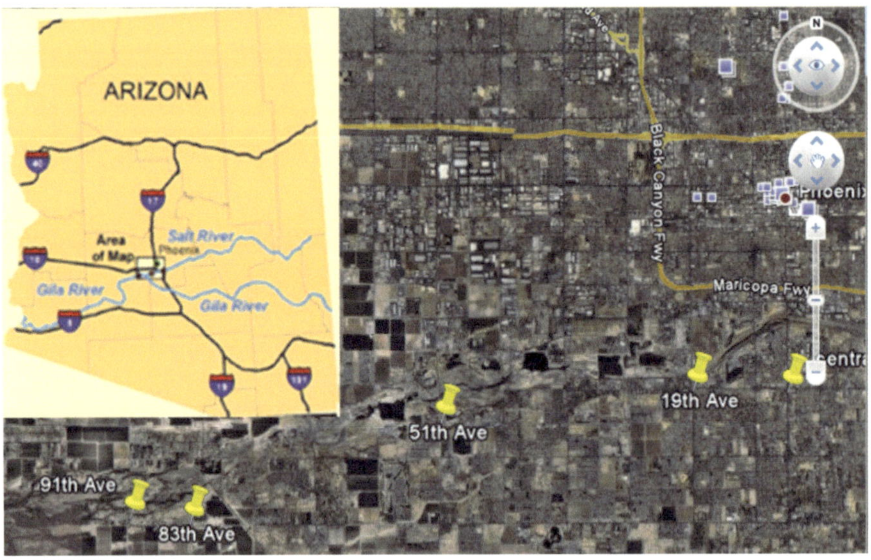

Fig. 4. Oeste reach of the Rio Salado, Salt River, Phoenix, Arizona

This study was designed to understand the impact of gravel mining on the flood zone coverage and channel geomorphology of the Oeste reach of the Salt River. The Oeste reach (study reach) is approximately 9.5 miles of the Salt River extending from 19th Avenue on the east to 91st Avenue on the west. The study area is within the boundaries of the City of Phoenix, Arizona. Previous sediment transport analyses have showed that sediment dynamics was more significant in the proximity of mining operations (USACE, 2005). Sand gravel mining has been going on in the Salt River for generations. There are numerous large mining pits in the Salt River around the city of Phoenix and the demands of more mining pits are growing day by day (shown in Fig. 4). Sand and gravel mining operations can cause changes in channel geomorphology as well as hydrology. This study was designed to understand the impacts of deep instream gravel mining pits (with a depth of 10 meters) on inundated area and riverbed change of the Oeste reach.

5.2 Bed material gradation

Bed material gradations along the channel were obtained through a combination of Pebble Count Method and Sampling Method at five accessible locations (Chen et al, 2007). Samples were analyzed at the soil laboratory of Desert Research Institute to obtain the grain size distribution. American Society for Testing and Materials (ASTM) procedures were followed for sieving analysis as described under Standard Test Method for Particle-Size Analysis of Soils. All samples were oven dried at 105°C for 24 hours, then placed in a series of ASTM approved sieves and positioned on a mechanical shaker for a minimum of 10 minutes each. The resulting size distributions of bed surface material are plotted in Fig. 5. Based on Fig. 5, less than 0.6% surface material consists of wash load (clay and silt) whose size is less than 0.0625 mm. About 2% bed material belonged to fine sand at 19th and 91st Avenues. Bed surface material mean size D_{50} ranges from 20mm to 40mm. More than 90% of the bed surface material consists of very coarse sand and gravels. Materials at the 51st Avenue are coarser than the bed material at other locations.

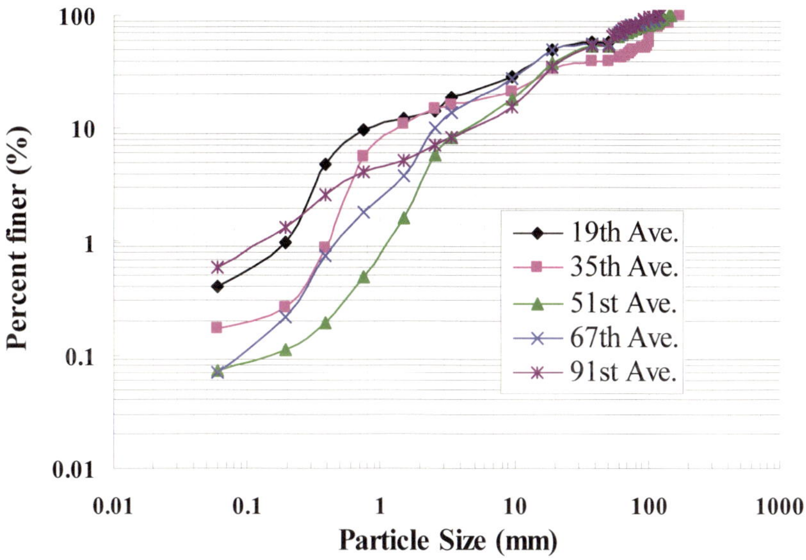

Fig. 5. Bed Surface Material Gradation in the Channel

5.3 Generation of 2-D computational mesh

A two-dimensional computational mesh was generated using the cross section data from HEC-RAS and channel topography data extracted from a digital contour map. The 2D mesh covered the entire study reach of the study site. Total of 77 cross sections extracted from the digital contour map were used to specify the bed elevations. Additional cross sections were interpolated based on these cross sections. The computational mesh had 678 cross sections with 120 computational nodes at each cross section. Fig 6 shows a part of computational mesh which includes a mining pit.

5.4 Sediment transport analysis

The sediment transport model treats suspended bed load and bed materials as mixed, grain-sized sediments and divides bed load into ten groups based on quantity and fluvial characteristics. The Wu, Wang and Jia's (2000) Formula and size distribution of substrate material are adopted in the present study. There is no field measurement of suspended load and bed load in the study reach. In the present study, we simulate sediment transport by assuming an equilibrium sediment supply at the inlet boundary which is calculated by using HEC-6T which uses Yang's (1984) Equation. In the present, we choose $\alpha = 1$ as suggested by Han et al. (1980) and Wu and Li (1992) and $L = 1000$ m (about the distance between alternative bars) based on personal discussion with NCCHE.

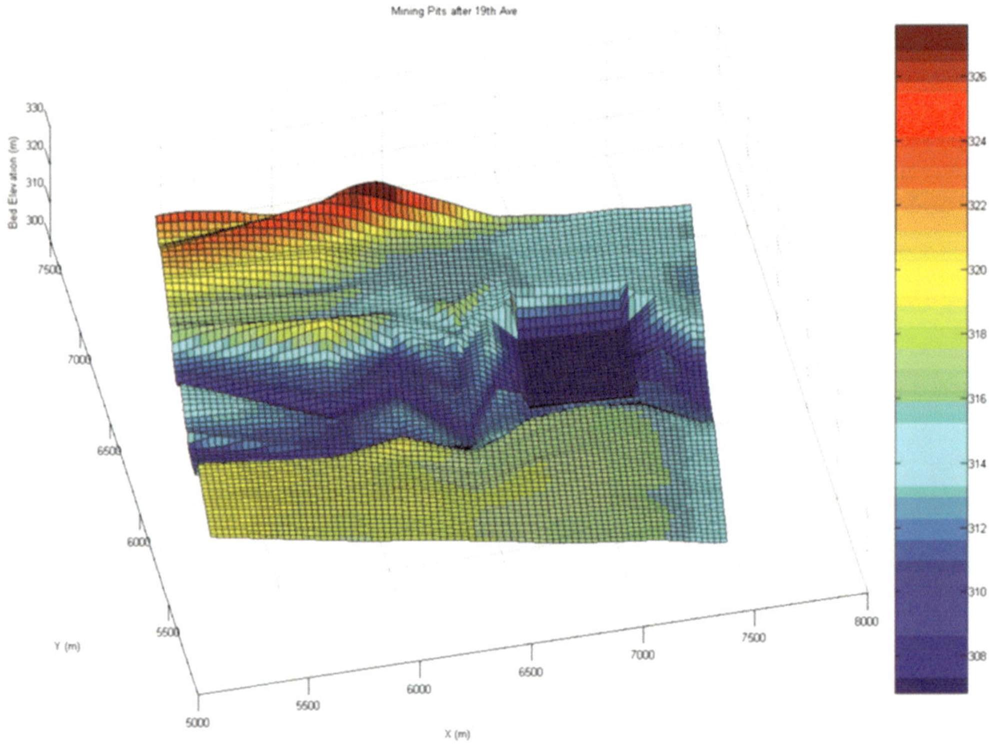

Fig. 6. Part of computational mesh including a mining pit

5.5 Hydrodynamic simulation

The effect of mining pits on inundated areas during different discharges was simulated in this study. Fig.7 provides an example of the calculated inundated areas with and without the mining pits at 500-year event. Based on Fig. 7, HEC-RAS and CCHE2D have similar calculated results which indicate that the inundated area has been reduced around the pits. However, this modification is minor comparing to the whole inundated area because the mining area only occupied a small part in the whole domain. Similar conclusions were drawn under other discharges which are not included in the report. However, our results also suggested that mining pits altered the local flow direction and magnitude. It may cause channel instability if those mining pits are close to riverbanks (Chen et al., 2008; 2007). The maximum riverbank shift has been as much as one-half mile in some locations through the area between 19th and 91st Avenues, Phoenix (USACE, 2005). However, the quantitative relation between bank erosion and gravel mining activities has not been formulated yet.

5.6 Simulation of bed elevation change

The effect of mining pits on bed elevation change was also simulated in this study. Fig. 8 and 9 show the bed topography change induced by gravel mining operation at 500-year flood event with HEC-RAS and CCHE2D models, respectively. Based on the simulation results, there was substantial erosion occurring upstream and downstream of the pits, however, the modeling results of HEC-RAS and CCHE2D model differed with each other. HEC-RAS results indicated serious head-cutting occurred upstream the pits, while CCHE2D results suggested "downstream erosion" was more noticeable. Modeling results of HEC-RAS showed the maximum bed degradation were 3.6 meters and 2.9 meters upstream of the Pit #1 and Pit #3, respectively. However, modeling results of CCHE2D exhibited the maximum bed degradation was 3.3 meters which occurred downstream the Pit #1. Besides, modeling results of CCHE2D indicated most sediment deposition was occurred in the upstream end of the mining pits. The maximum bed aggradations were 1.7 meters and 1.9 meters in the Pit #1 and Pit #2, respectively. Based on author's previous modeling study (Chen et al., 2008), the two-dimensional model was more robust in simulating flood zone coverage, non-uniform sediment sorting, and channel geomorphologic changes.

6. Conclusions

Impacts of instream gravel mining on flood zone coverage and riverbed change of the Rio Salado, Salt River, Phoenix, Arizona were simulated using HEC-RAS and CCHE2D in the present study. The capability of CCHE2D model in simulating bed changes due to mining was verified by two laboratorial cases. The following conclusions can be obtained:

a. Presence of mining pits insignificantly reduced the inundate area in the study reach, however, those pits changed the local flow directions and magnitudes which may accelerate stream bank erosion.
b. Calculated results of bed elevations of the two models differ with each other. There was substantial erosion occurring upstream and downstream of the pits, however, HEC-RAS results indicated serious head-cutting occurred upstream the pits, while CCHE2D results showed "downstream erosion" was more noticeable.
c. CCHE2D results indicated most sediment was deposited in the upstream end of the pits.

7. Discussion

The two-dimensional model was more robust in simulating the impacts of mining on flood zone coverage, non-uniform sediment sorting, and channel geomorphologic changes since one dimensional model hardly consider the information along the cross-sectional direction. The overestimation of the head-cutting by HEC-RAS was most likely caused by the improper calculation of the cross-sectional velocity upstream the pits, which is totally determined by the local water surface elevation in the one-dimensional calculation. In Figure 8b, the calculated water surface curve dropped dramatically upstream the pit #3, which is not realistic since the location of the pit is offset from the high velocity zone.

Flow structure around a deep mining pits present evident three-dimensional characteristics, however, three-dimensional sediment transport and bed deformation models are either complicate or time-consuming in preparing input data and calculation. The two dimensional model (CCHE2D) is able to examine the mining-induced bed change after appropriately determining non-equilibrium adaptation parameters. One-dimensional or vertical two-dimensional models may be applicable when the mining pit covers most area of the main flow zone.

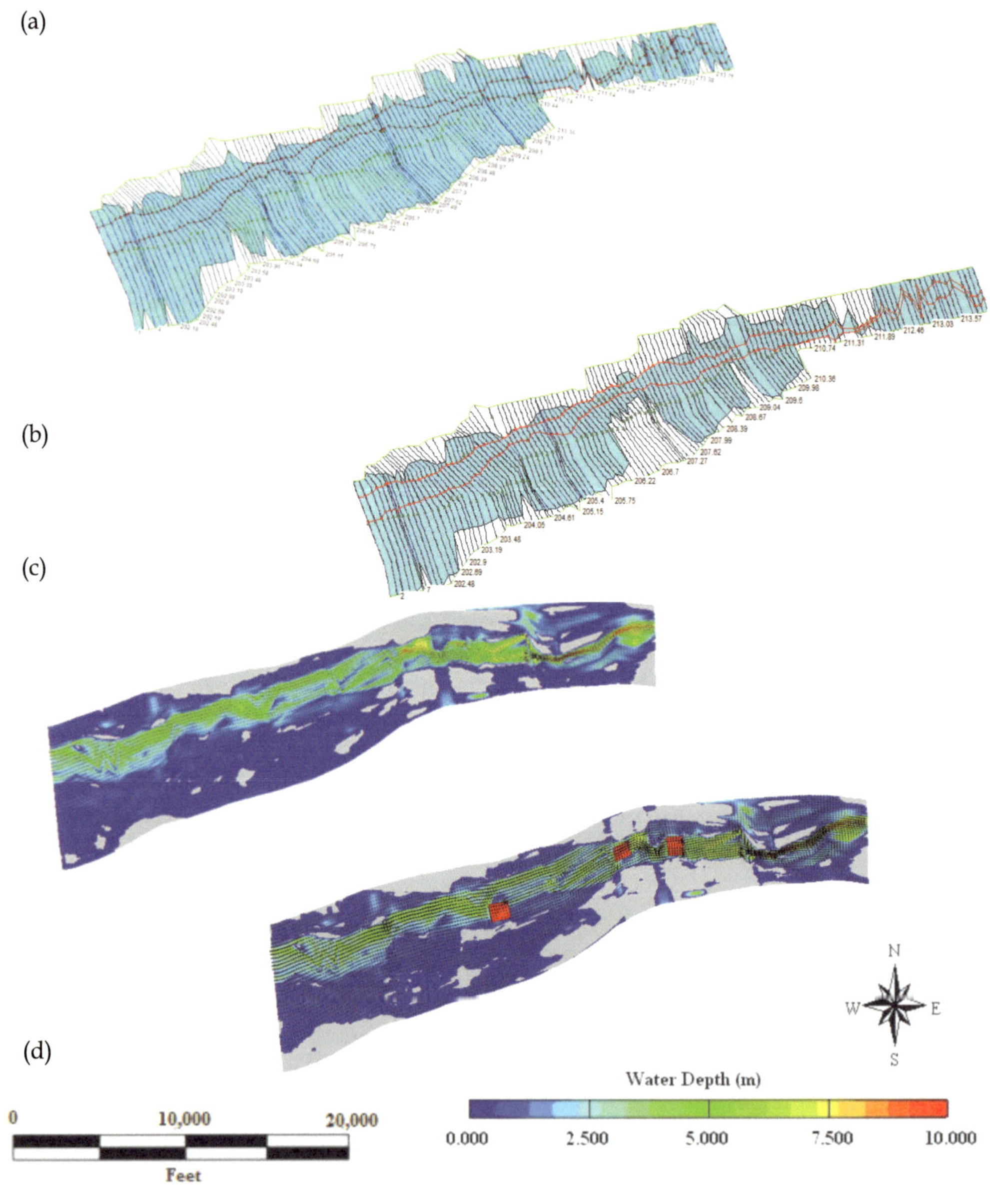

Fig. 7. Calculated inundated areas at 500-year event (a) HEC-RAS without mining pits; (b) HEC-RAS with three 10m pits; (c) CCHE2D without mining pits; (d) CCHE2D with three 10m pits

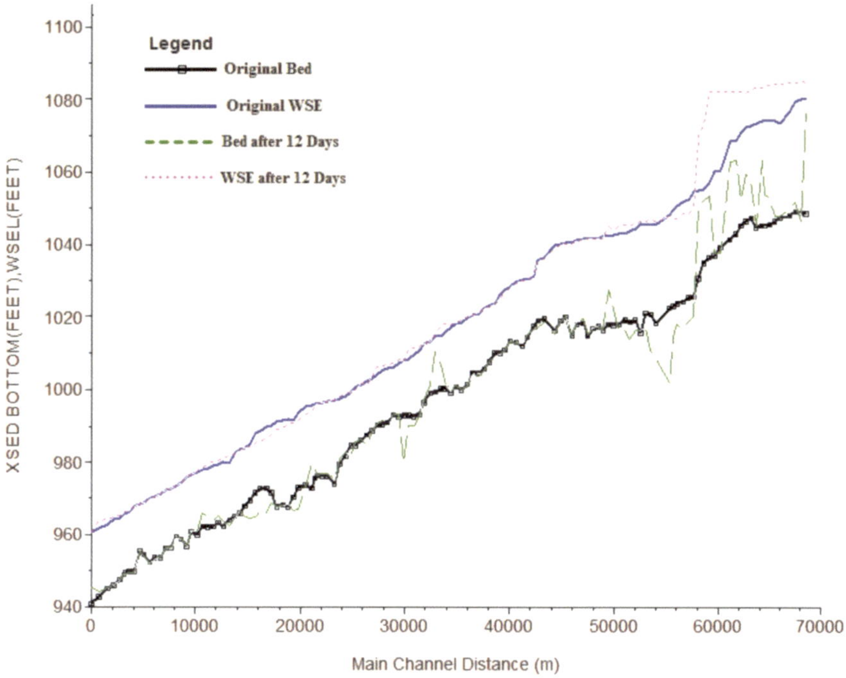

(a)

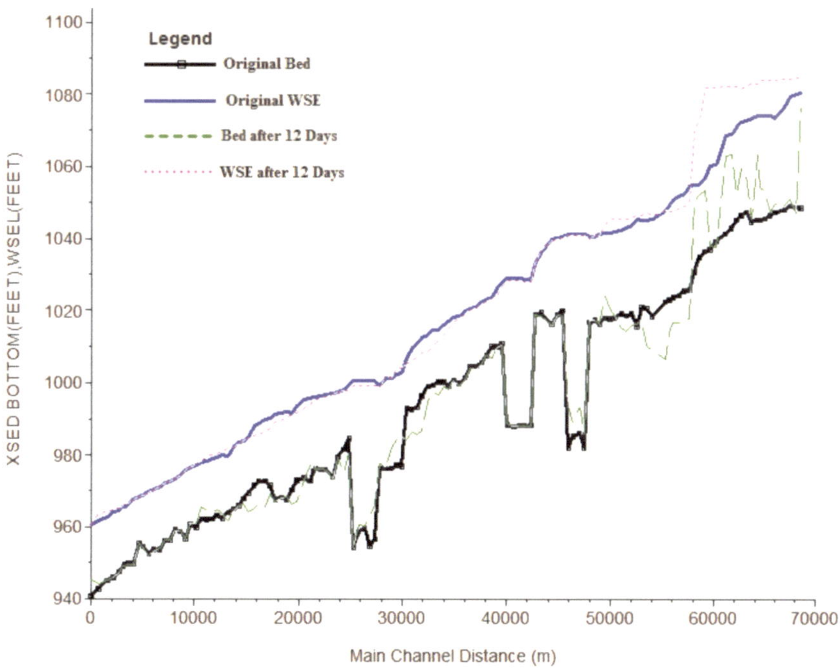

(b)

Fig. 8. Calculated bed topography change with HEC-RAS (a) without mining pits; (b) with three 10m mining pits (500-year event).

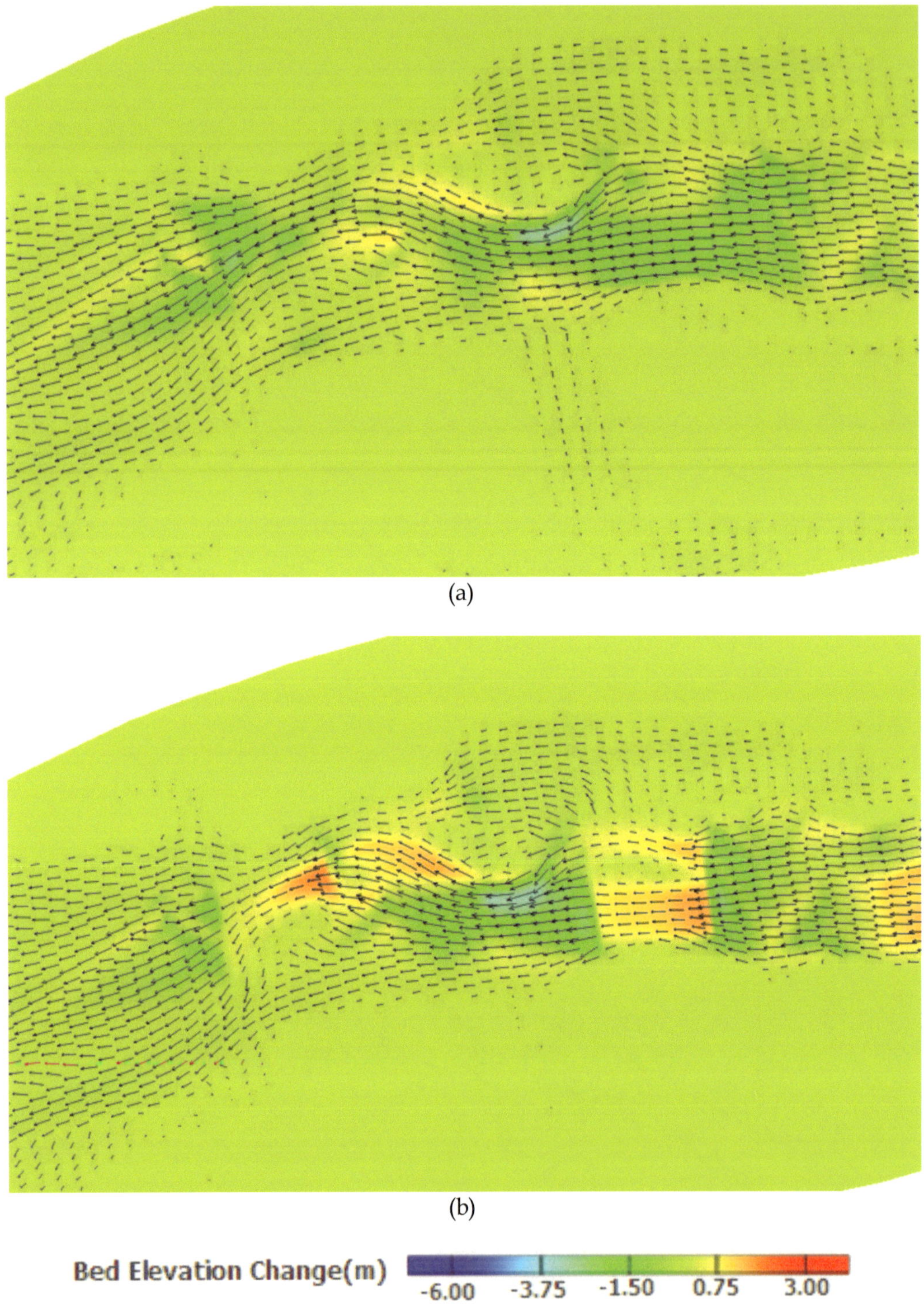

Fig. 9. Calculated bed topography change with CCHE2D (a) without mining pits; (b) with three 10m mining pits (500-year event).

8. Acknowledgement

This study was funded by US Corps of Army Engineers (USACE), Department of Defense under Urban Flood Demonstration Program of Arid and Semi-Arid Regions (UFDP). We thank the personnel of Flood Control District of Maricopa County and Phoenix district office of the USACE for their assistance in data collection and providing over view of the sites. We would also like to acknowledge the National Center for Computational Hydro-science and Engineering at the University of Mississippi for CCHE2D model.

9. References

Armanini, A,. and Di Silvio G. (1988). "A one-dimensional model for the transport of a sediment mixture in non-equilibrium conditions," Journal of Hydraulic Research, 26(3), 275-292.

Bell, R. G., and Sutherland, A. J. (1983). "Nonequilibrium bedload transport by steady flow." J. of Hydr. Engrg,, ASCE, 109(3), 351-367.

Cao, Z. and Pender, G. (2004). "Numerical modeling of alluvial rivers subject to interactive sediment mining and feeding." Advances in Water Resources, 27, 533-546.

Chang. H. H. (1987). "Modeling fluvial processes in streams with gravel mining." Sediment Transport in Gravel Bed Rivers, ed. by C. R. Thorne, J.C. Bathurst, and R.D. Hey, John Wiley & Sons, Ltd., 977-988.

Chen, D.; Acharya, K.; Stone, M. (2010). Sensitivity analysis of non-equilibrium adaptation parameters for modeling mining-pit migration. *J. of Hydraulic Engineering*, 136(10), 806-811

Chen, D.; Liu, M. (2009). One- and Two-dimensional modeling on deep gravel mining in the Rio Salado. *World Water & Environmental Resources Congress, ASCE*, Kansas City, Missouri.

Chen, D., Acharya, K., and Stone, M. (2008). Modeling of Gravel Mining in the Rio Salado, Arizona. *World Water & Environmental Resources Congress, ASCE* 2008. Honolulu, Hawaii.

Chen, D., Zhang, Y., Duan, J., Stone, M., and Acharya, K. (2007). 2D Simulation of hydrodynamic and sediment transport in a gravel bed channel: the Salt River. *World Water & Environmental Resources Congress, ASCE* 2007. Temple, Florida.

Cotton, G. K., and Ottozawa-Chatupron, V. (1990). "Longitudinal channel response due to in-stream mining." Proc., Specialty Conf. of Hydr. Engrg., ASCE, New York, N. Y.

Fredsøe, J. (1978). "Sedimentation of river navigation channels." J. Hydr. Div., ASCE, 104(2), 223-236.

Galappatti, G. and Vreugdenhil, C. B. (1985). "A depth-integrated model for suspended sediment transport. " J. Hydraul. Res., 23(4), 359-377.

Gill, M.A. (1994). "Hydrodynamics of mining pits in erodible bed under steady flow." J. of Hydr. Engrg,, ASCE, 120(11), 1337-1348.

Guo, Q.C and Jin, Y.C. (1999). "Modeling sediment transport using depth-averaged and moment equations." J. of Hydr. Engrg,, ASCE, 125(12), 1262-1269.

Han, Q. (1980). "A Study on the Non-Equilibrium Transport of Suspended Load," Proceedings of the International Symposium on River Sedimentation, Beijing, China, 793-802 (in Chinese).

Jia, Y. F., and Wang, S.S.Y. (2001). CCHE2D: Two-dimensional Hydrodynamic and Sediment Transport Model for Unsteady Open Channel Flow Over Loose Bed. Technical Report of National Center for Computational Hydroscience and Engineering, The University of Mississippi. NCCHE-TR-2001-01.

Kondolf, G. M. (1997). "Hungry water: effects of dams and gravel mining on river channels." Environmental Management, 21(4), 533-551.

Kondolf, G. M. (1994). "Geomorphic and environmental effects of instream gravel mining." Landscape and Urban Planning, 28, 225-243.

Kornis, S., and Laczay, I.A. (1988). "Effects of extensive dredging on the river regime." Int. conf. on River Regime, Hydraulic Research Ltd., John Wiley and Sons, New York, N. Y.

Lee, H. Y. and Chen, S.C. (1996). "Migration of rectangular mining pit composed of nonuniform sediment." Journal of the Chinese Institute of Engineers, 19(2), 255-264.

Lee, H. Y., Fu, D.T., and Song, M.H. (1993). "Migration of rectangular mining pit composed of uniform sediments." J. of Hydr. Engrg,, ASCE, 119(1), 64-80.

NCCHE(2011). website - http://www.ncche.olemiss.edu/

Neyshabouri, S.A., Farhadzadeh, A., and Amini, A. (2002). "Experimental and field study on mining-pit migration." International Journal of Sediment Research, 17(4), 2002, 323-331.

Phillips, B.C. and Sutherland, A.J. (1989). "Spatial lag effects in bed load sediment transport." J. Hydraul. Res. 27(1), 115-133.

Rahuel, J. L., Holly, F. M., Chollet, J. P., Belleudy, P.J., and Yang, G. (1989). "Modeling of riverbed evolution for bedload sediment mixtures." J. of Hydr. Engrg,, ASCE, 115(11), 1521-1542.

Rinaldi, M., Wyżga, B., and Surian, N. (2005). "Sediment mining in alluvial channels: physical effects and management respectives." River Res. Applic. 21, 805-828.

Rovira, A., Batalla, R.J. and Sala, M., 2005. Response of a river sediment budget after historical gravel mining (the Lower Tordera, NE Spain). River Research and Applications, 21(7), 829-847.

Soni, J. P. (1981). "Unsteady sediment transport law and prediction of aggradation parameters." Water Resources Research, 17(1), 33-40.

US Army Corps of Engineers (2005). *Rio Salado Oeste Interim Feasibility Report* (Alternative Formulation Briefing). 63 pp

US Army Corps of Engineers (2002). *HEC-RAS River Analysis System*, Hydraulic Reference Manual

van Rijn, L.C. (1986). "Sedimentation of dredged channels by currents and waves." J. Wtrway, Port, Coast, and Oc. Engrg., 112(5), 541-549.

Wang. Z.Y. (1999). "Experimental study on scour rate and river bed inertia." J. Hydraul. Res. 37(1), 17-37.

Wu, W. M. (2001). CCHE2D Sediment Transport Model (Version 2.1). Technical Report of National Center for Computational Hydroscience and Engineering, The University of Mississippi. NCCHE-TR-2001-03.

Wu, W.M. and Li, Y. (1992). "One-and two-dimensional nesting model for iver flow and sedimentation." Proc. 5th Int. Symp. on River Sedimentation, Karlsruhe, Germany.

Wu, W. M., Wang S.S.Y., and Jia, Y. F. (2000). "Nonuniform Sediment Transport in Alluvial Rivers". J. Hydraul. Res. 38(6), 427-434.

Wu, W. M. (2008). Computational River Dynamics. Taylor & Francis Group, London, UK. 463p.

Yang, C.T. (1984). "Unit stream power equation for gravel." J. of Hydr. Engrg,, ASCE, 110(12), 1783-1797.

Yue, J., and Anderson, J. L. (1990). "Numerical modeling of a trench fill and scour." Proc., Hydr. Engrg. Nat. Conf., ASCE, 487-492.

7

Modeling of Sediment Transport in Surface Flow with a Grass Strip

Takahiro Shiono[1] and Kuniaki Miyamoto[2]
[1]*National Institute for Rural Engineering, NARO*
[2]*Graduate School of Life and Environmental Sciences*
University of Tsukuba
Japan

1. Introduction

Reddish sediment runoff from land areas during rainfall causes environmental problems in coastal areas of the Okinawa region, Japan. Sediment delivered to the coastal areas causes water pollution, sedimentation and degrades the coastal ecosystems and fisheries resources. Agricultural fields are a major source of sediment runoff in the region (Yoshinaga & Onaga, 1993; Minami et al., 2002). Nakasonoe et al. (1998) reported that sediment from agricultural fields accounts for 70% of the total sediment runoff in the region. Countermeasures for runoff in agricultural fields are promoted as an important issue in the Okinawa region.

Countermeasures so far proposed include terrace work, drainage canals, sediment ponds, grass strips, cover cropping, mulch farming, contour farming, deep tillage, crop rotation and green manure (Hudson, 1995; Morgan, 1995). The grass strip countermeasure involves installing grass bands at the downstream end of an agricultural field to reduce the amount of non-point source pollutants, such as sediment and nutrients from an agricultural field into the stream (Dillaha et al., 1989). Grass strips are currently installed by prefectural governments assisted by the Ministry of Agriculture, Forestry and Fisheries as public work projects for water conservation, and also as a measure to help conserve agricultural land, water and the environment in rural areas. Grass strips can also be installed by farmers themselves as one of agricultural activities.

There are various factors affecting how effectively grass strips reduce sediment runoff, such as flow rate of inflowing water, sediment properties, slope conditions and features of the grass (Haan et al., 1994). When installing grass strips as a countermeasure for sediment runoff, various conditions must be taken into consideration. In designing grass strips, it is necessary to know the quantitative relationships between the various factors influencing grass strips and their effects on reduction of sediment runoff, and to determine appropriate and reasonable parameters for installing grass strips.

In previous studies, Sugawara et al. (2001), Osawa et al. (2005) and Shiono et al. (2005) conducted field experiments in Ishigaki Island and the northern part of the Okinawa Main Island, and reported their results on the effects of grass strips for reducing reddish sediment runoff. Shiono et al. (2007) also reported the influence that the length of grass strips in the direction of flow and particle sizes of sediment flowing into grass strips have on the ability of grass strips to reduce reddish sediment runoff. These results were obtained from field experiments conducted in the northern part of the Okinawa Main Island. However, since the effects that grass strips have on reducing reddish sediment runoff presented in these studies were obtained under limited conditions, these effects do not fully reveal the relationship between various installation conditions of grass strips conceivable at a site and their effects on reducing the runoff of reddish sediment.

Therefore, this study aims to establish a mathematical model to represent the sediment trapping process of grass strips as part of efforts to present, through the analysis of a mathematical model, the relationship between various installation conditions of grass strips and their effects on reducing the runoff of reddish sediment. First, flume experiments were performed using a simulated grass model and then a grass model developed to experimentally simulate the sediment-trapping process of grass strips. The results of these experiments allowed us to obtain the necessary characteristics. Next, a mathematical model was established for simulating the sediment trapping processes of the grass strips used in the flume experiments, and the model was verified.

2. Flume experiment

2.1 Flume experiment with bamboo rod model

Flume experiments with a bamboo rod model simulating grass strips were conducted to clarify the process of surface flow and sediment transport in and around a grass strip. The experiments used an acrylic flume with a rectangular cross section, total length 12.0 m, width 0.145 m, depth 0.150 m and gradient of 0.02. The rod model was made of a polyvinyl chloride plate with a width of 0.145 m and thickness of 3 mm, where 2.5 mm-diameter and 0.15 m-long bamboo rods were arranged vertically to form grids of 40,000 rods/m^2. The rod model was installed as a section extending 1.50 m from the downstream end of the flume. To prevent the occurrence of steps on the flume floor when installing this model, a plywood board with a thickness of 9 mm was placed on the flume floor immediately under the model, and another board with a thickness of 12 mm placed upstream of the model. Both boards were coated with varnish.

Four types of sediment were used for the experiments: GK-9, NK-9, Goto Clay (these three were made by Kumamoto Silicasand Industry Co., Ltd.) and Kunigami Maji. The last soil was collected from the Arashiyama field in Nago on the northern part of the Okinawa Main Island (26°38′N, 127°59.5′E). The particle densities of GK-9, NK-9, Goto Clay and Kunigami Maji were 2.63, 2.61, 2.71 and 2.71 Mg/m^3, respectively. Figure 1 shows the particle size distribution of the test sediments used.

Figure 2 shows an outline of the flume experiment with the rod model, and Table 1 shows the experimental conditions. In these experiments, clean water with a constant flow rate was supplied from the upstream end of the flume. Then, a sediment-water mixture was supplied at a constant rate into the running water at a point 2.0 m downstream from the upstream end, so that sediment-laden water with a constant concentration of sediment flowed at a constant rate. The flow rate was set to form a surpercritical flow so as to prevent sediment from being deposited upstream of the flume. The flow rate was measured by means of an HS flume installed at the downstream end. The sediment-water mixture was prepared in advance in a bucket, which was mixed with a mixer during the experiments. The flow rate of the mixture supplied was regulated using a meter pump (Masterflex Inc, 7523-60). The experiments continued for durations of 150 to 390 minutes. While the water was running, water levels were measured at intervals of 0.2 to 0.5 m along the length of the flume with a point gage (KENEK, PH-340) to determine the water surface profile. Two hundred ml of water was sampled from the running water both upstream and downstream of the rod model every 30 minutes and sediment concentrations in the running water measured by the oven-drying method. The upstream sampling point was set immediately upstream of the hydraulic jump due to the influence of the rod model. After each experiment, the bed surface profile was measured along the length of the flume with the point gage.

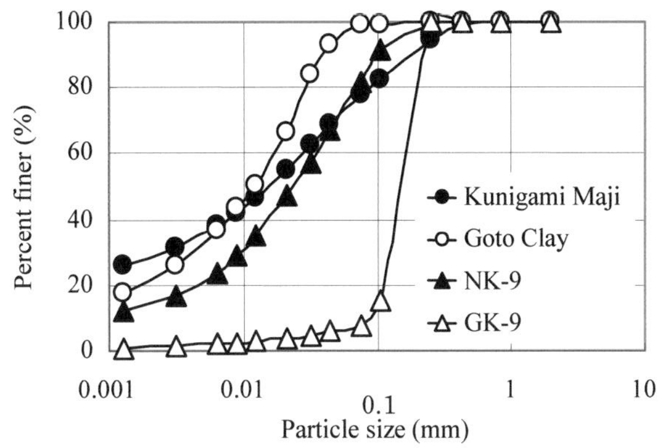

Fig. 1. Particle size distribution of test sediments

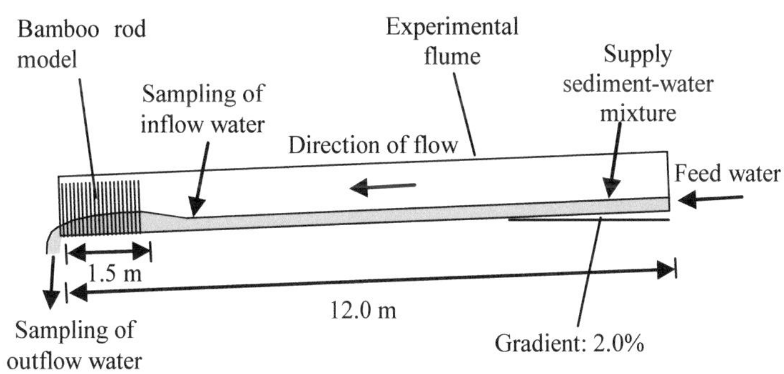

Fig. 2. Outline of flume experiment with a bamboo rod model

2.2 Flume experiment with grass model

Flume experiments with a grass model simulating grass strips were also conducted to clarify the characteristics of sediment removal by grass strips. Centipede grass (*Eremochloa ophiuroides* (Munro) Hack.), a turf grass, was used for the grass model. Shiono et al. (2007) reported that grass strips made from centipede grass were effective for reducing reddish sediment loads under typical farmland conditions. Dead grass was substituted for living grass in these experiments because the shape and the stiffness of the dead grass was almost the same as the living grass and multiple flume experiments could be conducted under the same grass conditions. The turf grass was grown at a field in the National Agricultural

Research Center for Kyushu Okinawa Region in Kumamoto, Japan. When collected, the grass was 10 cm high, the vegetation cover ratio was 100% and the surface portion was dead. The grass bodies collected also covered the root section from the surface down to a depth of 10 cm with soil attached. After collection, only the surface portion was selected, and the roots and soil were removed. These plant bodies were then attached to a polyvinyl chloride plate with a width of 0.145 m and thickness of 3 mm to form the grass model used for the experiments.

Case	Sediment	Flow rate Q ($m^3 s^{-1}$)	Water flowing time (min)	Water depth on upstream side h_0 (mm)	Froude number Fr_0	Average sediment concentration on upstream side C_{in} ($m^3 m^{-3}$)	Average sediment concentration on downstream side C_{out} ($m^3 m^{-3}$)	Sediment reduction rate (%)
A-1	GK-9	2.07×10^{-4}	285	4.33	1.60	5.93×10^{-4}	1.41×10^{-5}	97.6
A-2	NK-9	2.02×10^{-4}	285	3.59	2.07	9.89×10^{-4}	3.91×10^{-4}	60.5
A-3	NK-9	2.13×10^{-4}	205	4.69	1.46	2.26×10^{-3}	8.47×10^{-4}	62.5
A-4	NK-9	1.03×10^{-4}	390	3.38	1.15	8.43×10^{-4}	3.28×10^{-4}	61.1
A-5	NK-9	1.10×10^{-4}	390	3.06	1.43	4.75×10^{-4}	9.96×10^{-5}	79.0
A-6	Goto Clay	2.07×10^{-4}	390	4.34	1.60	1.07×10^{-3}	6.33×10^{-4}	40.8
A-7	Goto Clay	2.02×10^{-4}	195	4.73	1.37	2.66×10^{-3}	1.50×10^{-3}	43.6
A-8	Goto Clay	1.07×10^{-4}	390	3.47	1.16	1.13×10^{-3}	4.89×10^{-4}	56.7
A-9	Goto Clay	1.10×10^{-4}	375	3.23	1.32	4.59×10^{-4}	1.66×10^{-4}	63.8
A-10	Kunigami Maji	2.07×10^{-4}	270	5.09	1.26	1.68×10^{-3}	7.52×10^{-4}	55.2
A-11	Kunigami Maji	1.10×10^{-4}	240	3.41	1.22	9.15×10^{-4}	4.19×10^{-4}	54.2
A-12	Kunigami Maji	3.04×10^{-4}	150	5.50	1.64	1.55×10^{-3}	3.96×10^{-4}	74.4

Table 1. Summary of experimental conditions and results

The flume and grass models were installed in the same way as in 2.1, and the above-mentioned Kunigami Maji sediment was used in the experiments. The experiments were conducted in a manner similar to that in 2.1, that is, clean water was supplied from the upstream end of the flume. Then at a point 2.0 m downstream from the upstream end, a sediment-water mixture was supplied into the running water, so that sediment-laden water with a constant concentration of sediment flowed at a constant rate. Each experiment continued for 60 minutes. While the water was running, water level was measured for the water surface profile along the length of the flume as in 2.1. Also water was sampled every 10 minutes to determine the sediment concentration in the upstream and downstream flow of the grass model. The experimental conditions are given in Table 2. In Cases B-1 to B-6, the length of the model in the longitudinal direction was fixed at 1.5 m, while the flow rate was set at three steps within the range of 1.01×10^{-4} to 2.95×10^{-4} $m^3 \cdot s^{-1}$ and the sediment concentration was set at two steps of 7×10^{-4} and 1.4×10^{-3} $m^3 \cdot m^{-3}$. On the other hand, in Cases C-1 to C-4, the experiments were performed with the model length set at four steps within the range of 0.5 to 5.9 m. The experiments also included sampling of running water in the upstream and downstream flows of the grass model to determine the distribution of equivalent sizes of sediment particles contained in the running water. The distribution of equivalent particle size was obtained by the pipette method. The equivalent particle size is defined here as the diameter of a spherical particle having a sediment particle density and a falling velocity that are equal to the target sediment particles or aggregates under consideration. For this purpose, only distilled water was used for dispersion of the sediment samples in the process of the pipette method.

Case	Model length (m)	Flow rate Q ($m^3 s^{-1}$)	Water flowing time (min)	Water depth on upstream side h_0 (mm)	Froude number Fr_0	Average sediment concentration on upstream side C_{in} ($m^3 m^{-3}$)	Average sediment concentration on downstream side C_{out} ($m^3 m^{-3}$)	Sediment reduction rate (%)
B-1	1.5	1.01×10^{-4}	60	2.39	1.96	1.48×10^{-3}	4.65×10^{-4}	68.6
B-2	1.5	1.01×10^{-4}	60	2.39	1.96	6.16×10^{-4}	1.47×10^{-4}	76.1
B-3	1.5	1.98×10^{-4}	60	4.04	1.70	1.13×10^{-3}	5.35×10^{-4}	52.7
B-4	1.5	1.98×10^{-4}	60	4.04	1.70	5.20×10^{-4}	1.93×10^{-4}	62.9
B-5	1.5	2.95×10^{-4}	60	4.80	1.97	1.53×10^{-3}	7.75×10^{-4}	49.3
B-6	1.5	2.95×10^{-4}	60	4.80	1.97	7.82×10^{-4}	3.45×10^{-4}	55.9
C-1	0.5	1.98×10^{-4}	60	3.89	1.80	1.35×10^{-3}	7.27×10^{-4}	46.1
C-2	1.5	1.98×10^{-4}	60	3.86	1.82	1.50×10^{-3}	6.64×10^{-4}	55.7
C-3	3.1	1.98×10^{-4}	60	3.95	1.76	1.45×10^{-3}	6.31×10^{-4}	56.5
C-4	5.9	1.98×10^{-4}	60	3.85	1.83	1.29×10^{-3}	5.09×10^{-4}	60.5

Table 2. Summary of experimental conditions and result

2.3 Experiments on flow conditions in grass and rod models

Pipe channel experiments were performed to identify flow conditions within the grass models used in the above mentioned experiments. These experiments adopted an acrylic pipe channel with a rectangular cross section, total length 1.0 m, width 0.145 m and depth 0.10 m, where the grass model was placed along the entire channel. Then, as indicated in Fig. 3, the channel was filled with clear water at a constant flow rate to determine the relationship between the flow rate and the hydraulic gradient. The flow rate was Q = 6.94×10^{-5} to 2.03×10^{-3} m·s^{-3}, and the average flow velocity in the cross section u = 1.45×10^{-2} to 1.76×10^{-1} m·s^{-1}. The experiments provided data on the flow conditions corresponding to 9 to 15 steps within the above parameter ranges, where measurements were made using a bucket. The hydraulic gradient was obtained from the water level differences between the two acrylic pipes installed in the pipe channel and the distance (0.90 m) between them. The heights of the grass model were set at four steps: 3.2, 5.5, 7.7 and 10.0 cm. First, the experiment was conducted for the 10.0 cm-high grass model, and then the top was cut to 7.7 cm. Subsequently, other heights were also set in a similar way. For grass heights other than 10.0 cm, plywood boards were placed beneath the grass model to adjust the channel floor height and ensure that all water flowing into the channel passed through the grass.

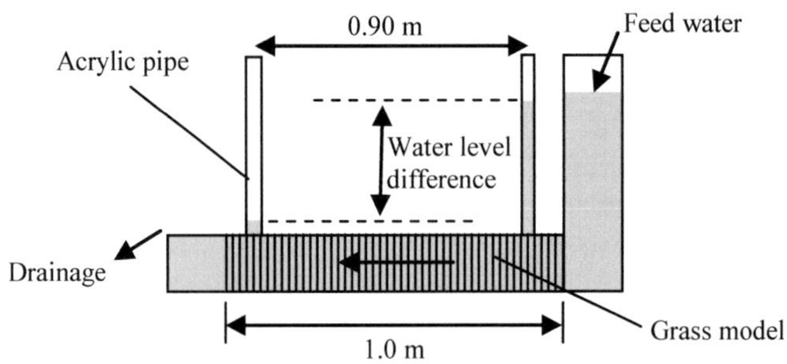

Fig. 3. Outline of experiment on flow conditions in grass

Experiments on the flow characteristics of the rod model were also performed in a similar manner, but the acrylic pipe channel used had a rectangular cross section with a total length of 1.0 m, a width of 0.145 m and a depth of 0.16 m, while the model height was set at one step only of 0.15 m. The flow rate given was Q = 2.62×10^{-4} to 3.18×10^{-3} m·s^{-3}, while the average flow velocity in the cross section was u = 1.39×10^{-2} to 1.69×10^{-1} m·s^{-1}. The experiments provided data on the flow conditions corresponding to 17 steps within the above parameter ranges.

3. Results and discussion for flume experiments

3.1 Flume experiments with bamboo rod model

In the experiments for Cases A-1 to A-12, the running water flowed upstream of the flume as a supercritical flow, resulting in the generation of a hydraulic jump upstream of the rod model. Downstream of the hydraulic jump the flow velocity decreased so the water ran at a subcritical flow. At the start of the experiments, the hydraulic jump was located 0.7 to 2.1 m upstream of the model upstream end. However, during the experiments, the jump location gradually shifted upstream due to sediment being deposited on the flume bed, and by the time the experiments were finished the location was 1.4 to 4.0 m upstream of the model upstream end.

The sediment contained in the running water did not deposit on the flume bed in the section upstream of the hydraulic jump, but instead flowed as bed load or suspended load sediment. In the section downstream of the hydraulic jump, that is, in the backwater section or the rod model section, some sediment was deposited on the flume bed due to a decrease in flow velocity, while other sediment passed with the running water through the rod model section. The depth of sediment deposited peaked in the backwater section, while the depth gradually decreased along the downstream direction in the model section. During the experiments, the sediment deposition depth and the length of the deposition section gradually increased, while the sediment concentration on the model downstream side remained almost unchanged.

As indicated in Table 1, the sediment reduction rates for the rod model obtained from the time average of sediment concentrations in the upstream and downstream flows of the model ranged from 40.8 to 97.6%. It is evident that the sediment reduction rates vary depending on the conditions set. Particularly, in experiment Case A-1 using sediment GK-9 with course particles, the rate was as high as 97.6%.

3.2 Flume experiments with the grass model

The sediment reduction rates for the grass model varied depending on the flow rate and concentration of sediment in the upstream side. As indicated in Table 2, the sediment reduction rates obtained from the time average of sediment concentrations in the upstream and downstream flows of the grass model for Cases B-1 to B-6 ranged from 49.3 to 76.1%. An overview of the relationship between the sediment reduction rate and measurements of the flow rate and the upstream-side sediment concentrations suggests the sediment reduction rate was smaller for increased flow rates, while the rate decreased to some extent for higher upstream sediment concentrations.

The sediment reduction rates of the grass model also varied in accordance with the model length. As indicated in Table 2, the sediment reduction rates obtained from the time average of sediment concentrations in the upstream and downstream flows of the grass model for Cases C-1 to C-4 ranged from 46.1 to 60.5%. In other words, the reduction rate was larger when model length increased.

The results of the analysis for particle sizes of the sediment collected on the upstream side of the grass model in each experiment showed that the average fractions of equivalent particle size for 0 to 0.002, 0.002 to 0.02, 0.02 to 0.2 and 0.2 to 0.5 mm were 13%, 44%, 39% and 4%, respectively. This evidences the sediment with a particle size equivalent to silt and fine sand is predominant, while the fraction of particles equivalent in size to clay and course sand is small.

The effect of the grass model on reducing sediment runoff varies depending on the equivalent particle size. Figure 4 shows the time-averaged sediment loads for each equivalent particle size class in the upstream and downstream flows of the grass model of experiment C-3. As is shown here, the sediment loads for particle size classes of 0.02 to 0.2 and 0.2 to 0.5 mm on the downstream side are remarkably small compared with those on the upstream side, while those of the 0 to 0.002 and 0.002 to 0.02 mm particle size classes on the downstream side are somewhat smaller than on the upstream side. The sediment reduction rates in all cases with an equivalent particle size of 0 to 0.002, 0.002 to 0.02, 0.02 to 0.2 and 0.2 to 0.5 mm were 0 to 68, 2 to 64, 88 to 100 and 100%, respectively. Sediment with an equivalent particle size of larger than 0.02 mm was mostly trapped by the grass model, while only some sediment with particle sizes smaller than 0.02 mm was trapped. Such characteristics of the sediment trapping by grass depending on particle sizes were also reported by Shiono et al. (2007).

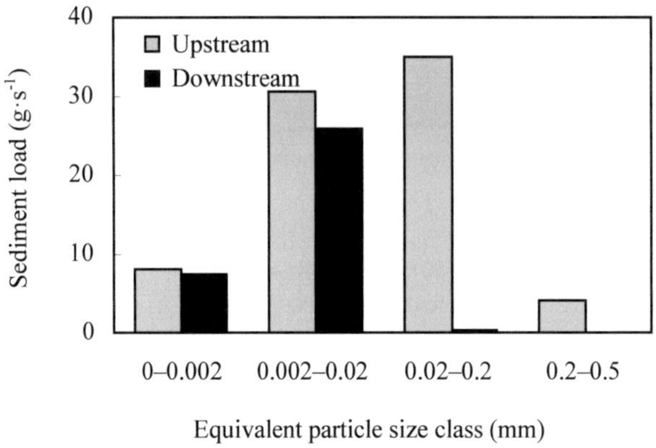

Fig. 4. Sediment load in equivalent particle size class (C-3)

3.3 Experiment on flow conditions in grass and rod models

Table 3 shows the values of parameters a and b of the empirical model $I = a{\cdot}u^b$ determined by measuring the average flow velocity in the cross section u ($m{\cdot}s^{-1}$) and the hydraulic gradient I obtained by the experiments on flow characteristics in the grass model and the rod model. Parameter b in the grass model was 1.17 to 1.32 while that in the rod model was 1.36. The above results indicate that the flows in the grass model and the rod model are in a state of transition between laminar flow and turbulent flow. If flow is turbulent the square of the velocity is proportional to the hydraulic gradient and if flow is laminar the velocity is proportional to the hydraulic gradient (Nezu and Tominaga, 2000). The above observation suggests that sediment deposited in the flume bed is less likely to be picked up as suspending sediment, because flows in the model are not in a fully developed turbulent state for flow in the model of the experiments performed in 2.1, and 2.2.

4. Numerical model calculation

A numerical model was constructed for simulating surface flow and sediment transport processes in and around the rod model and the grass model.

Model type	Model height (m)	Parameter		Coefficient of determination
		a	b	
Grass model	0.10	1.60	1.32	0.99
Grass model	0.077	2.52	1.31	0.99
Grass model	0.055	3.71	1.31	1.00
Grass model	0.032	2.93	1.17	0.99
Bamboo rod model	0.15	2.18	1.36	0.99

Table 3. Value of parameter a and b in empirical model $I = a{\cdot}u^b$

4.1 Basic model equations

When simulating the sediment trapping process in the flume experiments with the rod model, the continuity equation (1) and the momentum conservation equation (2) were employed for calculating flows in the flume, while calculation of sediment transport used the continuity equation (3).

$$\frac{\partial B\theta h}{\partial t}+\frac{\partial Q}{\partial x}=0 \tag{1}$$

$$\frac{\partial Q}{\partial t}+\frac{\partial}{\partial x}\left(\frac{gB\theta h^2}{2}+\frac{Q^2}{B\theta h}\right)=-gB\theta h\left\{\frac{\partial z}{\partial x}+\frac{n^2Q^2}{B^2h^{10/3}}+\frac{1}{2g}NdC_d\left(\frac{Q}{B\theta h}\right)^2\right\} \tag{2}$$

$$\frac{\partial z}{\partial t}+\frac{1}{1-\lambda}\left\{\frac{\partial\sum(q_{bi})}{\partial x}+\sum(q_{sui}-w_{0i}C_i)\right\}=0 \tag{3}$$

where; B: flume width, θ: volumetric water content in the water flowing cross section (θ= 0.80 for the rod model and θ= 0.85 for the grass model), h: flow depth, t: time, Q: flow rate, x: distance from the lower end of the flume, g: acceleration of gravity, z: flume bed level, n: Manning's roughness coefficient (n = 0.013), N: number of rods per unit area, d: diameter of rod, C_d: drag coefficient of rod (C_d = 2.5), λ: porosity of deposited sediment (λ= 0.42 for GK-9, λ= 0.40 for other than GK-9), q_{bi}: bed load transport rate per unit width of sediment in an ith size class, q_{sui}: pick-up rate of suspended load per unit area of sediment in an ith size class, C_i: sediment load concentration in an ith size class, and w_{oi}: settling velocity of sediment in an ith size class. q_{bi} is calculated by the Meyer-Peter Müller formula, while q_{sui} for the flume section is calculated by using the Itakura and Kishi formula (1980), wherein the parameter K = 0.0001 (Shimada et al., 2005). On the basis of the experimental results in 3.3, q_{sui} for the model section for both models is set at zero. The Rubey equation is used to calculate w_{oi}.

The above equations (1) and (3) are also used for experiments with the grass model, but equation (4) given below is used instead of (2) as the equation of conservation of momentum for calculating flows in the flume.

$$\frac{\partial Q}{\partial t}+\frac{\partial}{\partial x}\left(\frac{gB\theta h^2}{2}+\frac{Q^2}{B\theta h}\right)=-gB\theta h\left\{\frac{\partial z}{\partial x}+\frac{n^2Q^2}{B^2h^{10/3}}+\frac{1}{K_s^{\,2}}\left(\frac{Q}{B\theta h}\right)^2\right\} \tag{4}$$

where K_s is the grass permeability coefficient (Shimizu et al., 1991). For K_s (m·s^{-1}) an empirical formula (K_s = 0.0011exp(57.9·h)+0.313) obtained from the results of the flume experiment performed with a grass model was used, where h is the water depth (m).

4.2 Calculation method

The numerical calculation based on the above equations used the MacCormack method (Okabe, 1992). The difference distance interval Δx and difference time interval Δt were set at 0.02 m and 0.005 sec, respectively. Sediment particle sizes in terms of the equivalent particle sizes were divided into four classes: 0 to 0.002, 0.002 to 0.02, 0.02 to 0.2 and 0.2 to 0.5 mm. The first two classes were used for calculations as suspended sediment while the last two were used as bed load sediment. The upstream end in the calculation was set at a point 1 to 2 m on the upstream side from the hydraulic jump location in the experiments. At this point, the boundary conditions were set for Q, h and C_i, that is, the flow rate obtained in the experiments was taken as Q, the normal depth of the flow condition was taken as h, while C_i was calculated on the basis of sediment concentration and composition of particle sizes on the upstream side obtained in the experiment. The downstream end in the calculation was set at the downstream end of the flume, and the critical depth of the flow condition was given for h as a boundary condition.

First, calculations used only the flow equations under the initial conditions, and at a time when the steady state condition was reached, C_i was introduced at the upstream end as the sediment concentration to start the calculation. Thereafter, calculations were performed for the experiment duration for each case.

5. Results and discussion for numerical model simulation

5.1 Numerical simulation for rod model experiments

The measurements of the experiments using the bamboo rod model and the numerical simulation were compared. Figure 5 shows the measured and calculated values for the water surface and the flume bed levels in and around the rod model in the flume upon completion of the experiments on Cases A-1, A-2, A-7 and A-10. This figure indicates the calculations for the water surface and the bed level profile along the length of the flume are generally in good agreement with the experimental results. Also in terms of the water surface profile, the water level at each point including the hydraulic jump is generally simulated. Certain differences appear between the measured and calculated values of the bed level profile in the flume in the vicinity of the upstream end of the rod model. The sediment deposition depth peaks in the backwater section, while in the model section, the depth gradually becomes smaller towards the downstream end. These experimental features of sediment deposition were also simulated. The simulations also showed that the sediment concentration at the downstream end had almost no change over time.

Figure 6 indicates the comparison of the calculated and measured values for the time average of the sediment concentration at the downstream end of the flume. Calculated values of sediment concentration are similar to the experimental values, with the magnitude of relative error being 0 to 58% and mean magnitude of relative error being 23%. This shows the simulation generally reproduced the time average for the measurements of sediment concentrations obtained in each experiment.

The above comparisons suggest the numerical model established in this study can effectively simulate the sediment transport process in and around the rod model and also can estimate the time average of the concentration of sediment passing through the rod model.

5.2 Numerical simulation for grass model experiments

Then, the measurement results of the flume experiments with the grass model and the numerical simulation results were compared. Figure 7 shows the comparison of the calculated and measured values of the time average of the sediment concentration at the downstream end obtained in the experiments of Cases B-1 to B-6, and C-1 to C-4. Although two calculated values of B-1 and B-2 are larger than their measured values, in the other eight cases, there is good agreement between the calculated and measured values. The plotted points are above the 1:1 straight line, that is, calculated values tend to be larger than the measured values. Magnitude of relative error was in the range of 0 to 130%. More specifically, 72% and 130% for Cases B-1 and B-2, respectively, and 0 to 45% for the other cases, with the overall mean magnitude of relative error being 29%.

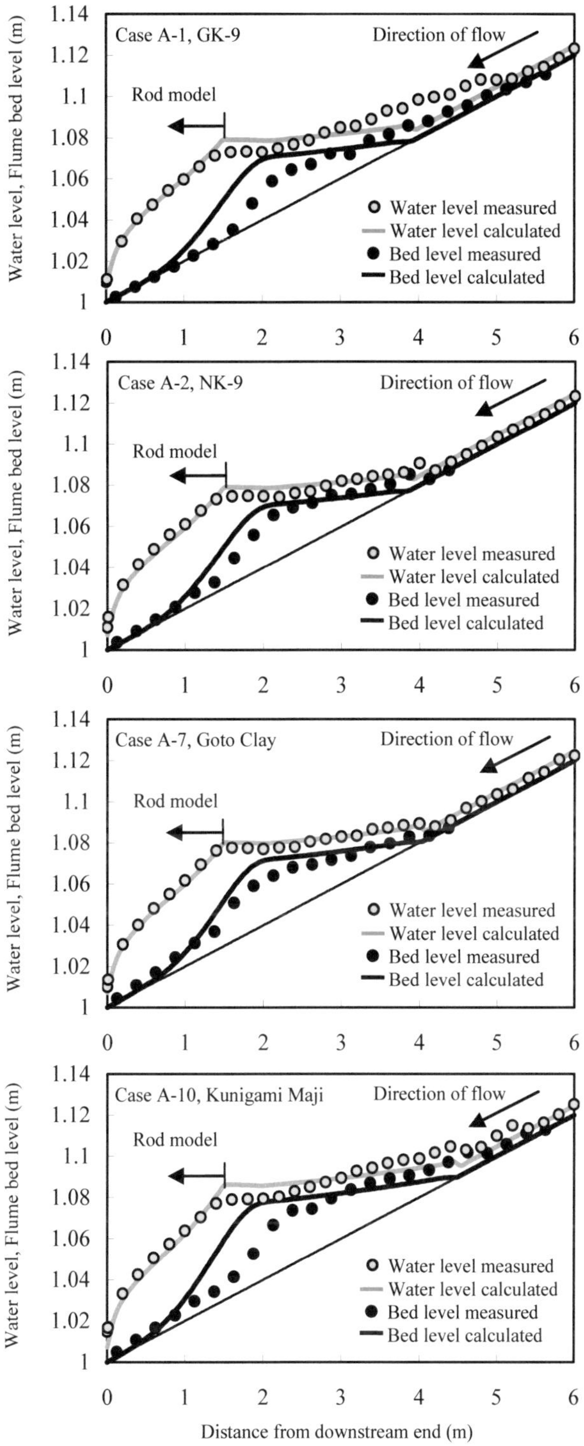

Fig. 5. Measured and calculated water and bed surface profiles in and around the rod model

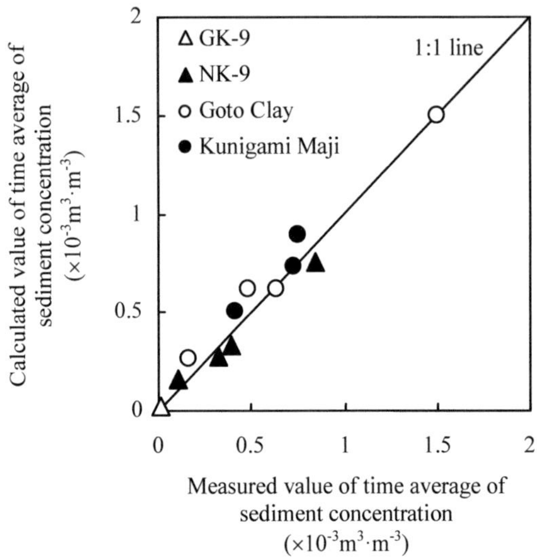

Fig. 6. Relationship between measured and calculated value of the time average of sediment concentration at the downstream end of the flume (A-1 to A-12)

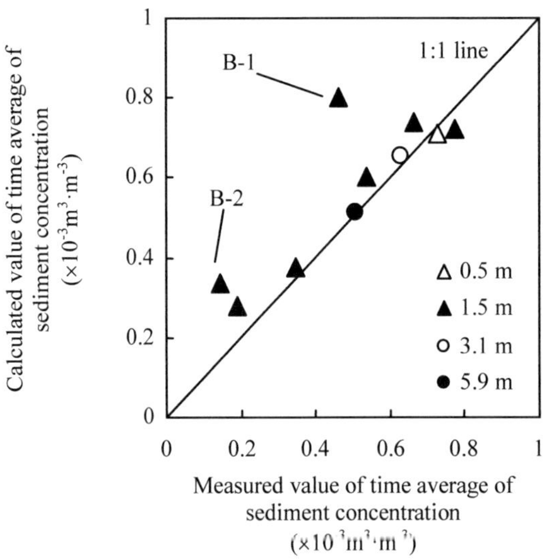

Fig. 7. Relationship between measured and calculated values of the time average of sediment concentration at the downstream end of the flume (B-1 to B-6, C-1 to C-4)

Figure 8 shows the comparison of the concentration of sediment having equivalent particle sizes of 0 to 0.002 mm, where all the calculated values are above the measured ones. The calculated values of Cases B-1, B-2 and B-4 are fairly large compared with the measured values, and in the other seven cases, the calculated values are somewhat larger than the measured ones. Magnitude of relative error was in the range of 12 to 147%. More specifically, the values were 137%, 147% and 109% for Cases B-1, B-2 and B-4, respectively, and 12 to 51% for other cases, with the overall mean magnitude of relative error being 62%.

The comparisons of the concentration of sediment having an equivalent particle size of 0.002 to 0.02 mm are given in Fig. 9. The calculated value of Case B-2 is fairly large compared with the measured value, but in the other nine cases the calculated values are similar to the measured ones. Magnitude of relative error was in the range of 3 to 124%. More specifically, the values were 124% for B-2 and 3 to 57% for the other cases, with the overall mean magnitude of relative error being 27%.

For the two equivalent particle size classes larger than 0.02 mm, all the time averages of the calculated results of C_{out} were 0.00 $m^3 \cdot m^{-3}$. Comparison with the experimental results presented in 3.2 shows that the calculated values well reproduce the measured values of each experiment.

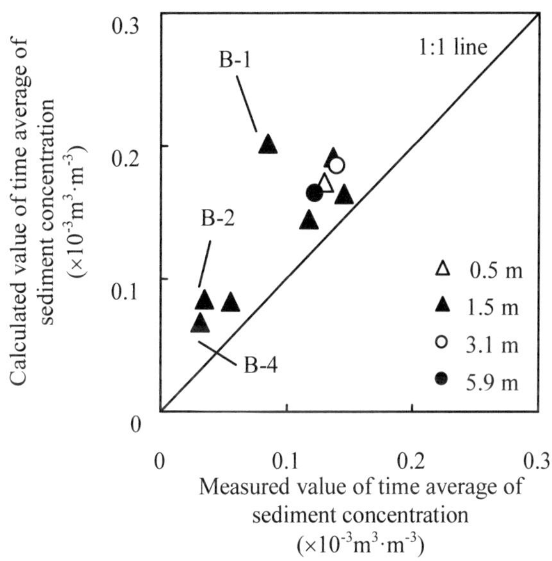

Fig. 8. Relationship between measured and calculated value of the time average of sediment concentration of 0–0.002 mm-size particles at the downstream end of the flume (B-1 to B-6, C-1 to C-4)

Comparing the time averages of the sediment concentrations for the grass model flume experiments with the numerical simulation showed good agreement for the equivalent particle size classes other than the 0 to 0.002 mm class, the calculated sediment concentration of 0 to 0.002 mm class was large. Since the ratio of the equivalent particle size class 0 to 0.002 mm was only 13% of all the sediment used in the experiments, the influence of over estimation for this size class in the model simulation does not severely affect the overall estimation. Therefore, the numerical model established in this study is valid for determining concentrations of sediment passing through the grass. The validation results indicate the model is useful to evaluate grass strips for reduction of sediment runoff.

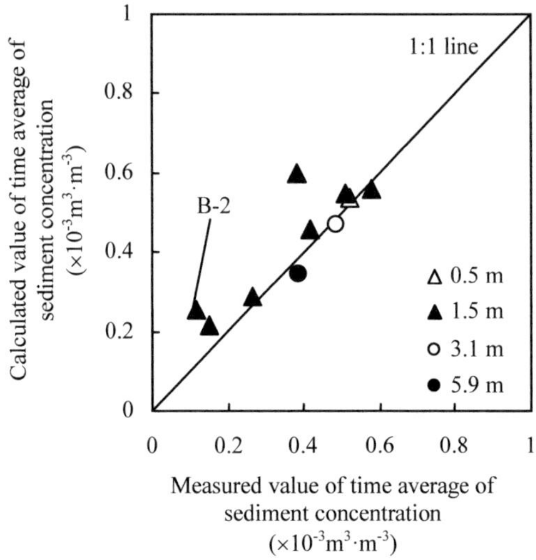

Fig. 9. Relationship between measured and calculated value of the time average of sediment concentrations of the 0.002–0.02 mm-size class at the downstream end of the flume (B-1 to B-6, C-1 to C-4)

6. Conclusion

This study provides an understanding of the sediment transport process in and around grass strips as part of an effort to accurately evaluate the effect of grass strips to reduce the runoff of reddish sediments to prevent runoff of reddish sediment in the Okinawa region of Japan. Based on this understanding, a mathematical model has been established to simulate the processes.

That is, flume experiments performed to simulate sediment transport in and around grass strips demonstrated (i) the features of deposition depth distribution of sediment trapped by the rod model, (ii) the relationship between the reduction ratio of runoff of reddish sediment in a grass model and the inflow rate, the length of the grass model and particle size classes, and (iii) flows in the grass are in a state of transition between laminar and turbulent flow.

Calculations from the numerical model for the rod model flume experiments verified that the model generally simulated (i) the water and the bed level profiles of the flume used in the experiments where the rod model was applied, and (ii) the concentration of sediment on the downstream side of the rod model.

Calculations with the numerical model generally simulated the concentrations of sediment for the downstream end of the grass model that were measured in the flume experiments. Although the concentration of sediment with an equivalent particle size of 0 to 0.002 mm was fairly large, the concentration of sediment with other particle sizes and the overall concentrations were in good agreement. Therefore, the numerical model established in this study can effectively estimate the concentration of sediment passing through the grass model and also the concentration of sediment of each equivalent particle size class demonstrated in this study.

Further studies are required on applicability of the numerical model under field conditions, followed by evaluation of the sediment trapping effect of grass strips with the numerical model simulation.

7. Acknowledgements

This study was partly granted by KAKENHI (No: 20580271). In conducting the flume experiments, Mr. Kazuya Ishizaka of the National Agricultural Research Center for Kyushu Okinawa Region offered assistance. We would like to express our sincere thanks for his efforts.

8. References

Dillaha, T.A., Reneau, R.B., Mostaghimi, S. and Lee, D. (1989). Vegetative Filter Strips for Agricultural Nonpoint Source Pollution Control. *Trans. ASAE,* Vol.32, No.2, pp. 513-519, ISSN 0001-2351

Haan, C.T., Barfield, B.J. & Hayes, J.C. (1994). Sediment control structures, In: *Design Hydrology and Sedimentology for Small Catchments,* Haan, C.T., Barfield, B.J. & Hayes, J.C., pp. 359-375, Academic Press, ISBN 0-12-312340-2, San Diego

Hudson, N. (1995). *Soil conservation,* Bastford, ISBN 0-7134-7353-3, London

Itakura, T. & Kishi, T. (1980). Open Channel Flow with Suspended Sediment. *Proc. ASCE,* Vol.106, No.HY8, pp. 1325-1343, ISSN 0733-9429

Minami, N., Yamada, T., Nakano, M., Tomisaka, M., Tokunaga, T. & Yamashiro O. (2002). The characteristics of red sediment discharge at different stages of cultivation of pineapple fields (in Japanese with English abstract). *J. JSECE,* Vol.54, No.5, pp. 30-38, ISSN 0286-8385

Morgan, R.P.C. (1995). *Soil erosion and conservation,* Longman, ISBN 0-582-24492-7, Essex, England

Nakasone, K., Higa, E., Mitsumoto, H. & Omija (1998). Estimation of soil loss in Okinawa prefecture (II) (in Japanese). *Ann. Rep. Okinawa Pref. Inst. Health Environ.,* No.32, pp. 67-72, ISSN 1341-0636

Nezu, I. & Tominaga, A. (2000). *Hydraulics* (in Japanese), Asakura Publishing, ISBN4-254-26139-X, Tokyo

Okabe, T. (1992). Improved model for bed-level changes in mountain rivers. *Proc. of the Int. Symposium on Erosion, Debris Flows and Environment in Mountain Regions,* pp. 139-146, ISBN 0-947571-38-8, Chengdu, China, July 1992

Osawa, K., Yamaguchi, S., Ikeda, S. & Takamuku, K. (2005). Field observation of sediment runoff reduction methods on farmland (in Japanese with English abstract). *Ann. J. Hydraul. Eng., JSCE,* No.49, p.p. 1099-1104, ISSN 0916-7374

Shimada, T., Yoshikawa, Y. & Watanabe, Y. (2005). Sediment transport in the Nibutani dam reservoir at 2003 flood of the Saru river (in Japanese with English abstract). *Ann. J. Hydraul. Eng., JSCE,* No.49, pp. 913-918, ISSN 0916-7374

Shimizu, Y., Tsujimoto, T., Nakagawa, H. & Kitamura, T. (1991). Experimental study on flow over rigid vegetation simulated by cylinders with equi-spacing (in Japanese with English abstract). *Proc. JSCE,* No.438/II-17, pp. 31-40, ISSN 0289-7806

Shiono, T., Nakamura, H., Haraguchi, N., Taruya, H. & Miyamoto T. (2005). Effectiveness of Vegetative Filter Strips for Sediment Removal under Field Conditions. *J. Agric. Meteorol.,* Vol.60, No.5, pp. 1021-1024, ISSN 0021-8588

Shiono, T., Yamamoto, N., Haraguchi, N. & Yoshinaga, A. (2007). Performance of Grass Strips for Sediment Control in Okinawa. *JARQ*, Vol.41, No.4, pp. 291-297, ISSN 0021-3551

Sugawara, K., Ohwaki, Y. & Banzai, K. (2001). Erosion Control in Pineapple Fields on the Island of Ishigaki. *JARQ*, Vol.35, No.2, pp. 91-96, ISSN 0021-3551

Yoshinaga, A. & Onaga, K. (1993). Suspended soil from reclaimed farmland of Kunigami-Maji area in Okinawa (in Japanese with English abstract). *Trans. JSIDRE*, No.168, pp. 105-110, ISSN 0387-2335

8

Clear-Water Scour at Labyrinth Side Weir Intersection Along the Bend

M. Emin Emiroglu
Firat University
Turkey

1. Introduction

Weirs are among the oldest and simplest hydraulic structures, and have been used for centuries by hydraulic and environmental engineers for flow measurement, flow diversion, regulation of flow depth, flood passage, and other means. Although the definition of many different kinds of weirs is very simple and similar, the application and hydraulic behavior of each is quite different (Borghei *et al.*, 1999). Weirs are normally placed perpendicular to the direction of flow of the main channel. The most significant parameters in determining the capacity of a weir are its height relative to the upstream depth, the crest shape, and the crest length. Here, capacity refers to the flow rate or discharge for a given depth of flow over the crest of the weir. Of these parameters, the crest length has the greatest influence on the discharge capacity. The crest length can be increased further and can still keep the downstream dimension small by folding the weir into several sections. One implementation of this idea is the duckbill spillway (Falvey, 2003). A Labyrinth weir is an overflow weir folded in plan view to provide a longer total effective length for a given overall weir width. A Labyrinth weir has advantages compared to the straight overflow weir and the standard ogee crest. The total length of the labyrinth weir is typically three to five times the weir width. Its capacity varies with head and is typically about twice that of a standard weir or overflow crest of the same width (Tullis *et al.*, 1995). Labyrinth weirs have long been used as the spillway of a dam reservoir. Emiroglu *et al.* (2010) used the labyrinth weirs as a side weir structure. If the crest is placed parallel with the channel centerline, it is called a side weir. A side weir is an overflow weir framed in the side of a channel, which allows lateral overflow when the surface of the liquid in the channel rises above the weir crest. Side weirs, also known as a lateral weir, are control structures, widely used in canal systems and for storm water overflow from urban drainage systems. Like normal weirs, side weirs may be sharp, broad or rounded-crested. In addition, flow in the main channel along a side weir may be subcritical or supercritical.

The flow over a side weir falls within the category of spatially varied flow. The existing studies of side weir flow deal mainly with the application of the energy principle. The energy principle assumes that the longitudinal component of velocity vector of spill flow at any section is equal to the average velocity of flow in the channel. Therefore the total energy per unit mass of water remaining in the channel is unaffected by the spill flow occurring and, apart from frictional losses, the total energy of the flow in the main channel remains constant. The concept of constant specific energy (De Marchi, 1934) is often adopted when studying the flow characteristics of these weirs. The discharge over the side weir per unit length, q, is assumed as:

$$q = -\frac{dQ}{ds} = C_d\sqrt{2g}\left[h-p\right]^{3/2} \tag{1}$$

where Q is the discharge in the main channel, s is the distance from the beginning of the side weir, dQ/ds (or q) is the spill discharge per unit length of the side opening, g is the acceleration due to gravity, p is the crest height of the side weir, h is the depth of flow at the section s (at s=0: h=h_1 and Q=Q_1), (h-p) is the pressure head on the weir, and C_d is the discharge coefficient (De Marchi coefficient) of the side weir (see Fig. 1). In which C_d is dimensionless, g is m/s^2 and, h and p are in meters. Thus, the discharge within the main channel at the downstream end of the side weir can be written as:

$$Q_2 = Q_1 - Q_w \tag{2}$$

where Q_2 is the discharge within the main channel at the downstream end of the side weir and Q_1 is the total discharge within the main channel at the upstream end of the side weir.

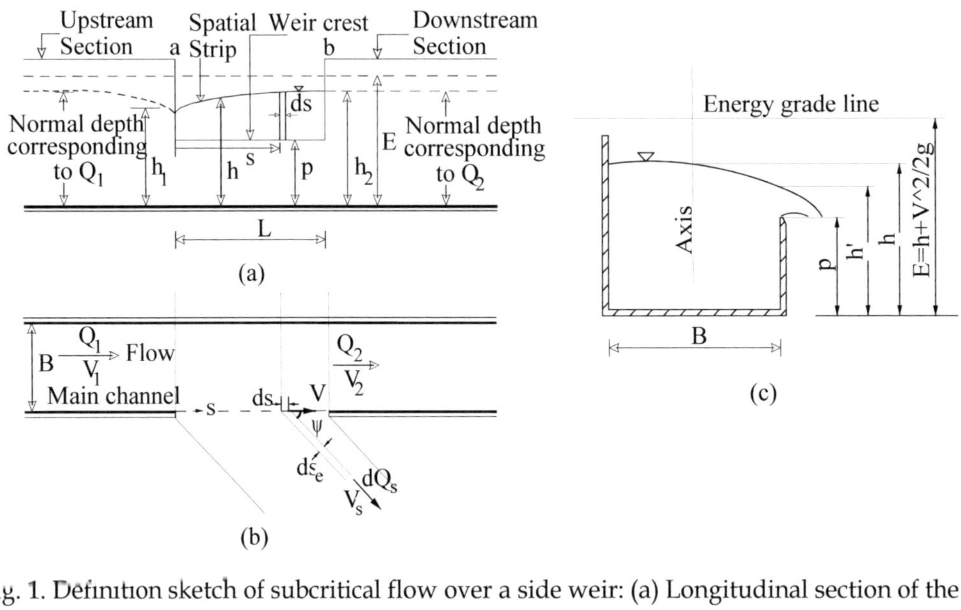

Fig. 1. Definition sketch of subcritical flow over a side weir: (a) Longitudinal section of the side weir, (b) Plan view of the side weir, (c) Front view of the side weir.

2. Experimental set-up and experiments

Labyrinth side weir experiments were conducted at the Hydraulic Laboratory of Firat University, Elazig, Turkey (Emiroglu *et al.*, 2007). A schematic representation of the experimental set-up is shown in Fig. 2. The experimental set-up consisted of a main channel and a discharge collection channel. The main channel was 12 m long and the bed had a rectangular cross-section. The main channel was 0.5 m wide, 0.5 m deep, with a 0.001 bed slope. The channel was constructed from a smooth, horizontal, well-painted steel bed with vertical glass sidewalls. A sluice gate was fitted at the end of the main channel in order to control the depth of flow. The collection channel was 0.5 m wide and 0.7 m deep, and was situated parallel to the main channel. The width of the collection channel across the side weir was 1.3 m and constructed in a circular shape, to provide free overflow conditions. A rectangular weir was placed at the end of the collection channel, in order to measure the discharge of the side weir. A Mitutoyo digital point gauge with ±0.01 mm sensitivity was fixed at a location 0.4 m from the weir. Trapezoidal labyrinth side weirs were produced from steel plates, which had sharp edges and were fully aerated. These were installed flush with the main channel wall.

Experiments were conducted at subcritical flow, stable flow conditions and free overflow conditions. Surface tension σ is an important parameter for small nape heights. Novak & Cabelka (1981) reported that minimum nape height over side weirs should not be less than 30 mm because of the surface tension over the weir crest. As mentioned above, the minimum nape height over the side weir is taken as 30 mm. Therefore, the influence of surface tension can be ignored.

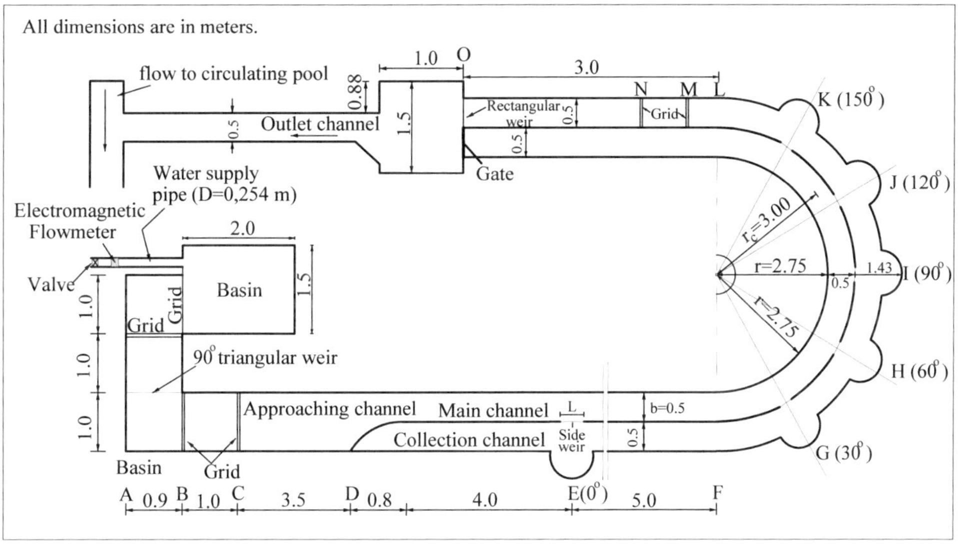

Fig. 2. Experimental arrangement.

3. Discharge coefficient for labyrinth side weirs

3.1 Triangular labyrinth side weir

Figure 3 shows schematic and plan view of a triangular labyrinth side weir with one cycle. Emiroglu *et al.* (2010) studied the discharge coefficient of sharp-crested triangular labyrinth side weirs on a straight channel. They obtained the following results regarding the labyrinth side weir discharge coefficient: (1) Discharge coefficient of the labyrinth side weir is 1.5 to 4.5 times higher than that of the rectangular side weir. (2) The discharge coefficient C_d increases when L/B ratio increases. A decrease in the labyrinth weir included angle θ causes a considerable increase in C_d, due to increasing the overflow length. A labyrinth side weir with θ=45° was shown to have the greatest C_d values of the weir included angles that were tested.

Emiroglu *et al.* (in press) gave the Eq. (3) for the triangular labyrinth side weirs. This equation was developed for $(h_1\text{-}p)\geq 30$ mm. In which, $(h_1\text{-}p)$ is nap height (i.e., water head above the weir). Moreover, Eq. (3) is valid for $45°\leq\theta\leq 150°$ in subcritical flow regime.

$$C_d = 0.4 + \left[\begin{array}{l} -2.62 + 0.634\left(\frac{L}{B}\right)^{0.254} + 3.214\left(\frac{L}{\ell}\right)^{-0.122} - \\ -0.684\left(\frac{p}{h_1}\right)^{-0.4} + 0.122\left(\sin\left(\frac{\theta}{4}\right)\right)^{1.982} + 0.22\mathrm{F}_1^{2.458} \end{array} \right]^{3.857} \quad (3)$$

where the weir width L, the channel width B, the weir crest length ℓ, the height of weir crest p are in meters, the labyrinth side weir included angle θ is in degrees and Froude number F_1 is dimensionless. Determination coefficient of this equation (R^2) is 0.94. It should be noted that Eq. (3) is valid for labyrinth side weir with one cycle.

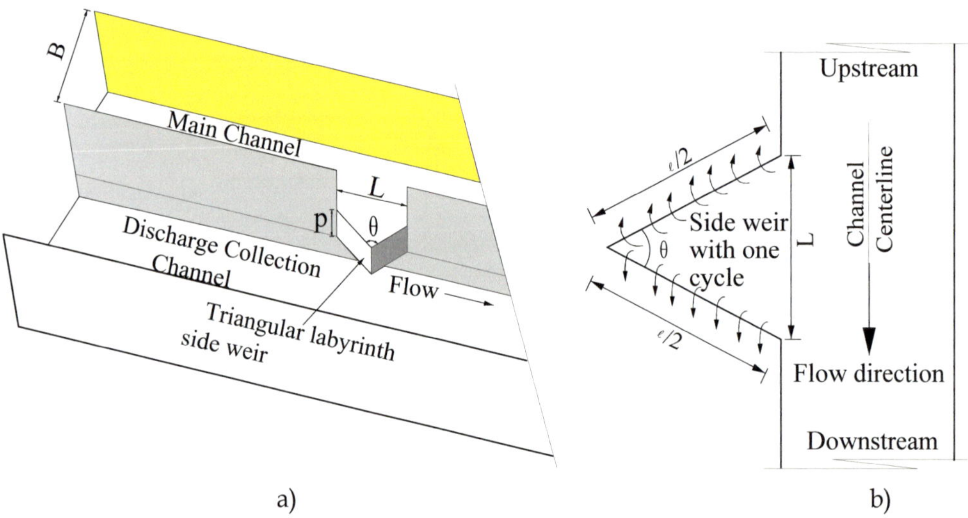

Fig. 3. Triangular labyrinth side weir: (a) Schematic view, (b) Plan view.

Fig. 4 shows the variation of Q_w/Q versus F_1 for constant crest height (p=0.20 m) and weir included angle (θ=45°). In which, Q_w is total flow over the side weir and Q is the discharge in the main channel before beginning of side weir. The ratio of Q_w/Q decreases with increasing Froude number. The primary reason is that the increase in the Q_w is less that that of Q. For example; Q_w is 14.62 L/s for Q=125 L/s, F_1=0.765, θ=45°, p=20 cm and L=50 cm.

Figure 5 shows that C_d is plotted against F_1. Obtained values for triangular labyrinth side-weirs having θ=45° for L/B=1.5 located on a straight channel were compared with the classical rectangular side-weir values. It is observed from Figure 5 that discharge coefficients of labyrinth side weirs have much higher values than those of rectangular side weirs. The crest length of labyrinth side-weir is always longer than that of classical rectangular side-weir. The main reason of labyrinth side-weir is to occur higher length of the crest and more severe secondary flow.

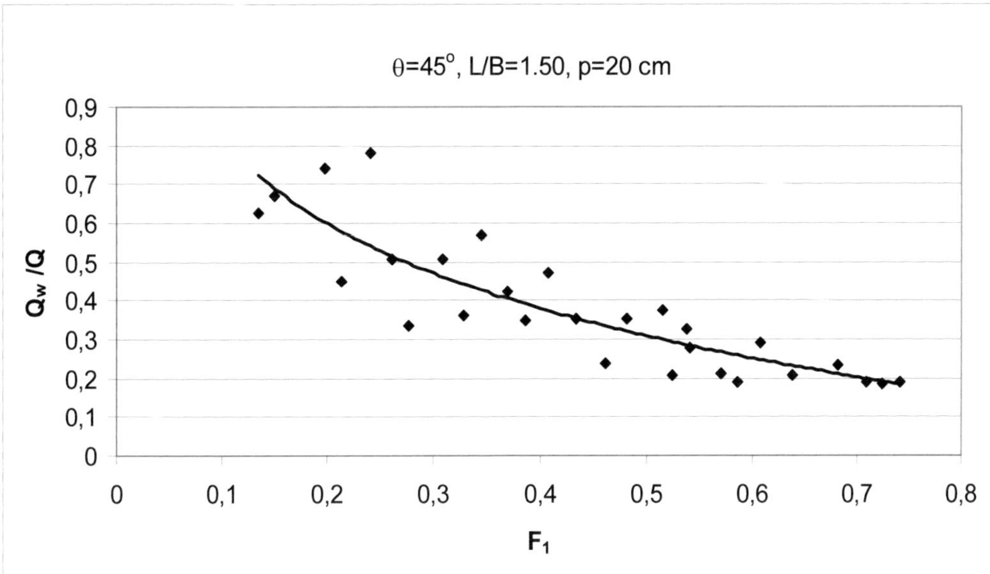

Fig. 4. Variation of Q_w/Q versus F_1 for θ=45°.

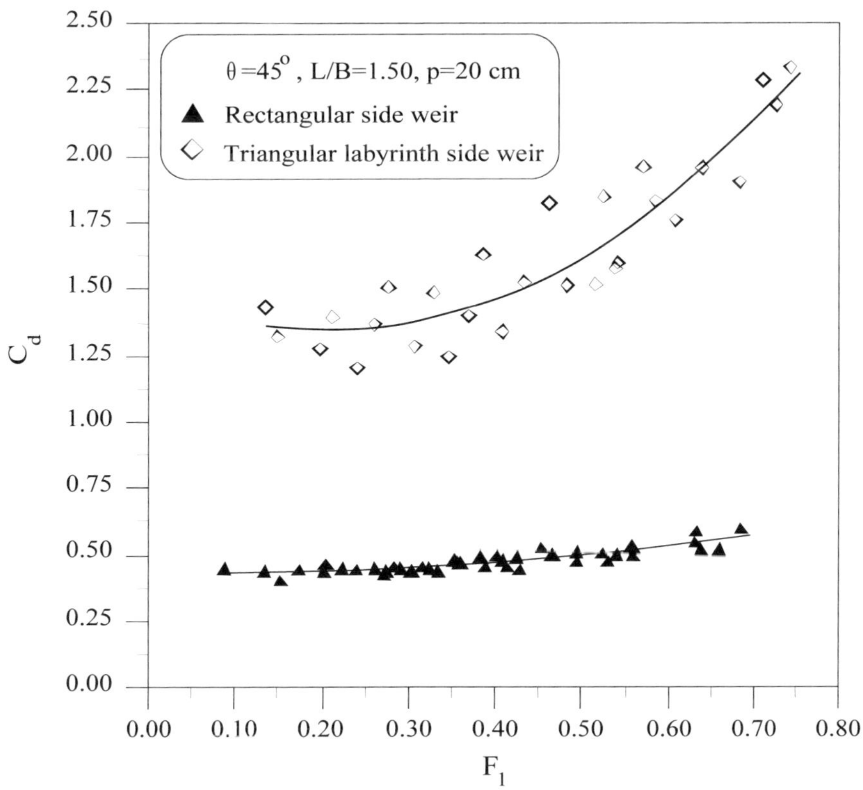

Fig. 5. Variations of C_d versus F_1

As it can be seen in Fig. 6 that as h_1-p/p ratio increases C_d values decrease. This situation is well known by hydraulic engineers. Tullis *et al.* (1995) and Falvey (2003) also stated that as nap height increase, discharge coefficient values decrease. Therefore, the designers should consider this important situation. It should be noted that the decreasing tendency in the discharge coefficient for the low weir included angles is higher than that of high weir included angles.

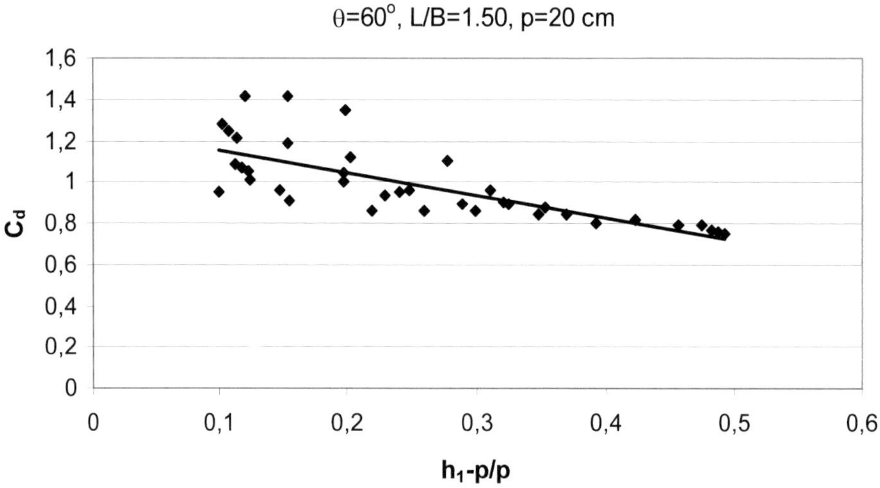

Fig. 6. Variation of C_d versus h_1-p/p for θ=60°

3.2 Trapezoidal labyrinth side weir

Figure 7 show schematic and plan view of a trapezoidal labyrinth side weir with one cycle. Laboratory experiments were carried out by Emiroglu & Kaya (2011) using a trapezoidal labyrinth side weir located on a straight channel. They obtained the following results regarding the labyrinth side weir discharge coefficient: (1) The discharge coefficients of trapezoidal labyrinth side weirs have much higher values than those of rectangular side weirs, but slightly higher than that of a triangular labyrinth side weir. The discharge coefficient C_d increases with reduction in sidewall angle. The primary reason for this may be the increased crest length. The intensity of secondary motion created by lateral flow increases with increasing overflow length. (2) The discharge coefficient of the trapezoidal labyrinth side weir is 1.5 to 5.0 times higher than for a rectangular side weir. (3) The values of the discharge coefficient C_d increase with an increase in p/h_1 and F_1 values. Moreover, the discharge coefficient C_d increases with increasing L/B ratio.

A nonlinear equation has been developed for the discharge coefficient of the trapezoidal labyrinth side weir. Minimum nap height at this equation is 30 mm.

$$C_d = \frac{\left[0.27 + 0.89\mathrm{F}_1^{6.28} + 0.95\left(\frac{L}{B}\right)^{0.21}\right]^{2.42}}{\left[0.09 + 0.16\left(\frac{L}{\ell}\right)^{-1.66} - 6.86\left(\frac{p}{h_1}\right)^{0.75} + 14.22(\sin\alpha)^{0.62}\right]^{0.60}} \tag{4}$$

where Froude number $F_1(=V_1/(g.h_1)^{1/2})$ is dimensionless, the discharge coefficient C_d is dimensionless, the crest height of the side weir p, the width of the side weir $L(=2(2c+d.\sin\alpha))$, the width of the main channel B, the overflow length of the side weir ℓ (=2d+4c), the depth of flow at the upstream end of the side weir in the main channel centerline h_1 are in meters, and the sidewall angle α is degrees.

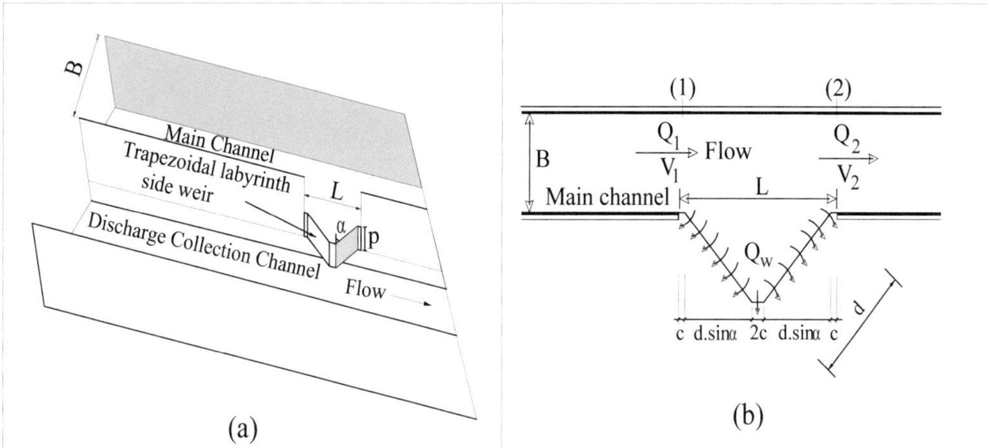

Fig. 7. Trapezoidal labyrinth side weir: (a) Schematic view, (b) Plan view

Equation 4 is valid for $16° \leq \alpha \leq 30°$. Correlation coefficient (R) of Eq. (4) is 0.97. It should be noted that the discharge coefficient of the trapezoidal labyrinth side weir is calculated for a single (i.e. a control section) cycle. The C_d that is calculated is substituted into Eq. (1). The depth of flow (h) in Eq. (1) is taken as the flow depth at upstream end of side weir at channel center (h_1).

Fig. 8 shows the variation of Q_w/Q versus F_1 for the crest heights (p=0.12, 0.16 and 0.20 m) and weir included angle (θ=45°). As mentioned above, the ratio of Q_w/Q generally decreases with increasing Froude number.

The values of C_d were plotted against F_1, as shown in Fig. 9. The values obtained for trapezoidal labyrinth side weirs having sidewall angle of α=30°, L/B=1.5 and the ratio of weir crest height to upstream flow depth [p/B=0.24, 0.32 and, 0.40] located on a straight channel were compared with values for the classic rectangular side weir. The discharge coefficients of trapezoidal labyrinth side weirs were found to have much higher values than those of rectangular side weirs. In other words, the discharge over trapezoidal labyrinth side weirs is greater than that over an equivalent straight weir. The overflow length of the trapezoidal labyrinth side weir is always longer than that of a classic rectangular side weir. The main reason is the greater length of the crest and the more severe secondary flow in trapezoidal labyrinth side weirs.

As h_1-p/p ratio increases C_d values decrease, as illustrated in Fig. 10. As mentioned above, this situation is very important for low sidewall angles because decreasing tendency in discharge coefficient for low sidewall angles is higher than that of high sidewall angles. This situation should be considered in the hydraulic design. To reduce the nap height, the labyrinth side weir can be selected as multi-cycles (see Fig. 11).

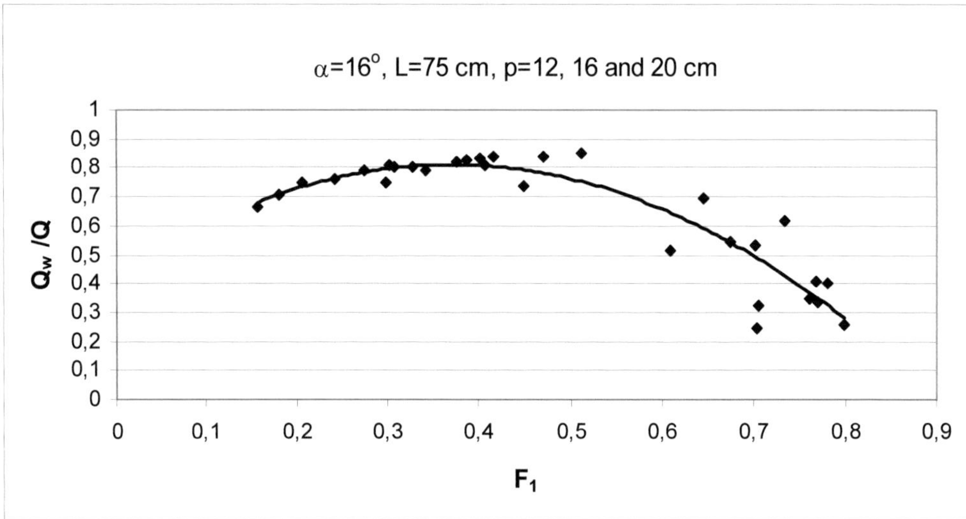

Fig. 8. Variation of Q_w/Q versus F_1 for α=16°.

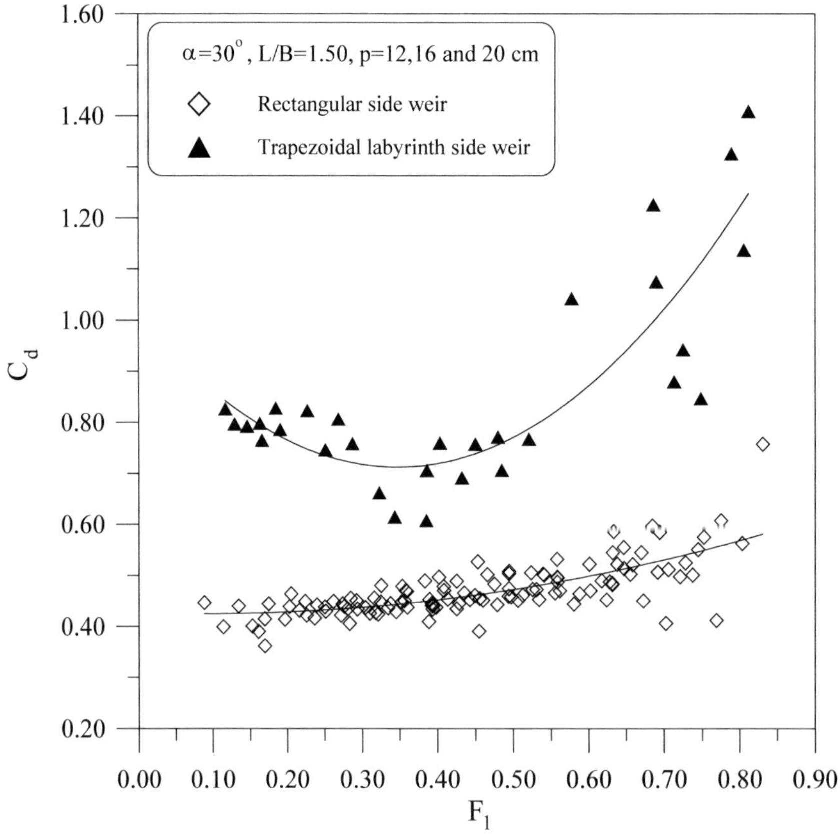

Fig. 9. Variations of C_d versus F_1.

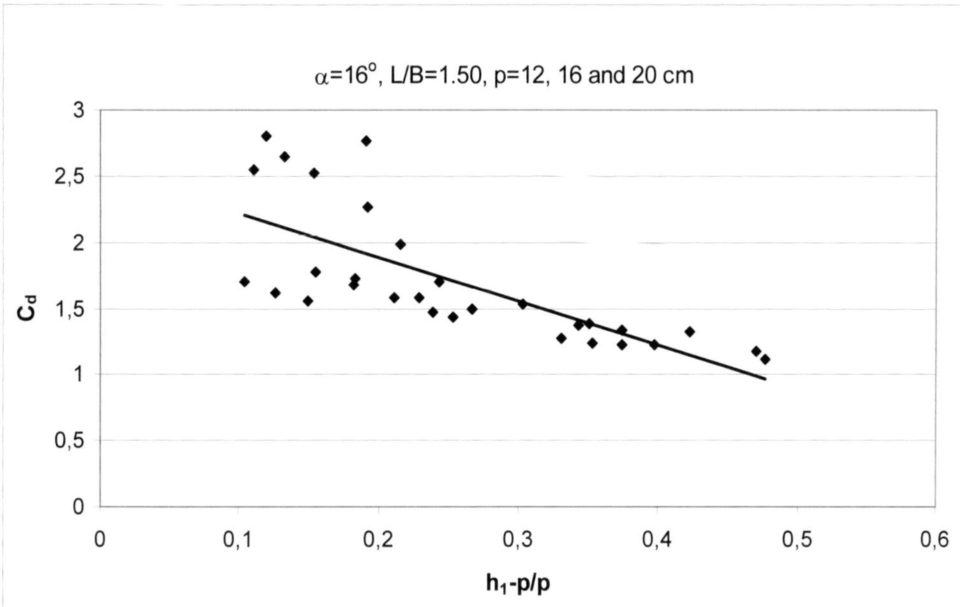

Fig. 10. Variation of C_d versus h_1-p/p for α=16°.

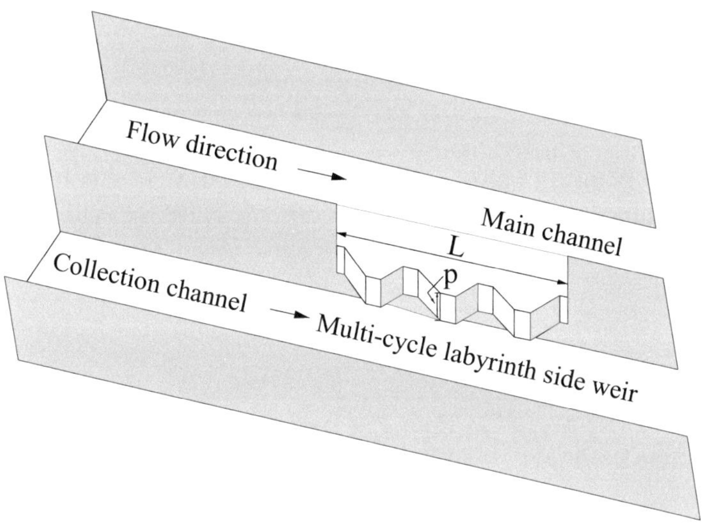

Fig. 11. Trapezoidal labyrinth side weir with multi-cycle.

3.3 Semi-elliptical labyrinth side weir

Figure 12 show schematic and plan view of a semi-elliptical labyrinth side weir with one cycle. Laboratory experiments using a semi-elliptical side weir located on a straight channel were carried out, in order to investigate the effect of the dimensionless parameters F_1, p/h_1, L/B, L/ℓ and b/a on the discharge coefficient (Kaya *et al.*, 2011). They obtained the following results regarding the labyrinth side weir discharge coefficient: (1) As a result of dimensional analysis, the following dimensionless parameters were found; $C_d = f(F_1, L/B, L/\ell, p/h_1, b/a)$. (2) The discharge coefficient of the semi-elliptical side weir is higher than that of the rectangular side weir, but lower than that of the triangular and trapezoidal labyrinth side weirs. (3) The discharge coefficient C_d increases with increasing L/B ratio.

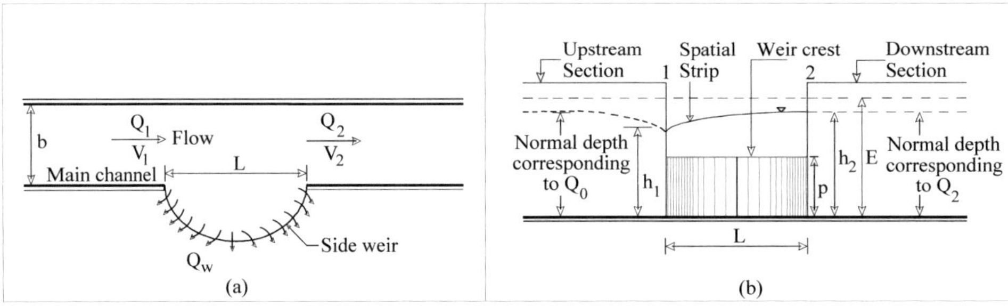

Fig. 12. Semi-elliptical side weir: (a) Plan view, (b) Longitudinal section.

An ellipse is the finite or bounded case of a conic section, the geometric shape that results from cutting a circular conical or cylindrical surface with an oblique plane. It is also the locus of all points of the plane whose distances to two fixed points add to the same constant. By using an appropriate coordinate system, the ellipse can be described by the canonical implicit equation

$$\frac{x^2}{a^2} + \frac{y^2}{b^2} = 1 \tag{5}$$

where, (x,y) are the point coordinates in the canonical system. In this system, the center is the origin (0,0) and the foci are $(-ea,0)$ and $(+ea,0)$. a is called the *major radius*, and b is the *minor radius*. The quantity $e = \sqrt{(1 - b^2/a^2)}$ is the *eccentricity* of the ellipse (Fig. 13).

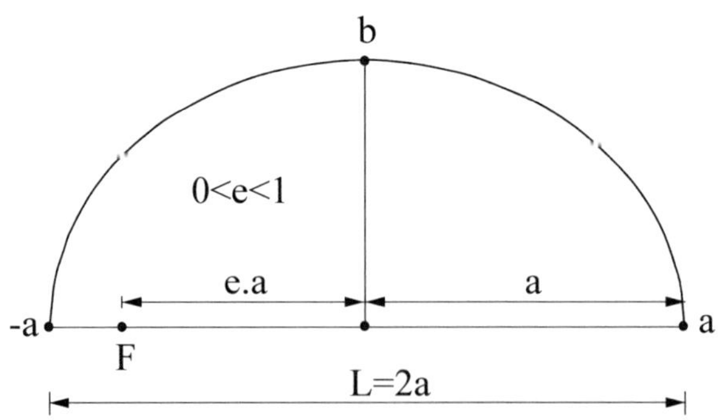

Fig. 13. The ellipse and some of its mathematical properties.

A nonlinear equation has been developed for the discharge coefficient of the semi-elliptical side weir. Minimum nap height was taken as 30 mm when Eq. (6) was developed.

$$C_d = \frac{\left[-0.025\mathrm{F}_1^{\;0.10} + 0.16\left(\frac{L}{B}\right)^{2.72}\right]^{0.05}}{\left[53.15\left(\frac{L}{\ell}\right)^{0.61} - 9.71\left(\frac{p}{h_1}\right)^{0.12} - 22.07\left(\frac{b}{a}\right)^{0.18}\right]^{0.17}} \qquad (6)$$

where Froude number F_1 is dimensionless, the discharge coefficient C_d is dimensionless, the crest height of the side weir p, the width (length) of the side weir L, the width of the main channel B, the major radius of ellipse a, the minor radius of ellipse b, the overflow length of side weir ℓ, and the depth of flow at the upstream end of the side weir in the main channel centerline h_1 are in meters.

Equation 6 is valid for $0.50 \leq b / a \leq 1.50$. It should be noted that the discharge coefficient of the semi-elliptical side weir is calculated for a single cycle. Correlation coefficient (R) of Eq. (6) is 0.93.

Figure 14 shows the variation of Q_w/Q versus F_1 for the crest heights (p=0.12, 0.16 and 0.20 m) and b/a=0.50. As mentioned above, the ratio of Q_w/Q generally decreases with increasing Froude number. This situation is similar to the other labyrinth side weirs.

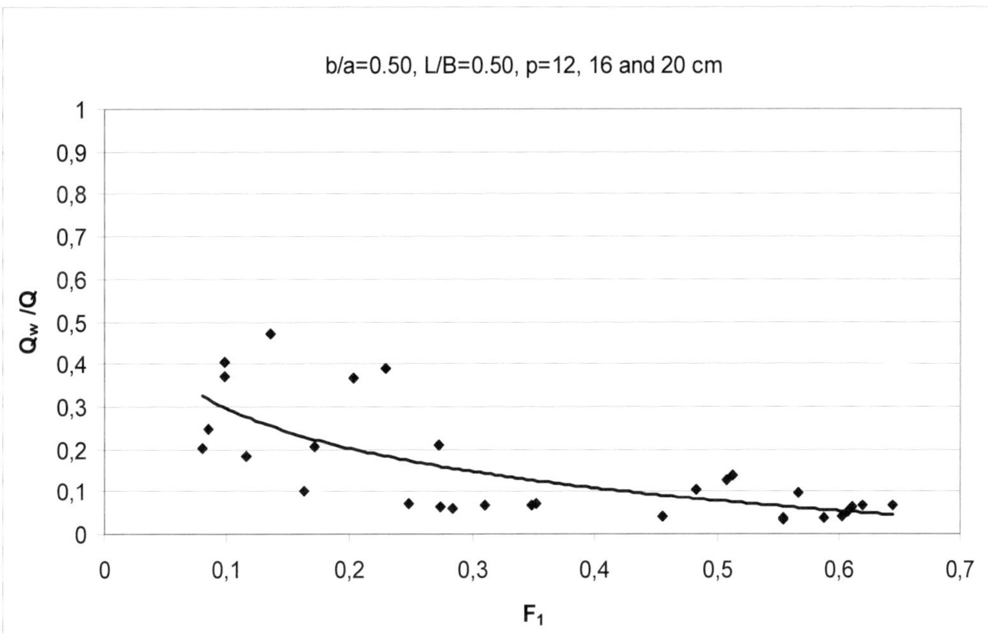

Fig. 14. Variation of Q_w/Q versus F_1 for b/a=0.50.

C_d is plotted against F_1, together with the values of the discharge coefficient of the rectangular side weir for the dimensionless ellipse radius b/a=1.50 and the dimensionless weir length L/B=1.50, as shown in Fig. 15. Figure 15 was plotted for p=12, 16 and 20 cm. The discharge coefficients of semi-elliptical side weirs have much higher values than those of rectangular side weirs. In other words, the discharge over semi-elliptical labyrinth side weirs is greater than that over an equivalent straight weir in the ratio of the weir sill lengths. In particular, the semi-elliptical side weir with b/a=1.50 has higher C_d values. The overflow length of the semi-elliptical side weir is always longer than that of a classical rectangular side weir. The main reason is the higher length of the crest and the more severe secondary flow in semi-elliptical side weirs.

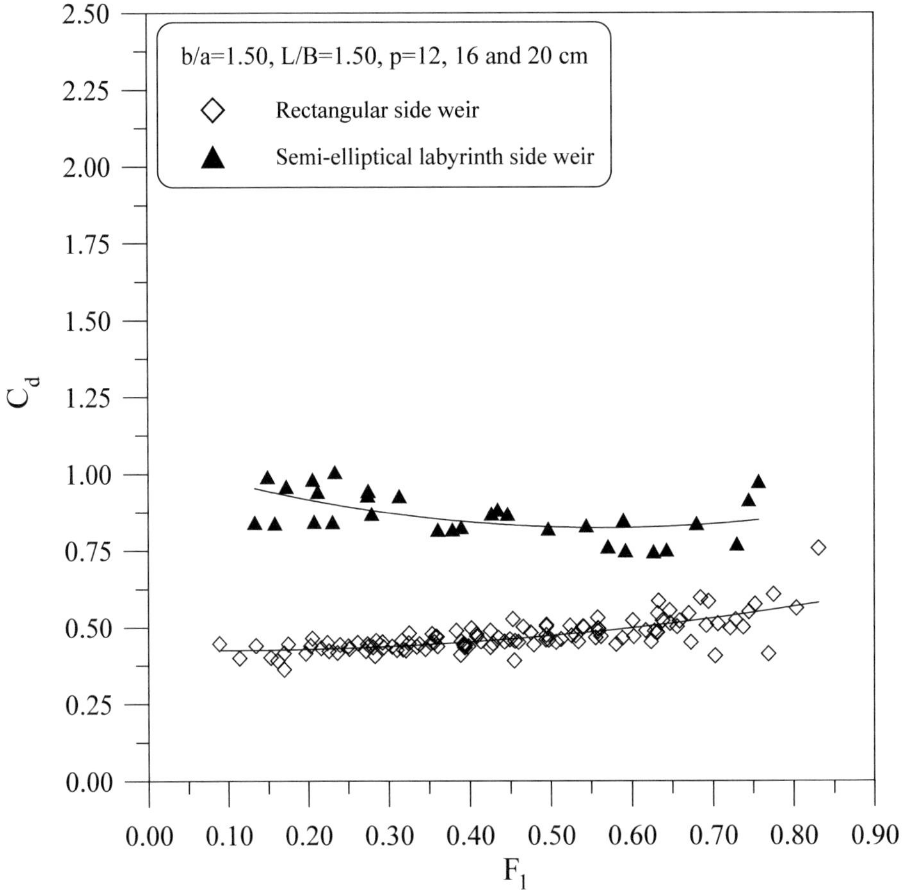

Fig. 15. Variations of C_d versus F_1.

Figure 16 shows the variation of C_d versus h_1-p/p ratio. The values of C_d decrease as h_1-p/p increases. It should be noted that decreasing tendency in Cd values is lower than that of triangular and trapezoidal labyrinth side weirs. This side weir also can be used as multi-cycles. Thus, the total crest length increases and the nap height over the weir decreases for a constant main channel discharge.

Consequently, to select trapezoidal labyrinth weir is more appropriate due to some reasons: For example, (1) Discharge coefficient of the trapezoidal labyrinth side weir is higher than that of the other labyrinth side weir types. (2) Its construction is easy in accordance with the other types. (3) The classical labyrinth weirs are also constructed as trapezoidal-shaped.

It should be noted that labyrinth side weirs can be constructed as multi-cycles (see Fig. 17). In this case, nap height decreases, discharge capacity over the weir increases.

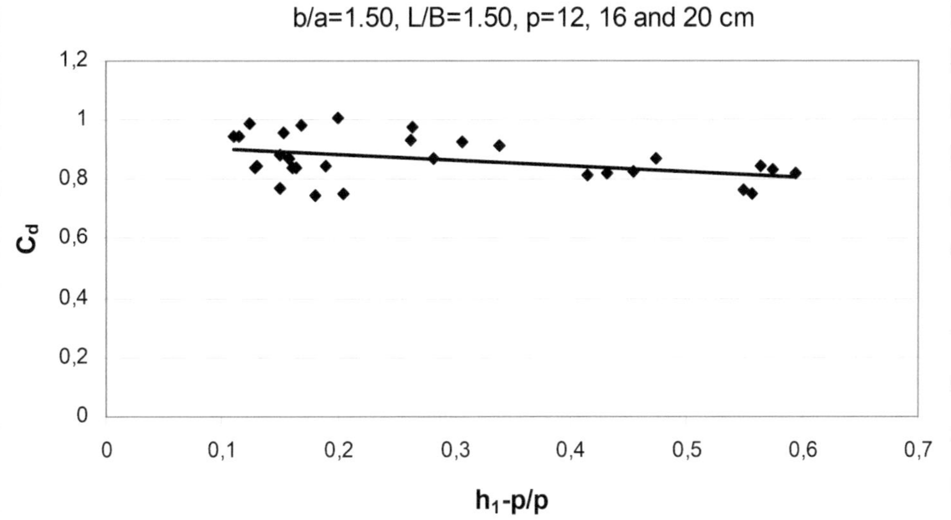

Fig. 16. Variation of C_d versus h_1-p/p for α=16°.

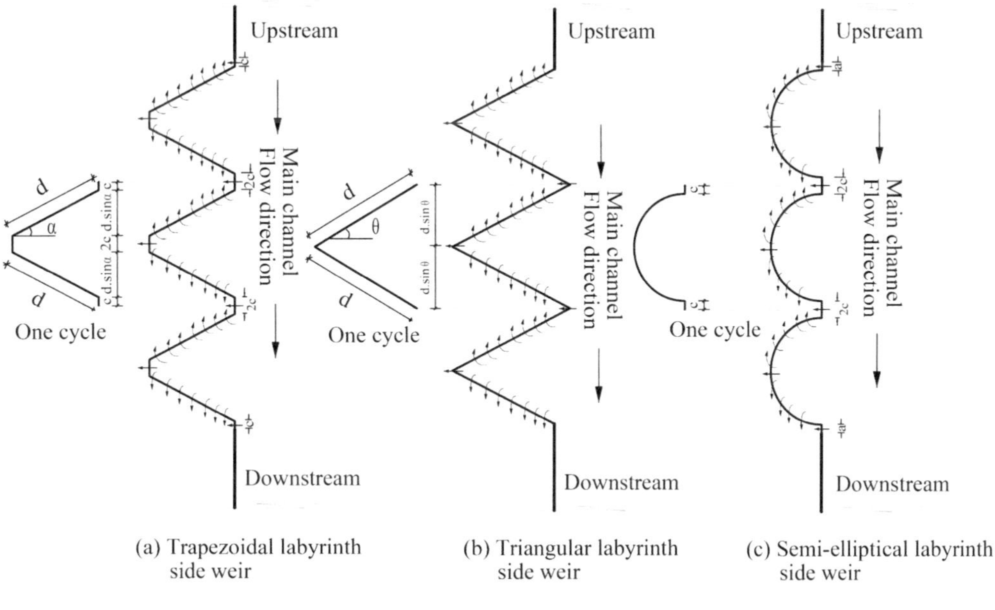

Fig. 17. Plan views of the labyrinth side weirs with multi-cycles.

4. Clear-water scour at labyrinth side weir intersection along the bend

Scour around hydraulic structures (for example; grade-control structures, pipe outlets, stilling basins, obstacles, bridge piers and lateral intakes) is a natural phenomenon by the flow of water in rivers and streams. Around these hydraulic structures, scouring introduces the possibilities of a reduction in the support given to foundations of these structures. Depths of scour can be seen to depend on the properties of the bend material in the stream (particle shape and size, grading, layering, alluvial or cohesive), the foundation geometries and the properties of flow (Raudkivi, 1986; Agaccioglu & Onen, 2005). Melville & Chiew (1999) reported that clear-water scour occurs for mean flow velocities, V, up to the threshold velocity for bed sediment entrainment, V_c, that is while live bed scour occurs for $V>V_c$. At the clear-water flow conditions, the local depth of scour in a uniform sediment increases almost linearly with flow intensity to a maximum value at the threshold velocity ($V=V_c$) (Agaccioglu & Onen, 2005).

A number of experiments for investigating the development of a scour hole geometry around a triangular labyrinth side weir are carried out with a model located on a rectangular plexiglas flume at the Hydraulic Laboratory of Firat University, Elazig, Turkey. The main and collection channels were located between two straight channels at a central angle of 180° with 3.00 and 3.50 m radius to the centerline, respectively. The bend is connected with an upstream straight reach of 13.30 m and a downstream straight reach of 3.00 m. The main channel is 0.50 m wide and 0.60 m deep with a well-painted bottom. The collection channel is 0.75 m deep, 0.50 m wide and situated parallel to the main channel. Details of the channel are shown in Fig. 2. Uniform bed materials are placed as a 20 cm thick layer in the main channel with a bed slope of 0.001. The bed material of the main channel was well mixed to prevent sediment grading and layering, and then flattened before each run. During the experiments, the upstream valve is adjusted slowly without causing any disturbance to the bed material until the desired flow conditions are obtained in the flume. No sediment inflow is allowed from upstream due to clear-water conditions. Triangular labyrinth side weirs were first placed at the 30° section of the bend centre and then at the 150° section of bend centre. Three different weir crest heights (0.07, 0.12 and 0.17 m) of the triangular labyrinth side weir were taken in this study.

The water depth was controlled by a sluice gate located at the end of the main channel to produce the uniform flow. Water was supplied from an overhead tank with a constant head and measured by an electromagnetic flowmeter (Siemens brand name). Moreover, the flow rate was also calibrated 90° V-notched weir (Q_1) at the beginning of the main channel. Side weir discharge at the end of the downstream of the collection channel (Q_w) was measured by calibrated a 90° V-notched weir. Thus, the downstream discharge (Q_2) in the main channel is obtained as:

$$Q_2 = Q_1 - Q_w \tag{7}$$

The measurements of water surfaces and bed levels are determined using digital point gauge with a capability of reading to the nearest 0.01 mm. In the present study, the mean particle size of the bed material (d_{50}) is 1.16 mm. The critical velocity (V_c) of motion of initiation is determined using a Shields diagram with a series of test runs which led to the following formula:

$$V_c = 0.1557.\log(4767.h_1) \tag{8}$$

where the flow depth at the upstream end of triangular labyrinth side weir at the channel centre h_1 is in meters and the critical velocity at the threshold conditions or initiation of motion V_c is in m.s^{-1}.

Figure 18 shows the development of clear-water scour with time in a curved channel when α=30°, θ=90°, L/B=0.50 and p/B=0.14. In other words; chute angle, weir included angle, dimensionless weir length and dimensionless crest height are constant in this figure, but approach flow intensity, V_1/V_c, is variable. Here, the weir opening length, L, is 25 cm, total crest length of the labyrinth side weir is 35.36 cm. Moreover, the crest height is 7 cm. As shown in Fig. (17), as V_1/V_c increases the depth of scour at the side-weir intersection, H, also increases. At the first 200 minutes, the increasing tendency in the scour depth is high. That is, the scour depth increases rapidly with time at the beginning. The increasing tendency in the scour depth is quietly low after from 600 minutes. Finally, the depth of clear-water scour approaches the equilibrium depth of scour, H_d, asymptotically with time, depending on flow intensity. In addition, there is no scour for approach flow intensities less than V_1/V_c=0.50 in the experimental study.

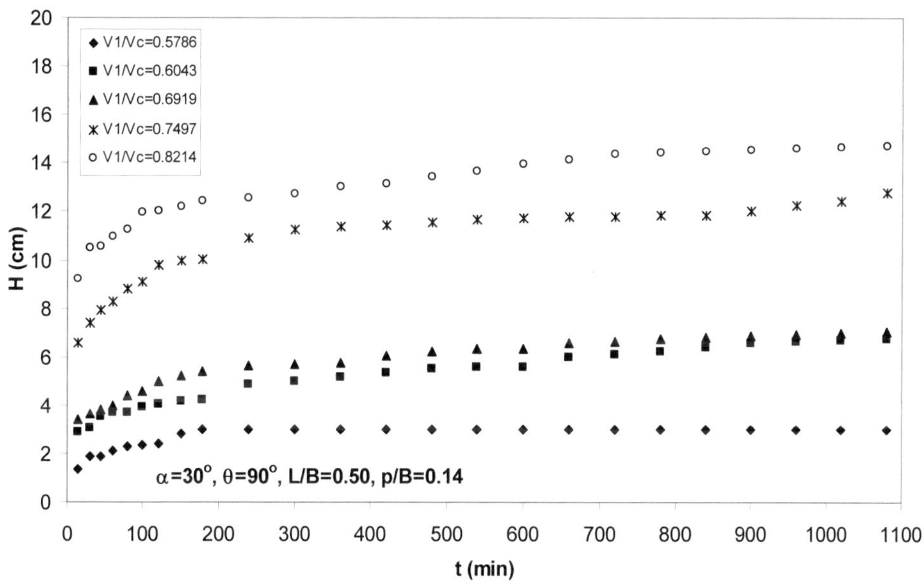

Fig. 18. The development of clear-water scour with time in a curved channel when α=30°, p=7 cm and θ=90°.

Figure 19 shows that the depth of scour at the side-weir intersection, H, is plotted against the temporal time, t, for α=150°. The increasing tendency of scour hole at the first 200 minutes for α=150° is higher than that of α=30°. Emiroglu *et al.* (2010) stated that the discharge capacity at the α=150° has the highest values according to the other angle of bend center (i.e., α=30, 60, 90 and 120°). Therefore, the intensity of the secondary motion is high at the angle of the bend center of 150°. Thus, higher scour holes generally were obtained at the angle of bend center of 150.

As is well known, the water intake structures in a curved river are generally located at the last part of the bend. Therefore, experiments are also conducted to determine the development of the depth of scour at the 150° section of the bend centre in order to understand the effect of time to scour depth for clear water. Figure 20 and Fig. 21 show the development of clear-water scour versus time. As shown in Figs. 20 and 21, the depth of

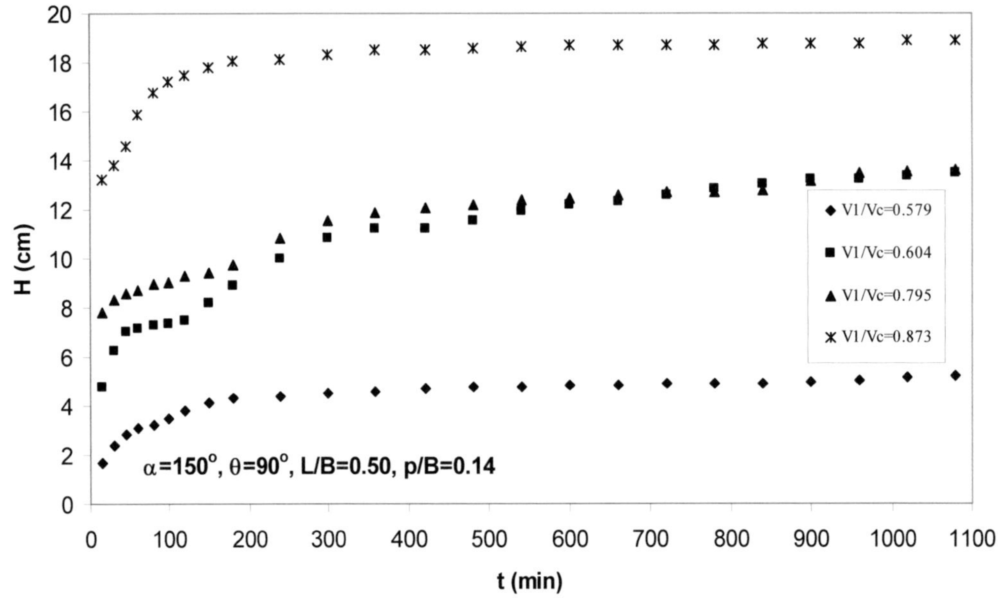

Fig. 19. The development of clear-water scour with time in a curved channel when α=150°, p=7 cm and θ=90°.

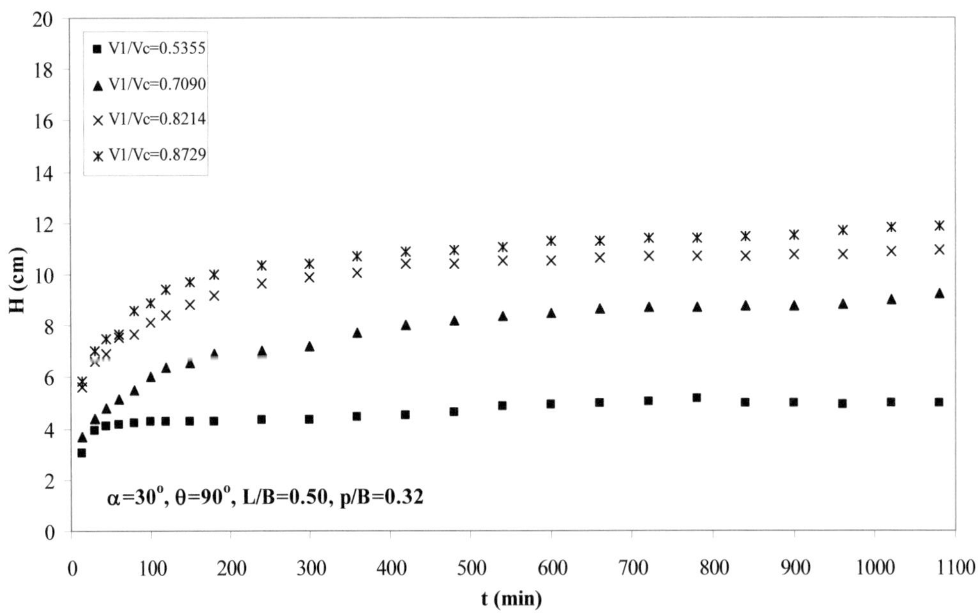

Fig. 20. The development of clear-water scour with time in a curved channel when α=30°, p=16 cm and θ=90°.

scour increases rapidly with time at the beginning, and then asymptotically approaches stable values. That means that the scour depth reaches an equilibrium condition. Time to equilibrium depends on the approach flow intensity and the dimensionless side weir crest height. When the flow intensity increases, time to the equilibrium depth of scour increases. As mentioned before, the intensity of the secondary circulation caused by the lateral flow increases with an increase of the overflow length because of the remaining momentum in the main channel. This also causes much scour around the side-weir intersection.

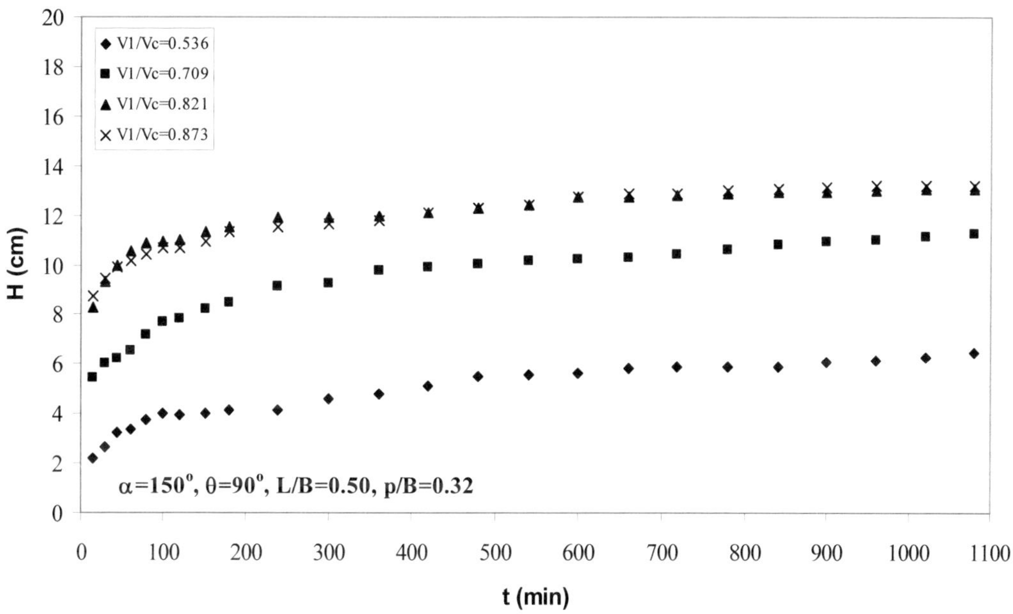

Fig. 21. The development of clear-water scour with time in a curved channel when α=150°, p=16 cm and θ=90°.

Figure 22 shows bed topography at a side weir intersection when V/V_c=0.7090, α=30°, θ=90°, L/B=0.50 and p/B=0.32. Figure 22 shows bed topography at a side-weir intersection when α=30°, θ=90°, V/V_c=0.8214, L/B=0.50 and p/B=0.32. As seen from Figures 22 and 23, maximum scour hole approximately occurs either at end the downstream of the triangular labyrinth side weir, or after from the end of the downstream of the triangular labyrinth side weir. V_1/V_c is a significant parameter for scour hole geometry and location of the maximum scour hole. As mentioned above, as V_1/V_c increase H also increase. At the downstream end of the triangular labyrinth side weir, a longitudinal bar in the middle of the channel and scour hole close the outer bank are formed because of the changes in shear stress field and velocities (see Fig. 22).

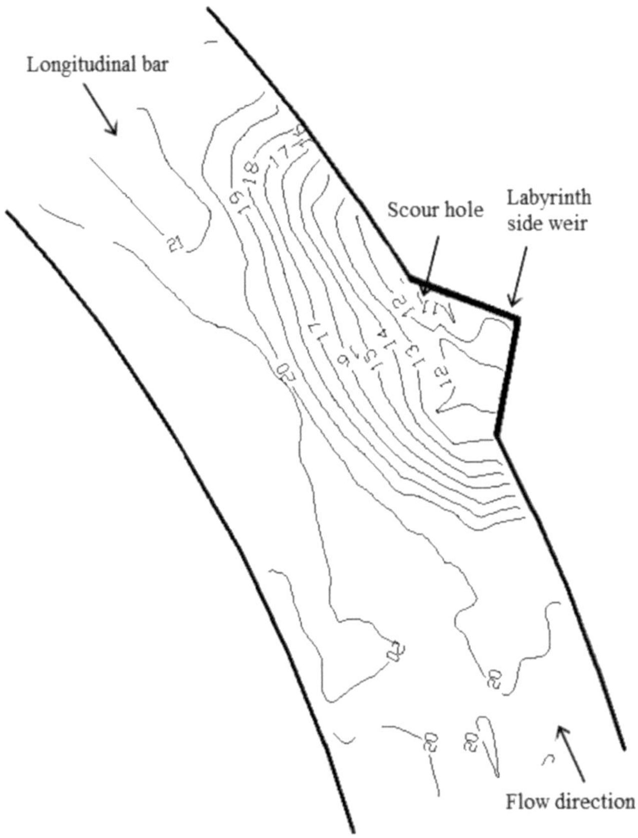

Fig. 22. Bed topography at a side weir intersection when V/V_c=0.7090, α=30°, θ=90°, L/B=0.50 and p/B=0.32 (*contours in cm and undisturbed bed level is 20 cm*).

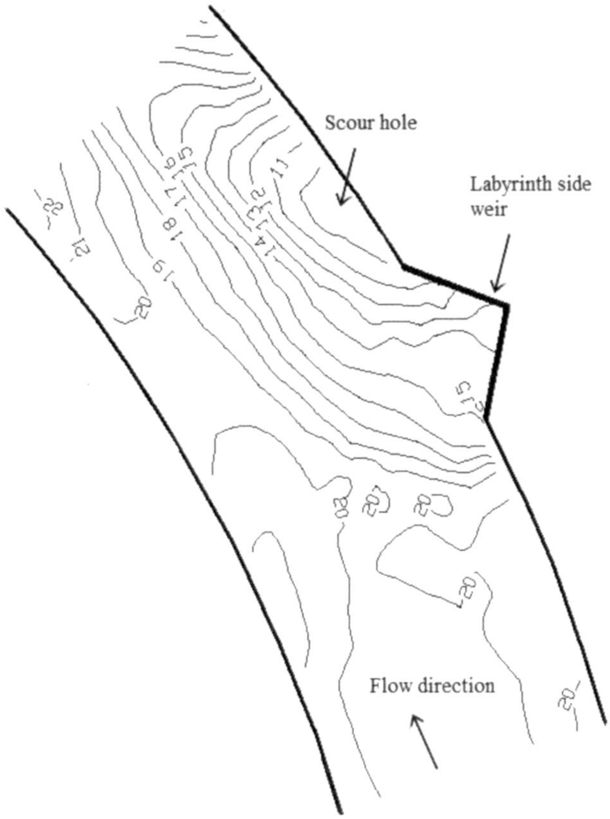

Fig. 23. Bed topography at a side weir intersection when α=30°, θ=90°, V/V_c=0.8214, L/B=0.50 and p/B=0.32 (*contours in cm and undisturbed bed level is 20 cm*).

5. Conclusions

Laboratory experiments, within the Tubitak Project (MAG 104M394), were carried out using triangular, trapezoidal and semi-elliptical labyrinth side weirs located on a straight and curved channel in order to investigate the discharge coefficient and the scour topography at the side-weir intersection area. Empirical correlations were submitted, predicting the discharge coefficient of these labyrinth side weirs. The following conclusions can be drawn from these findings:

- The discharge coefficients of labyrinth side weirs have much higher values than those of classical rectangular side weirs. The primary reason for this may be the increased crest length. The intensity of secondary motion created by lateral flow increases with increasing overflow length.
- The discharge coefficient of the labyrinth side weir is 1.5 to 5.0 times higher than for a rectangular side weir.
- The values of the discharge coefficient C_d increase with an increase in dimensionless crest height p/h_1 and Froude number F_1 values. Moreover, the discharge coefficient C_d increases with increasing dimensionless weir opening length L/B ratio.
- The flow intensity V_1/V_c is a significant parameter for scour hole geometry and location of the maximum scour hole.

- As the flow intensity V1/Vc increases the depth of scour at the side-weir intersection, H, also increases.
- The depth of scour increases rapidly with time at the beginning, and then asymptotically approaches stable values.
- Maximum scour hole approximately occurs either at end the downstream of the triangular labyrinth side weir or after from the end of the downstream of the triangular labyrinth side weir.
- At the downstream end of the triangular labyrinth side weir, a longitudinal bar in the middle of the channel and scour hole close the outer bank are formed because of the changes in shear stress field and velocities.

6. Acknowledgments

The author also thanks Prof. Hayrullah Agaccioglu for his helpful suggestions and contributions. The author also wishes to thank Dr. Nihat Kaya for his assistance in running the experiments. The author acknowledges the financial support of the Scientific and Technological Research Council of Turkey (TUBITAK) under Project No. MAG 104M394.

7. References

Agaccioglu H. & Onen F. (2005). Clear-water scour at a side-weir intersection along the bend. *Irrigation and Drainage*, Vol. 54, pp. 553-569.

Borghei M., Jalili M. R. & Ghodsian M. (1999). Discharge coefficient for sharp-crested side weir in subcritical flow. *J Hydraulic Eng ASCE*. Vol. 125, No. 10, pp. 1051-1056

De Marchi G. (1934). Saggio di teoria de funzionamente degli stramazzi letarali. *L'Energia Elettrica, Rome, Italy*, Vol. 11, No. 11, pp. 849-860

Emiroglu M. E., Kaya N. & Ozturk M. (2007). Investigation of Labyrinth Side Weir Flow and Scouring at the Lateral Intake Region in a Curved Channel. *The Scientific and Technological Research Council of Turkey (TUBITAK), Engineering Science Research Grant Group*, Project No: 104M394 [in Turkish]; 253p.

Emiroglu M.E., Kaya N. & Agaccioglu H. (2010). Discharge capacity of labyrinth side weir located on a straight channel. *J Irrig Drainage Eng ASCE*, Vol. 136, No. 1, pp. 37-46.

Emiroglu, M. E. & Kaya, N. (2011). Discharge coefficient for trapezoidal labyrinth side weir in subcritical flow. *Water Resources Management*, Vol. 25, pp. 1037-1058.

Emiroglu M. E., Kaya N. & Agaccioglu, H. Closure to "Discharge Capacity of Labyrinth Side-Weir Located on a Straight Channel. *J Irrig Drainage Eng ASCE*, in press.

Falvey H. T. (2003). Hydraulic design of labyrinth weirs. *American Society of Civil Engineers*, Virginia, 161p.

Kaya N., Emiroglu M. E. & Agaccioglu H. (2011). Discharge coefficient of semi-elliptical side weir in subcritical flow. *Flow Measurement and Instrumentation*, Vol. 22, pp. 25-32.

Melville B.W. & Chiew Y. M. (1999). Time scale for local scour at bridge scours. *Journal of Hydraulic Engineering ASCE*, Vol. 125, No. 1, pp. 59-65.

Novak P. & Cabelka J. (1981). Models in Hydraulic Engineering, *Pitman Publishing Limited*, London.

Raudkivi A. J. (1986). Functional trends of scour at bridge piers. *Journal of Hydraulic Engineering ASCE*, Vol. 112, No. 1, pp. 1-13.

Tullis J. P., Nosratollah A., Waldron D. (1995). Design of labyrinth spillways. *J Hydraulic Eng ASCE*, Vol. 121, No. 3, pp. 247-255.

9

On the Influence of the Nearbed Sediments in the Oxygen Budget of a Lagunar System: The Ria de Aveiro - Portugal

José Fortes Lopes
Universidade de Aveiro
Portugal

1. Introduction

The filtering capacity of an estuary for both heavy metals and organic micropollutants is largely due to sedimentation. Because sedimentation occurs only for the fraction of contaminant that is adsorbed to particulate matter, the process of sorption (adsorption and desorption) is an essential process. Sediments constitute by far the most important reservoir of nitrogen in shallow systems (Billen & Lancelot,1988). They play two key roles with respect to the plankton system: They act as regenerators of mineral nutrients for the water column, often supplying between 30-70% of nitrogen requirements by phytoplankton. On the other hand, being the only site in coastal marine ecosystems where anaerobic conditions exist, it is, therefore, the local of denitrification, allowing eliminating a large amount of the nitrogen mineralized in the benthos.

Sedimentary organic matter is composed of many different size fractions with different degradability (Wijsman, 1999; 2002). Due to the depletion of the degradable organic matter with depth and the preferential degradation of the labile compounds, benthic mineralization rate usually decreases rapidly with depth in the sediment. Consequently, most of the biological activity is restricted to the upper decimetres of the sediment column. Benthic macrofauna plays an important role in early digenesis. Filter feeding organisms, for example, can selectively filter food particles from the water column and thus increase the flux of high quality organic matter to the sediment (e.g. Heip et al., 1995). Additionally, bioturbating and bio-irrigating activity of the macrofauna can increase the transport processes within the sediment (Black, 1997). Conversely, the carbon loading and the effects of benthic mineralization, such as the production of reduced substances and depletion of oxygen, can influence the activity and composition of the benthic community (e.g. Pearson & Rosenberg, 1978).

Diagenensis characterises any chemical, physical, or biological changes undergone by sediments after their initial deposition and during and after their lithification (Wijsman, 1999; 2002).

Since only a limited number of mathematically formulated problems of organic matter and nutrient diagenesis in sediments can be solved analytically, there has been a shift over the years from a focus on analytical models to numerical models (Berg, 2003). These models are commonly based on mass conservation approaches expressing balances between vertical transport contributions of selected species and biogeochemical interactions between them.

Many researchers consider the classic book by Berner (1980) on early diagenesis to be a turning point in the use of modeling in the field. Berg (2003) has reviewed a number of state of the art of mathematically models focused on the organic matter and nutrient diagenesis processes in marine sediments. Examples of these models are the detailed transport-reaction models of Boudreau (1997), Dhakar & Burdige (1996), Soetaert et al (1996), Van Cappellen et al. (1993) and Van Cappellen & Wang (1996) based on a vertical one-dimensional description of the sediment column, accounting for molecular diffusion, burial, and bioturbation. Although similar, each model is particular, namely, in describing the carbon cycle. Meysman et al. (2003a,b) developed a new object-based software package for early diagenesis modeling. Several applications with simplifications or improvements of the above models have been performed since then (Boudreau, 1996, 1997; Herman et al., 2001 Luff et al., 2000; Archer et al., 2002; Wijsman et al., 2002). The bentic sediment diagenesis processes included in the *Mike3* eutrophication model (Mike3, 2005b) have a simple description of sediment release of nitrogen and phosphorus, which returns a fraction of the settled *P* to the sediment, back into the water column. The sediment module is, therefore, an add-on module to the main standard eutrophication module and, therefore, uses the state variables and some of the processes as input.
The main objective of this study is to assess the influence, of the main physical and biogeochemical processes occurring in the near bed sediments, on the oxygen conditions in Ria de Aveiro lagoon, focusing on the low trophic state of the water column, as well as on the organic matter and the nutrients processes associated to the sediments. It will be performed with the help of a couple hydrodynamic and ecological/eutrofication model, in order to simulate simultaneously, the hydrodynamic, the transport and dispersion as well as the Main biochemical processes within the water column. The eutrofication model (Mike3, 2005b) is focused on the lower trophic status of the water column, and it is able to describe the nutrient cycling, including, organic matter (detritus), organic and inorganic nutrients, phytoplankton and zooplankton growth, primary production and dissolved oxygen. It includes as well the nutrient diagenesis in the bottom sediments and the benthic vegetation processes. Although the model in not focused on the sediment transport, sedimentary processes for organic and inorganic matter are taken into account, namely, particles deposition into the bed.
Previous studies (Lopes & Silva, 2006) have shown that the dissolved oxygen (*DO)* distribution for the main areas of the lagoon is characterized by high concentration values (in general greater than 7 mg O_2 l^{-1}), which evidences the influence of the exchange of high oxygenated waters with the ocean, as well as the high productivity inside the lagoon. Hypoxic situations are not usual but reflect biogeochemical processes occurring at very limited areas, namely the far end and shallow areas, which are characterized by low flushing rates, high residence times and low residual currents. This study is focused on the relationship between the nitrogen processes, occurring in the nearbed sediments, and the biogeochemical condition within the water column. In particular, the study integrates the oxygen cycle associated to the phytoplankton processes, the nitrogen cycles, the fluxes between the water column and the sediments bed, and among other processes, nitrification and denitrification in the sediment processes.

2. The study area

Ria de Aveiro (Fig. 1) is a lagoon situated on the Northwest Atlantic coast of Portugal (40°38′ N, 8°45′ W), 45 km long and 10 km large, covering an area of 66-83 km^2, respectively, at low and high tide (Dias et al., 2000). It is characterised by several branching channels, the most important of which are S. Jacinto, Ovar, Espinheiro, Mira and Ilhavo that are directly connected to the lagoon mouth and the ocean by a single tidal channel. The central areas of the lagoon is constituted by a myriad of other small channels dominated by shallow and tidal flat areas, contributing to a strong damping of the currents and an increase of the phase delay of the tidal wave (Dias, 2001). The bathymetry (Fig. 1) shows that Ria de Aveiro is a very shallow lagoon (average depth of 1 m).

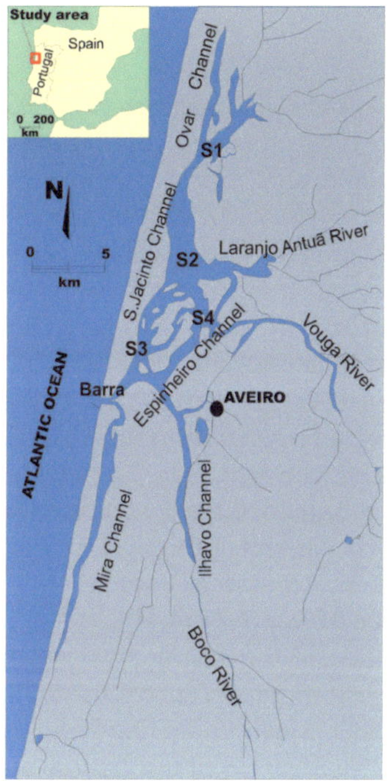

Fig. 1. The study area: The Ria de Aveiro Lagoon, with the location of the stations (**S**)

The deepest areas of the lagoon are confined to the inlet channel and to small areas close to the lagoon mouth, at the western boundary of the lagoon. In those areas, the depth may be of the order or greater than 10 m, whereas elsewhere, at the inner parts of the lagoon, the depth barely reaches values beyond 3 m. The hydrological characterisation of the lagoon (Dias et al., 1999; Dias et al., 2000; Dias, 2001) reveals that, excluding the areas close to the lagoon mouth, the salinity and the temperature vertical profiles have very well mixed structures. Tides, which are semi-diurnal, are the main forcing of the circulation in the Ria de Aveiro lagoon. The tidal range varies between 3.2 and 0.6 m, and the estimated freshwater input is very small (about 1.8 X 10^6 m^3 during a tidal cycle) when compared with the mean tidal prism at the mouth (about 70 X 10^6 m^3). In general, fluvial input to the lagoon occurs at small steady flow alternated by intense pulses due to the sudden increase of the river's tributaries flows. Two main rivers, the Antuã river (5 m^3 s^{-1} average flow) and the Vouga river (50 m^3 s^{-1}) located at eastern side of the lagoon, contribute with almost 90% of the freshwater input to the lagoon. Other small rivers situated at the far end of the lagoon, namely Caster located at the far end of the Ovar channel, may contribute substantially to the salinity patterns of the lagoon, during the rainy period. Therefore, in normal situation, as the lagoon is dominated by tides, the freshwater flows only affects restricted areas, close to the river mouth, along the main channels (Dias et al., 1999; Dias, 2001).

The salinity is an important physical variable for the understanding of the dynamic inside the lagoon, which is dominated by tides (Dias et al., 2001). During the dry season, generally coinciding with summer, the river's flow is very low, high salinity values (30-34) which are typical for the lagoon mouth, are observed inside the lagoon and even at its far ends, namely during the flood tide. These locations are close to the river mouths, present, in general, extremely low salinity values, between 0 and 5, during the wet season (autumn and spring).

From the biogeochemical point of view Ria de Aveiro lagoon may be considered as a mesotrophic shallow estuarine system (Almeida et al., 2005). It is characterised by a rich biodiversity as well as by an increasing pressure of the anthropogenic activities near its margins, namely building and land occupation, agricultural and industrial activities. This has resulted in a significant change of the lagoon morphology, and in a constant input of a large volume of anthropogenic nutrients and contaminant loads, with negative impact in the water circulation and the water quality of the lagoon. It is subjected to considerable inputs of industrial and domestic discharges which occur mainly at its periphery, currently receiving a mean annual influxes of total nitrogen (*N*) and total phosphorus (*P*) of 6118 t y^{-1} and 779 t y^{-1}, respectively, from its influent rivers (Silva et al, 2002). The assessment of the lagoon ecological state is given, for example, by assessing its dissolved oxygen distribution. *DO* (dissolved oxygen concentration) inside the lagoon vary in general between 8 to 12 mg O_2 l^{-1}. Given the salinity and the temperature distribution inside the lagoon, the last value correspond to a near oxygen saturation condition (Almeida et al., 2005). The far end of the lagoon, namely at the Ovar channel and near the Vouga river mouth, may show extreme *DO* values, the highest ones (12 mg O2 l^{-1} or greater), as well as the lowest ones (below 5 mg O_2 l^{-1}).

The bottom sediments distribution of the Ria de Aveiro consists of a mixture of mud and sand. The granulometric composition of sediments contains about 2 to 90% of sands, 10 to 80% of silts and 0 to 30% of clays. Meanwhile, there are significant differences in the bed composition between north and south areas of the lagoon. Fine cohesive particles and sands, mainly, compose the northern channels, whereas the southern channels are almost composed by coarse particles and sands (Lopes et al., 2006). A gradient of the bed composition has been observed at each channel of the Ria de Aveiro, with sand near the mouth and a mixture of sand and mud near its far extremity. Dias (2001) has calculated, from measured vertical profiles of currents, the bottom roughness length of different channels. Typical values, varying from 0.02 cm at the far end of the channels, corresponding to a bed predominantly composed by mud or a mixture of mud and sands, to 0.07 cm near the mouth, corresponding to a bed composed mainly by sand, have been found. The presence of a mixture of sands, gravel, shells and mud in the main channels of the Ria de Aveiro denotes the ocean influence in the lagoon. The bottom sediments distribution of the Ria de Aveiro lagoon bed has a cohesive behaviour. Following Torfs (Torfs, 1995; Mitchener & Torfs, 1996) the cohesiveness behaviour of a sediment bed depends depend on the ratio of non-cohesive and cohesive particles. Above a critical mud content of 4%, and a critical clay (Kaolinite) content of 3%, as the ripples and dunes are increasingly suppressed and the bed behaved cohesively. Dyer (1986;1995) gave a transition range for non-cohesive to cohesive behavior from 5-10% clay content by dry weight.

Ria de Aveiro is a tidal dominated system well connected to the ocean, which contributes to a constant oxygen renewal of its water (Dias & Lopes, 2003; 2006). Lopes et al., (2006; 2008) have characterized the lagoon in two main areas: (1) the central area under the influence of the tide, are characterized by strong currents, along the S. Jacinto and the Espinheiro channels, particularly close to the mouth where the intensity may reach values as high as 2 m s^{-1} (Fig. 3), which progressively decreases toward the inner areas of the channels; (2) the far end areas, which are mainly intertidal very shallow areas, (Fig. 1), where the currents may reach low values 0.1 m s^{-1} (Fig. 2), and the residence time two weeks (Dias et al., 2001, 2003).

3. The model description

3.1 The main features of the model

The physical model solves the hydrodynamic and the transport equations, and includes the main physical processes occurring in the water column, as well as the exchanges at the interface air-water between the atmosphere and the water system. The hydrodynamic model is the Hydrostatic Version of *Mike3 FLOW MODEL* (Mike3, 2005a), solves the continuity and the momentum equations. It is based on the solution of the three-dimensional incompressible Reynolds averaged Navier-Stokes equations, subjected to the assumptions of Boussinesq and hydrostatic pressure, therefore, the vertical accelerations are assumed to be negligible. The transport-diffusion equation is coupled to the hydrodynamic model, solves the temperature, T, and salinity, S, and any scalar quantity (biogeochemical variables), the turbulent kinetic energy per unit mass, k, and the dissipation of the turbulent kinetic energy, ε. It integrates radiatives exchanges between the atmosphere and the water system, including solar and non solar radiation (Gill, A.E., 1982). The hydrodynamic and the transport models are described in Dias & Lopes (2006) and Lopes et al., (2005; 2006). The model domain covers a rectangular area corresponding to 266 × 654 points, with a grid size of 60 m × 60 m. It makes use of the so-called *alternating direction implicit technique* (*ADI*) to integrate the mass and the momentum equations in the space-time domain. The equation matrices, which result for each direction and each individual grid, are resolved by a *Double Sweep* (*DS*) algorithm. It results in a third-order convective matricial equation that is solved for each direction and each grid line by a double sweep algorithm (Leendertse & Gritton, 1971; Abbott et al., 1981; Casuli, 1999). The time step for the hydrodynamic and the transport model has been chosen small, 2 seconds, in order to assure stable solutions for the transport equation. The transport model makes use of the *ULTIMATE-QUICKEST* scheme (Leonard, 1979) for the integration of the advection-diffusion equation in the space-time domain, extended to the third dimension. The scheme ensures mass conservation through the control volume formulation of the transport terms. It is simple and computationally fast and, thus, well suited for coupling to the three-dimensional hydrodynamic model in order to simulate density-driven flows. It results in a very efficient, stable and third-order accurate explicit algorithm (Vested et al., 1992). The validation of the *3D* hydrodynamic and transport models was performed with the help of the same data set and periods as those used to validate the *2D* models, as those described in Dias & Lopes (2006), Lopes & Silva (2006) and Lopes et al., (2008), the hydrodynamic model is setup by spinning up the lagoon from rest. The sea surface elevation is specified at the western open boundary at each time step. The lateral boundary conditions correspond to no slip velocity and insulation for temperature and salinity on the side-walls. Initial and boundaries conditions for the transport models were obtained from Almeida et al., (2005). During the simulation period, while the biogeochemical variables were kept constant at the river boundaries, the salinity

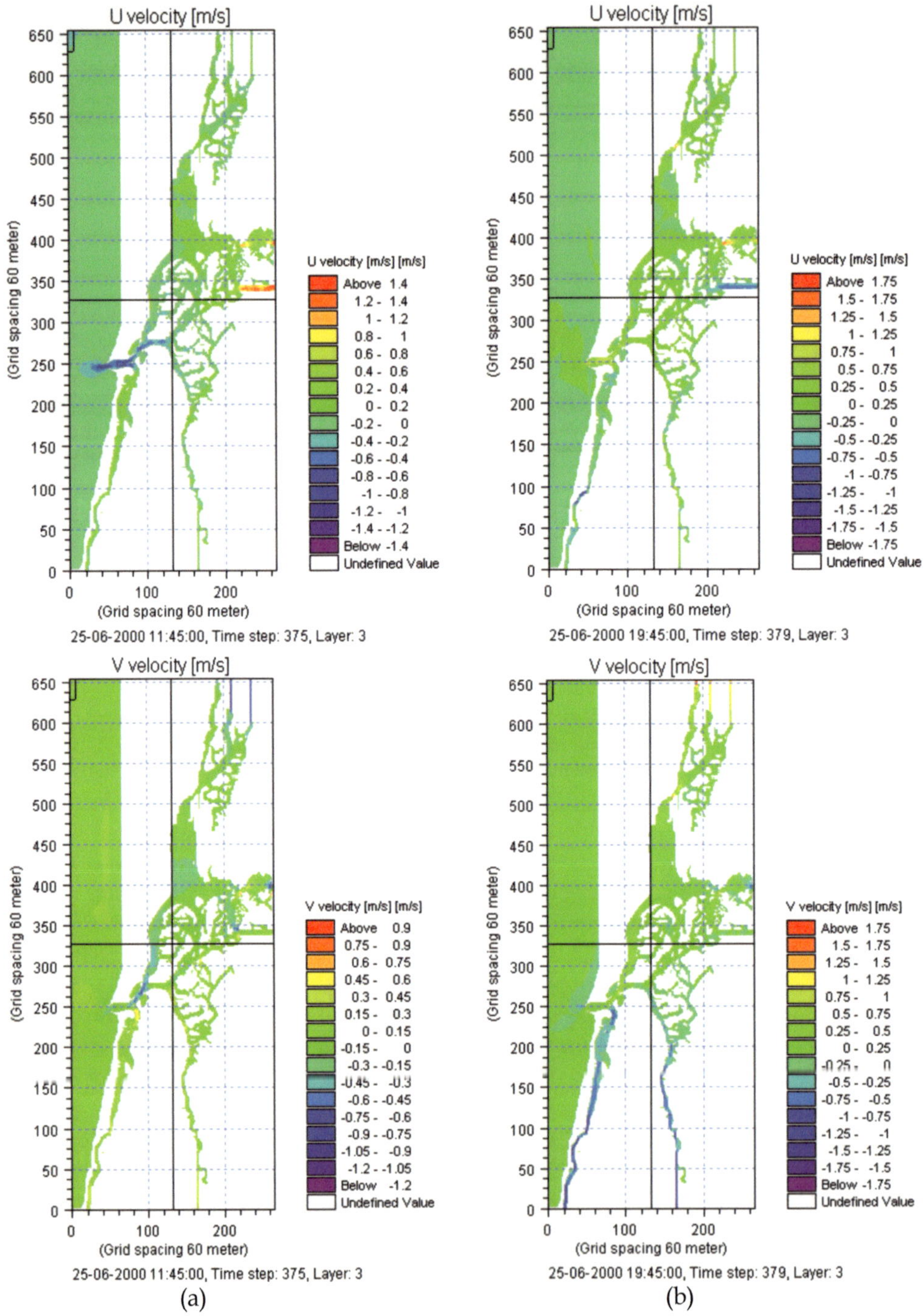

Fig. 2. Maps of U and V currents field, during flooding (a) and ebbing (b) tide at the mouth

and the temperature were allowed to vary at each time step, as a year time series were defined based on typical seasonality and historical data for these variables. Concerning the ocean open boundary, the von Newmann boundary conditions were imposed. The atmospheric parameters needed to compute the latent and sensible heat fluxes, as well as the radiation fluxes (relative humidity, air temperature, wind speed and cloudiness) were specified at each time step.
The ecological model, which is an eutrophication biogeochemical module, *Mike3-EU* (Mike3 (2005b), dealing with the basic aspects of the lower trophic status of the water column, has been applied to the study area (Lopes et al. 2010). The model includes a diagenesis sub-model, which deals with the nitrogen and the phosphorus cycles in the sediment bed.
It is coupled to the 3D hydrodynamic model described above, through the transport equation and require the definition of the concentrations at the model boundaries, flow and concentrations from pollution sources, water temperature and irradiance.

3.2 The ecological/eutrofication model

The ecological/eutrofication model describes the nutrient cycling, the phytoplankton and the zooplankton growth, the growth and the distribution of rooted vegetation and macroalgae. In addition it simulates the concentrations of phytoplankton, chlorophyll-a, zooplankton, organic matter (detritus), organic and inorganic nutrients, dissolved oxygen and the area-based biomass of benthic vegetation over time, as well as several derived variables, such as the primary production, the total nitrogen and phosphorus concentrations, the sediment oxygen demand and the secchi disc depth.
The model, which is described by Lessin & Raudsepp (2006) and Erichsen & Rasch (2001) consists of thirteen state variables represented by four functional groups (phytoplankton, zooplankton, benthic vegetation and detritus), nutrients and oxygen (Tab.1). Fig. 3 summarizes the conceptual diagram of the model (Mike3, 2005b). The model is based on nitrogen and phosphorus processes, but for the Ria de Aveiro lagoon, nitrogen is generally the only limiting nutrient (Almeida et al., 2005). Tab. 2 presents the parameter representing the main biochemical processes for the eutrophication model.
The mathematical formulations of the biological and chemical processes and transformations for each state variable are described by a 1st order ordinary differential equation. Phytoplankton is represented by one functional group and three state variables *PX* representing: phytoplankton carbon (*PC*), phytoplankton nitrogen (*PN*) and phytoplankton phosphorus (*PP*). The time evolution of each state variable *X* is related to its mass balance within the water column:

$$\frac{\partial X}{\partial t} = PRODUCTION - RESPIRATION - GRAZING - SEDIMENTATION - DEATH \quad (1)$$

Once the advection-dispersion terms as well as the sources and sinks terms of the transport equation are evaluated, the model makes an explicit time-integration of (1), then calculating the concentrations to the next time step, following a Runge-Kutta 4 integration method. Although each variable is characterized by specific processes, the mass balance is similar to (1). For instance, concerning *CH (Chlorophyll-a)* the following processes are considered: production, death and sedimentation, whereas for *ZC*, representing important grazers, such as copepods and various microzooplankters. Detritus are defined in the model as particles of dead organic material in the water, and its pool receives the dead primary producers, dead zooplankton grazers and unassimilated material left over after grazing. Sedimentation

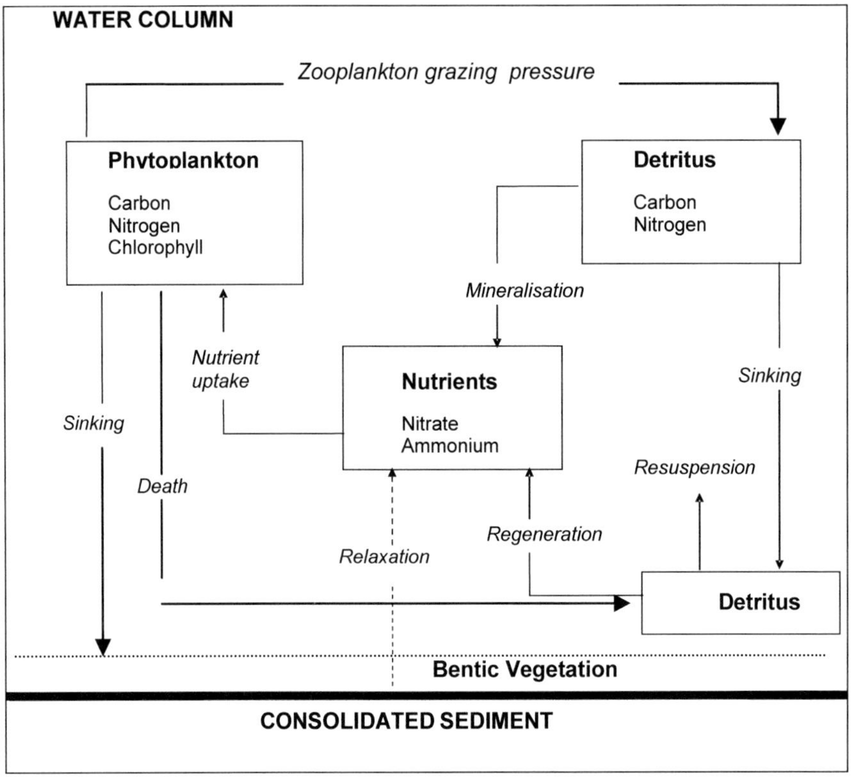

Fig. 3. The conceptual diagram for the ecological/eutrofication model

and mineralisation are the only processes draining the detritus pools. The mass balance for detritus (*DC, DN* and *DP*) includes instead, regeneration, mineralisation and sedimentation processes. The inorganic nitrogen is represented by two state variables, namely oxidised forms (sum of nitrate and nitrite) and reduced forms (sum of ammonia, *NH*, and urea). The main balance for the inorganic nitrogen and phosphorus includes the inputs from mineralisation and the uptake. The primary source of ammonia is the input from respiration and mineralisation processes, whereas the sinks include uptake by primary producers: phytoplankton and benthic vegetation, and nitrification from nitrate. The mineralization of *NH* is expressed as a fraction of the sedimentation of organic matter. Under anoxic conditions, the release of nutrients is not only a result of recently sedimented material, but also a zero order function, where large amounts of nutrient buried in the sediment will be released. Mineralization of organic matter is the main input of inorganic phosphorus (*IP*), and corresponds to the sum of mineralization of detritus, zooplankton and phytoplankton phosphorus as well as the release from the sediment, the last one being only relevant for the bottom layer. The benthic vegetation, *BC*, is assumed to be rooted and/or attached to stones etc. Fixed ratios of nitrogen to carbon and phosphorus to carbon are assumed. The mass balance for the benthic vegetation includes production and losses.

Hereafter the attention will be focused on the *DO* and *PC* related processes, as well as on the diagenesis processes, that is the nitrogen and the phosphorus cycles in the sediment bed.

The mass balances for *PN* and *PP* are similar to the mass balance of PC, as described in Lessin and Raudsepp (2006), and are not discussed here.

The net production of *PC* depends on light intensity, nutrient availability and ambient temperature. The gross production is computed with a multiplicative approach, considering the maximal rate of production (μ), the influence of light (*F(I))* (Jassby & Platt (1976)), a temperature Arrhenius function (*F(T)*) and the internal concentrations of nitrogen and phosphorus (*F(N,P))* calculated from the relative saturation of the internal *N* and *P* pools, following Droop (1973, 1975), Nyholm (1976), Mommaerts (1978), Tett et al. (1986) and Lancelot & Rousseau (1987)

Loss due to respiration is represented by a basal metabolic expenditure to maintain life processes, is considered a constant fraction of biomass *PC*, with a dependence on temperature represented by the Arrhenius function

Loss of phytoplankton due to grazing by zooplankton depends on zooplankton carbon, is regulated by zooplankton grazing function modified by a temperature function. The grazing rate of phytoplankton carbon by zooplankton is represented by a saturation equation relating food concentration to grazing rate, a threshold food concentration below which no grazing takes place (Kiørboe and Nielsen 1994), the Arrhenius temperature function and a function of dissolved oxygen suppression grazing at low *DO* (Roman et al. 1993)

The sedimentation rate of *PC* is described by a 1st order equation depending on the internal nutrient status and a 2nd order equation representing coagulation.

Natural mortality of phytoplankton, or autolysis, has been shown to be a significant phenomenon in the marine ecosystem (Jassby & Goldman, 1974) and this decay of blooms is partly mineralized in the water column (Lancelot et al., 1987). The mortality of phytoplankton is described by a 1st order equation, with a dependence on temperature represented by the Arrhenius function.

3.2.1 Oxygen processes in the water column

Dissolved oxygen mass balance plays a crucial role in the water column oxygen budget. It includes the photosynthesis, the respiration and the mineralization processes. The oxygen balance (*DO*) includes the production of oxygen by the primary producers and the benthic vegetation, the consumption oxygen and the exchange of oxygen between water and air i.e. re-aeration. The oxygen consumption is due to mineralization of organic matter in water and sediment, oxidation of ammonia (nitrification), respiration of zooplankton and phytoplankton, and mineralization of the part of the phytoplankton, which is mineralized immediately without entering the detritus pool. The oxygen balance (*DO*) follows a similar equation as (1) and includes the production of oxygen by primary producers and benthic vegetation, the consumption of oxygen and the exchange of oxygen between water and air. The production of phytoplankton is expressed by the same function as the gross production of PC. The oxygen consumption by respiration of zooplankton is considered a constant fraction of the zooplankton biomass and depends on temperature through an Arrhenius temperature dependence function for the phytoplankton respiration. The oxygen consumption in the mineralization of dead phytoplankton resulting from the natural mortality of phytoplankton, or autolysis, which is a significant phenomenon in the marine ecosystem is supposed proportional to the fraction of dead phytoplankton, undergoing immediate mineralization as well as on the phytoplankton concentration, and depends on temperature through an Arrhenius temperature dependence function for the phytoplankton respiration. The oxygen consumption involving the oxygen demand by the nitrification process, deals with transformation of ammonia into nitrate and nitrite into nitrate. These processes are expressed by a mathematical formula that expresses a proportionality on the ammonia and the nitrate concentrations. The sediment oxygen demand, *SOD*, is related to the carbon mineralization in the sediment which again is related to the sedimentation of organic matter (detritus and phytoplankton). It is supposed proportional to sedimentations rates of phytoplankton and detritus, on a temperature Arrhenius dependence function for the sediment oxygen demand and on a Michaelis-Menten oxyygen function for the sediment oxygen, Finaly the exchanges with the atmosphere, the reaeration, is supposed proportional to the difereence between the water oxygen concentration and the oxygen saturation concentration.

3.3 The nitrogen cycles in the sediment bed

The N cycle in the sediment is a dynamic proces (Williams, 1975). The diagenic model describes the changes and the fluxes between state variables, e.g. nitrification of NH_4 to NO_3 in the sediment. Fig. 4 summarizes the conceptual diagram of the diagenesis model (Mike3, 2005b), where the state variables and the processes for the nitrogen cycle are presented. As estuaries and lagoons tend to be more nitrogen limited than phosphorus, although its cycles and processes are as well included in the model, the phosphorus influence was not considered in the experiments.

Together with the state variables the processes may be regarded as the cornerstones in a dynamic model. The processes involved in the nitrogen cycle are described in connection with the state variables (Blackburn & Henriksen, 1983; Ruadij & Raaphorst,1995; Windolf et al., 1996). The nitrogen cycle in sediments consists of three state variables, the organic *N* (*SON*), the total NH_4 (*SNH*) and the NO_3 *(SN03)*. A sink of nitrogen is as well included in the model, as immobile nitrogen *(SNIM)*. The sedimentation of organic *N* or flux of NH_4 and NO_3 across the sediment surface connects the state variables to plankton *N*, detritus *N* and inorganic *N* in the water. The organic *N* in the sediment is mineralised producing NH_4, which enters the *SNH* pool. NH_4 in the sediment may either be exchanged with *IN*, in the water or nitrified into NO_3 in the uppermost layer of the sediment, with the help of O_2. (Sweerts 1990; Sweerts et al., 1989; Sweerts et al., 1991). The NO_3 entering the *SNO3* pool may either be denitrified or exchanged with inorganic *N* in the water.

3.3.1 The organic *N* in sediment (*SON*)

The input of new organic N to the sediment is mediated by sedimentation of living algae or dead organic mater from the overlaying water column. When the organic matter reaches the sediment surface, it often forms a loos layer of material, which is easily resuspended. Degradation of material, in this thin layer is fast compared to the underlying sediment layer. In the sediment, the organic material will degrade, releasing nitrogen as NH_4 to the pore water. However, as the *C: N* ratio in the remaining organic matter increases, the degradation decreases because the organic matter do not fulfil the needs of nitrogen for bacteria and other organisms involved in this mineralization. At a molar *C: N* ratio 1 of about 11, the net mineralization of NH_4 seems to stop (Blackburn, 1983).

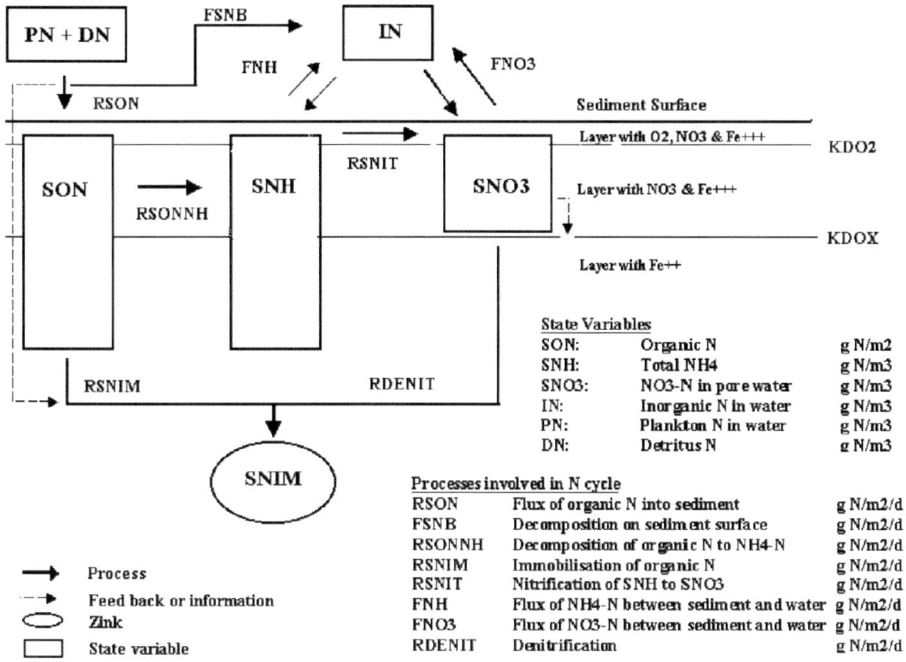

Fig. 4. The conceptual diagram for the N cycle in sediments, for the diagenesis model.

In the model the input of organic *N* to the sediment is calculated by the standard *EU* as sedimentation of algae *N* and detritus *N*. A fraction of the settled organic *N* is assumed to be degraded returning to inorganic *N* in the water. This process is temperature dependent and should account for the relative fast mineralization of organic material in the surface layer. The remaining organic *N*, *RSON*, is put into the pool of organic *N* in the sediment, *SON*.

A fraction of the settled nitrogen is assumed to be buried in the sediment. This fraction (*RSNIM*) is defined as the part of the settled nitrogen of the settled organic matter with a *C:N* ratio about 11. The pool of organic N is mineralised with NH_4 as a end product, this process (*RSONNH*) is set to be a temperature dependent fraction of *SON*.

The change with time of organic *N* in sediment is, therefore, calculated by adding all the processes together in a differential equation set up for *SON*:

$$\frac{d\ SON}{dt} = RSON - RSONNH - RSNIM \tag{2}$$

3.3.2 The total *NH4* in the sediment

A part, and sometime a major part, of this NH_4 is loosely sorbed to particles in the sediment. In the model it is assumed that fraction is available for nitrification and flux across the sediment surface. Three processes are connected to this state variable, mineralization of *SON*, nitrification, and flux of NH_4 across the sediment water surface. Bacteria in the sediment, mediate nitrification by oxidising NH_4 to NO_2 and NO_3, using O_2. In the model, nitrification is calculated by multiplying a potential nitrification capacity with a Monod kinetics for NH_4 in the sediment and a Monod kinetics for squared O_2 concentration in the water. As nitrification is an aerobic process, it will be restricted to the sediment layer with O_2. The flux of NH_4 between sediment and water is a process dependent on the concentration difference between water and sediment and the total diffusion for NH_4. In the standard *EU* model, NH_4 and NO_3 are lumped together as inorganic *N*, IN. It is, therefore, necessary to assume that the ratio of NH_4 and NO_3 in the water just above the sediment surface has the same ratio as in the sediment. The changes with time of the total NH_4 in the sediment, is calculated by including the fraction of organic *N, SON*, which is mineralised as NH_4, *RSONNH,* the nitrification of total NH_4 in sediment, *RSNIT*, the NH4 total flux between sediment and water, FNH, the mass (in grams) of dry matter of sediment per gram of water, *DM*, the specific gravity, *VF*, and the layer of O2 in the sediment layer, *KDO2*, as represented in Fig. 4 :

$$\frac{d\ SNH}{dt} = \frac{RSONNH - RSNIT - FNH}{(1 - DM) * VF * KDO2} \tag{3}$$

The *NO3* concentration in the sediment depends on three processes: nitrification, *RSNIT*, (previously defined under *SNH*), denitrification, *RDENIT*, and the flux of NO_3 across the sediment surface, *FNO3*, which is calculated in the same way as the flux of NH_4 (g $N/m^2/day$). The changes with time of the total NO_3 in the sediment, is calculated by adding all the processes together in the differential equation:

$$\frac{d\ SNO3}{dt} = \frac{RSNIT - RDENIT - FNO3}{(1 - DM) * VF * KDS} \tag{4}$$

The immobilisation of *N, SNIM*, may occur either as burial of slowly degradable organic N or as denitrification of NO_3. The changes with time of the total *SNIM* in the sediment is calculated by adding two processes together, previously described, the nitrification *RSNIT* and denitrification, *RDENIT,* in a differential equation set up for each state variable:

$$\frac{d\ SNIM}{dt} = RSNIM + RDENIT \tag{5}$$

Each one of the different terms presented in the above equations represent specific processes associated to the nitrogen cycles in the sediment, are described and detailed in Mike3 (2005b).

4. Results

In order to investigate the interplay between the biogeochemical conditions associated to the near bottom sediment processes and the dissolved oxygen mass balance in the water column, as well as on the organic matter and the nutrients processes associated to the sediments, the model described in the previous sections were applied to the Ria de Aveiro lagoon. In all the simulations it was considered a normal summer situation, corresponding to the mean river flow for this season (Lopes et al. 2010). For the simulation period, while the biogeochemical variables at the river boundaries were kept constant and their values chosen in accordance to the observations (Almeida et al. 2005), the salinity and the temperature were allowed to vary. Concerning the ocean open boundary, the von Newmann boundary conditions were imposed. The atmospheric parameters (relative humidity, air temperature, wind speed and cloudiness) needed to compute the latent and sensible heat fluxes, as well as the radiation fluxes were specified at each time step.

The next sections present the results concerning the influence of the near bottom sediment processes on the biogeochemical state of the water column. All the simulations are compared to the reference simulation, which was performed considering the reference value for all the parameters of the model and a low background sediment concentration, 5 mg l^{-1}, representing a typical summer simulations, as adopted by Lopes et al. (2010).

4.1 The influence of the suspended sediment concentration

Fig. 5 presents monthly time series of the main ecological model variables for two experiments: (1) the reference simulation (named *RS*) ; (2) simulation corresponding to a high background sediment concentration, 45 mg l^{-1} (named *SI1*). It can be observed that concerning *RS*, *DO*, although tidaly oscillate during the day between maximum and minimum values, remains quite high for all the stations. It varies within a range of 8-10 mg O_2 l^{-1}, excepting for station 1 where it falls below 8 mg O_2 l^{-1}.

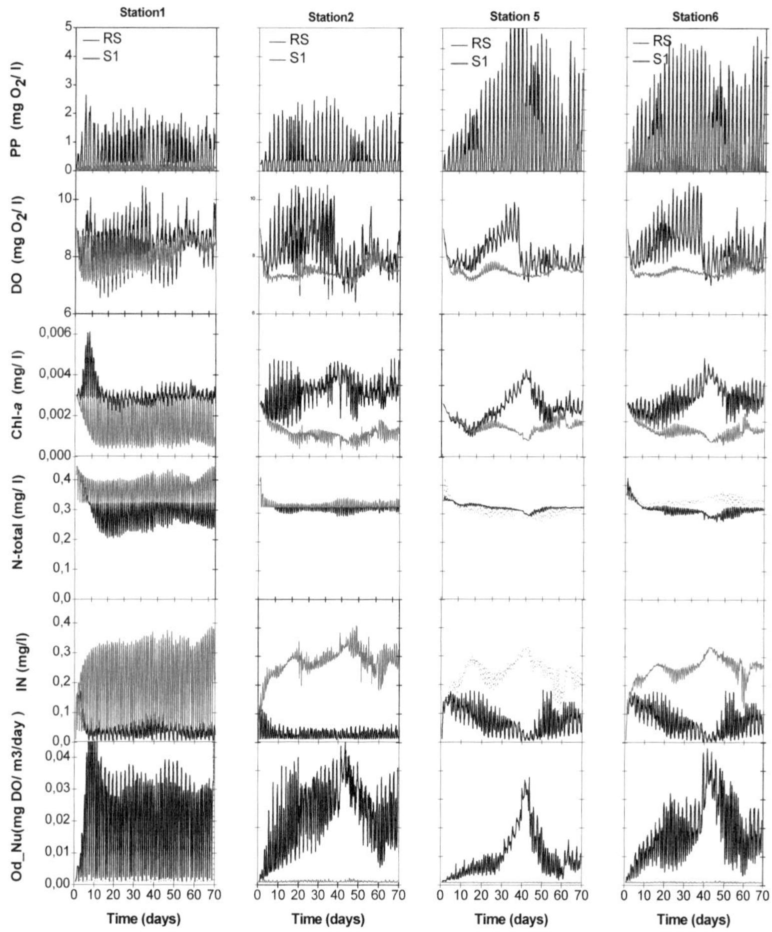

Fig. 5. The monthly time series for the main ecological model variables and for two experiments: (a) the reference simulation (*RS*); (b) the simulation (*SI1*) corresponding to a high background sediment concentration (45 mg l^{-1})

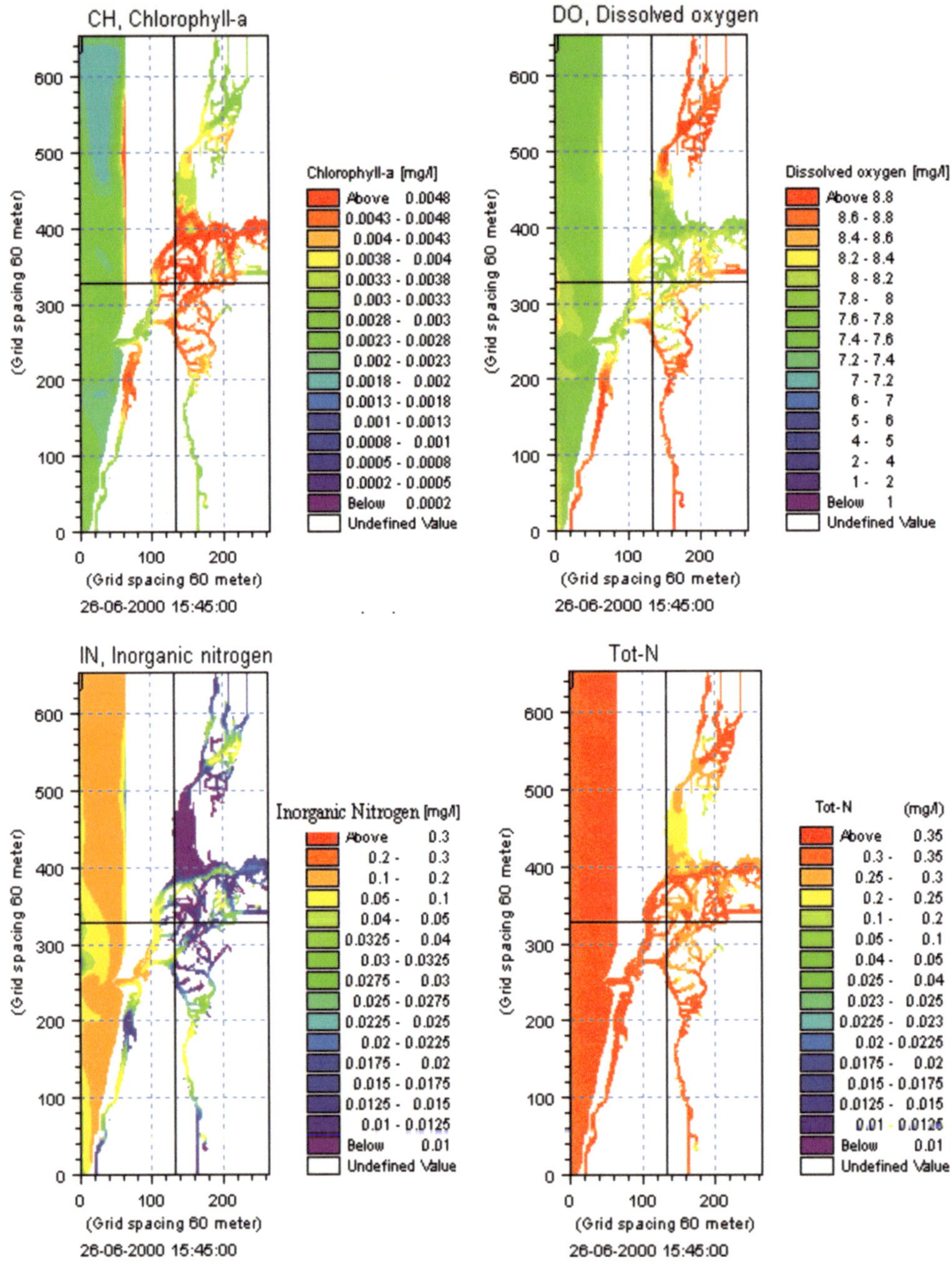

Fig. 6. The surface horizontal distributions for the main ecological model variables, for the reference simulation *(RS)*.

The semi-diurnal pattern for the time evolution, evidences the influence of the tide. On the other hand, *SI1* show significant low *DO* values, varying within a range of 7-8 mg O_2 l^{-1}. This situation, when compared to *RS*, evidences a significant consumption of oxygen within the water column. *Chl-a* and *PP* (Primary Production) present similar behaviour as *DO*.

While for *RS*, *Chl-a* concentrations tidaly oscillate around 3 μg l^{-1}, for *SI1* the concentrations fall below 2 μg l^{-1} and reach minimum values, close to 0.1 μg l^{-1}. *PP* values daily oscillate between the minimum value during night (0 g C m^2 day $^{-1}$) and the maximum values at noon. Excepting station 1 and 2, relatively high concentration values, greater than 2 g C m^2 day $^{-1}$, are observed during the day for RS. On the other hand, *SI1* always shows very low values for *PP*, well below 1 g C m^2 day $^{-1}$. *IN* presents relatively high concentrations values for *SI1*, reaching values as high as 0.3 mg l^{-1}, when compared to low values, below 0.05 mg l^{-1}, for *RS*. The strong values for *IN* concerning *SI1*, suggest a release of nitrogen into the water column. *N-total*, which includes organic and inorganic nitrogen, excepting station 3, shows greather concentration values for *SI1*, within the range 0.3-04 mg l^{-1}, even though of the same order of those for *RS*, Finally, *Od_Det* (the oxygen demand by detritus) shows extremely low concentration values, of the order of 0.04 mg O_2 l^{-1}, for *SI1*, when compared to the maximum values, of the order of 0.4 mg O_2 l^{-1}, for *RS*.

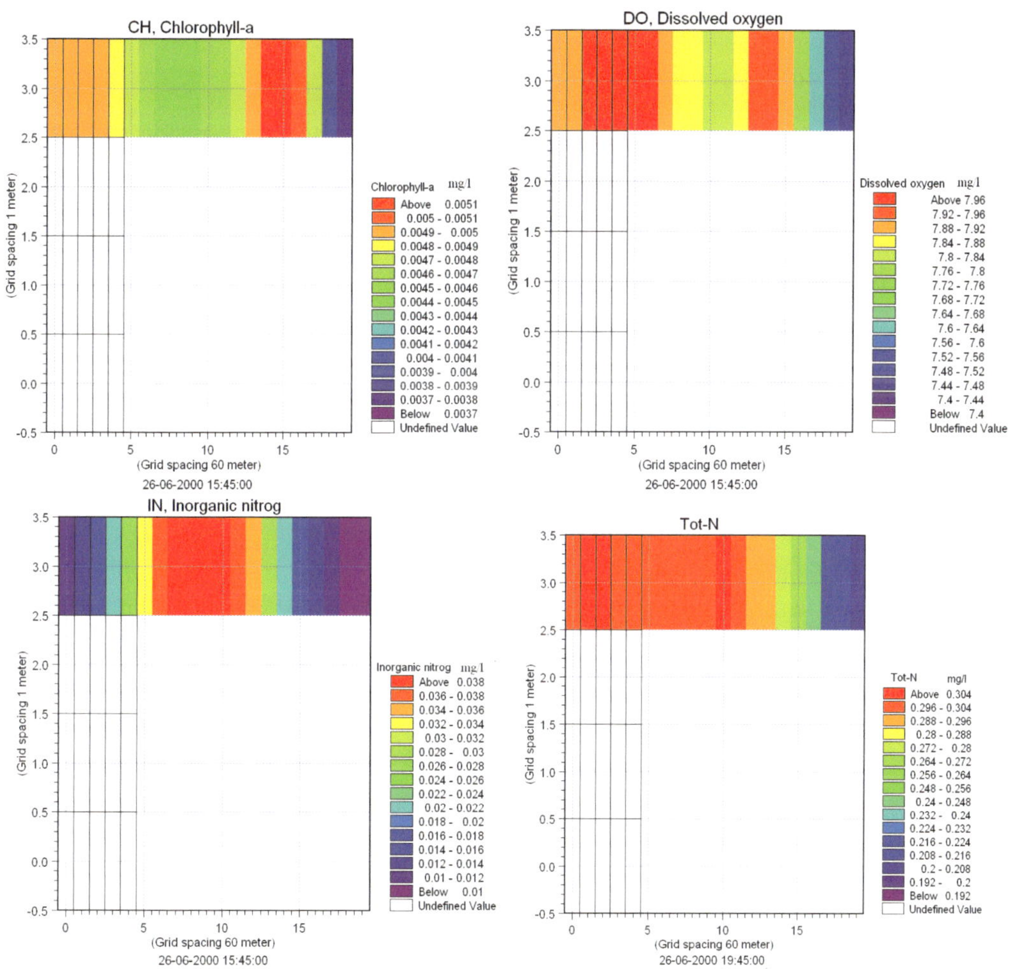

Fig. 7. Vertical distributions for the main ecological model variables, for the reference simulation, for station 2

The horizontal *DO* distributions for *RS* of Fig. 6, confirm that in the normal conditions, the concentration values remain quite high inside the lagoon, varying within the range of 8-9 mgl^{-1}. *DO* steadily increased from the early morning low values, below 8 mg l^{-1}, to maximum values at noon, of the order of 9 mg l^{-1}, and decreased again towards the minimum values, at the end of the day. The maximum values are observed at the far end shallow areas. In the inner areas of the lagoon, *DO* values are slightly above 8 mg l^{-1}, reflecting a condition of a strong influence of the ocean water. Between the central and the far end areas of the lagoon, namely those situated between the northernmost part of the S. Jacinto and the Ovar channel, *DO* shows maximum values (close 9 mg/l). On the other hand, the *DO* horizontal distributions for SI1 show significantly lower values, varying within the range of 7-8 mg l^{-1}. The central areas of the lagoon show, inclusively, values closely to 7 mg l^{-1}. Concerning *Chl-a,* the horizontal distribution for *RS*, shows maximum values of the order of 5 μg l^{-1} at the central area of the lagoon and at Laranjo, but not for the far end areas of Ovar, whereas for *SI1*, values below 2 μg l^{-1} are observed, namely at the far end shallow areas. The *IN* horizontal distributions for *SI1* show high concentration values, of the order of 0.3 mg l^{-1}, for the main lagoon areas, whereas *RS* shows values well below 0.1 mg l^{-1}. *N-total* distributions did not show significant differences between the two simulations. Fig. 7 presents the vertical profiles of the main model variables and for several sections for *RS*. Due to the shallowness of the lagoon, the distributions evidence a horizontal distribution, but no vertical stratification pattern. These results are in agreement with the physical characterisation of the lagoon by Dias et al. (1999), as a well mixed system.

4.2 The influence of the nearbed sediment processes

Fig. 8 presents monthly time series of the main ecological model variables for two experiments: a simulation corresponding to a high value of the sediment organic nitrogen mineralization rate 0.05 day^{-1} (named *SI2*) and the reference simulation (*RS*) for which the reference value for the parameter is 0.002 day^{-1}. Fig. 9 presents monthly time series of the main ecological model variables for two experiments: a simulation corresponding to a high value of the depth of the oxygen penetration into sediments, 0.012 m (named *SI3*), and the reference simulation (*RS*) for which the reference value for the parameter is 0.0012 m. As previously referred, the reference simulation was performed with the default value for all the model parameters, that is, their lowest values, as no interaction with the sediment processes (Lopes et al. 2010) were taken into consideration. The value of the nitrogen mineralization rate for the *SI2* is considerable lower than 0.075 day^{-1}, a value taken by Ambrose et al. (1993) and Zheng & Zhang (2004) for the Satilla River Estuary. The value for the depth of the oxygen penetration into sediments is in agreement with those referred by Revsbech (1983), Revsbech & Jorgensen (1986), Sorensen et al. (1984), Booij et al. (1994), Cai & Sayles (1996). According to the authors, the oxygen penetration of the sediment goes down to between few mm and few cm in estuarine and coastal sediments. It can be observed in Fig. 8 that the *DO* values for *SI2* remain within the range of 8-10 mg O_2 l^{-1}, but slightly increase from *RS*, reflecting an increase of *Chl-a*. *N-total,* shows some concentration increases for *SI2*, but the values remain within the range 0.3-04 mg l^{-1}, whereas *IN* slightly increases from *RS*, even though the values stay below 0.2 mg l^{-1}. Again, the increase of both *N-total* and *IN* for *SI2*, suggests a release of nitrogen into the water column. The *Chl-a* concentration increase may be attributed to the increase of nutrients availability within the water column. The most striking result corresponds the increase of *Od_Det*, from low values, below 0.05 mg O_2 l^{-1} day $^{-1}$, for *RS,* to extreme high values for *SI1*, reflecting an increase of the detritus within the water column. Concerning *SI3* it can be observed (Fig. 9) that *DO* values are, in general, lower than those for *RS,* namely at station 1, where the minimum values reach 6 mg O_2 l^{-1}. *N-total* and *IN* show small concentration increases, which suggests a small amount of the nitrogen being released into the water column. Likewise, *Chl-a* shows small increases. *Od_Det*, when compared to *RS,* shows extreme high values for *SI3*, varying within the range of 0.1-0.2 mg O_2 l^{-1} day $^{-1}$, reflecting, again, an increase of the detritus within the water column. In overall, the results show that the oxygen consumption due to the increase of the depth of the oxygen penetration layer and, therefore, the increase of the rate of the oxygen burial into sediment is only a small fraction of the oxygen budget within the water column. This lead to conclude that the DO distribution within the water column is dependent on the global transport, local production and advection by the currents, as well the exchanges with the atmosphere.

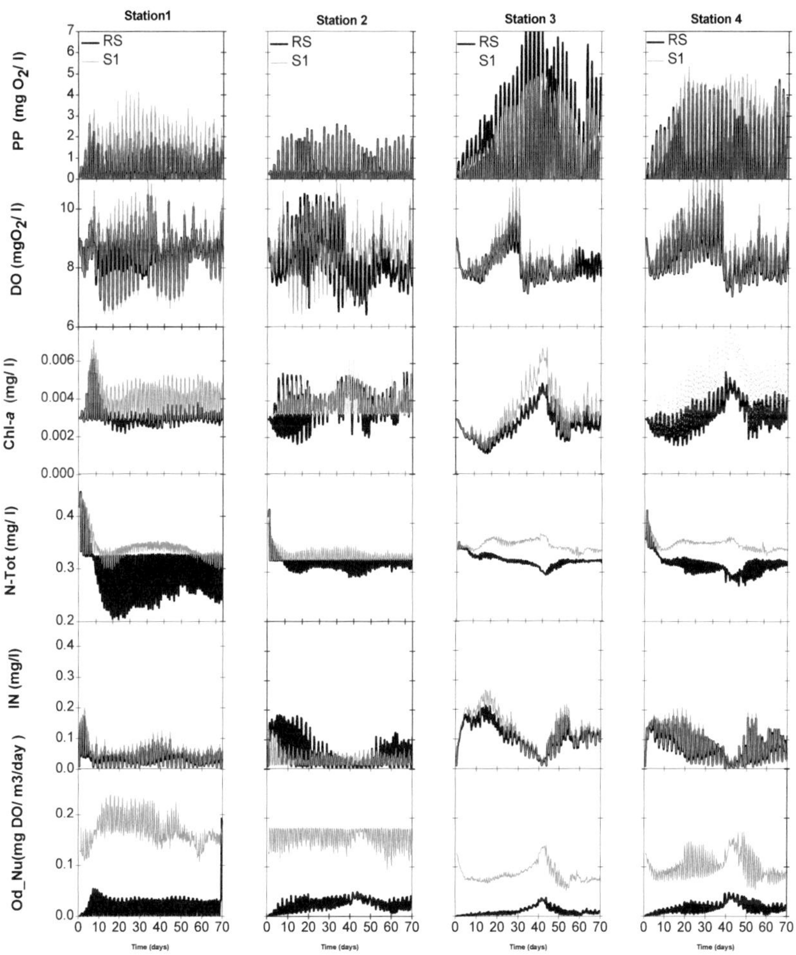

Fig. 8. The monthly time series for the main ecological model variables and for two different values of the sediment organic nitrogen mineralization rate: (a) 0.002 day^{-1} for the reference simulation (*RS*); (b) 0.05 day^{-1} for the simulation (*SI2*).

In the following section it will be discussed several other experiments, not presented here. A simulation corresponding to the proportionality factor for the sediment respiration of 2 (named *SI4*), compared to the reference simulation (*RS*) for which the value for the parameter is 1. The *DO* values for *SI4* are slightly sensitive to the sediment respiration, with values ranging within the interval, 7-10 mg O_2 l^{-1} . Stations 1 and 3 show the lowest minimum values, reflecting an increase of oxygen consumption by this process and the influence of shallow areas. Concerning the other variables, it was observed, for some stations, unexpected concentration changes when compared to RS, which may not be directly attributable to the specific process and location, but rather a consequence of the advective transport. *Od_Det* has shown similar results, namely, extreme high values for *SI5*, varying within the range of 0.1-0.2 mg O2 l^{-1} day -1, were observed, which reflects the increase of the detritus within the water column.

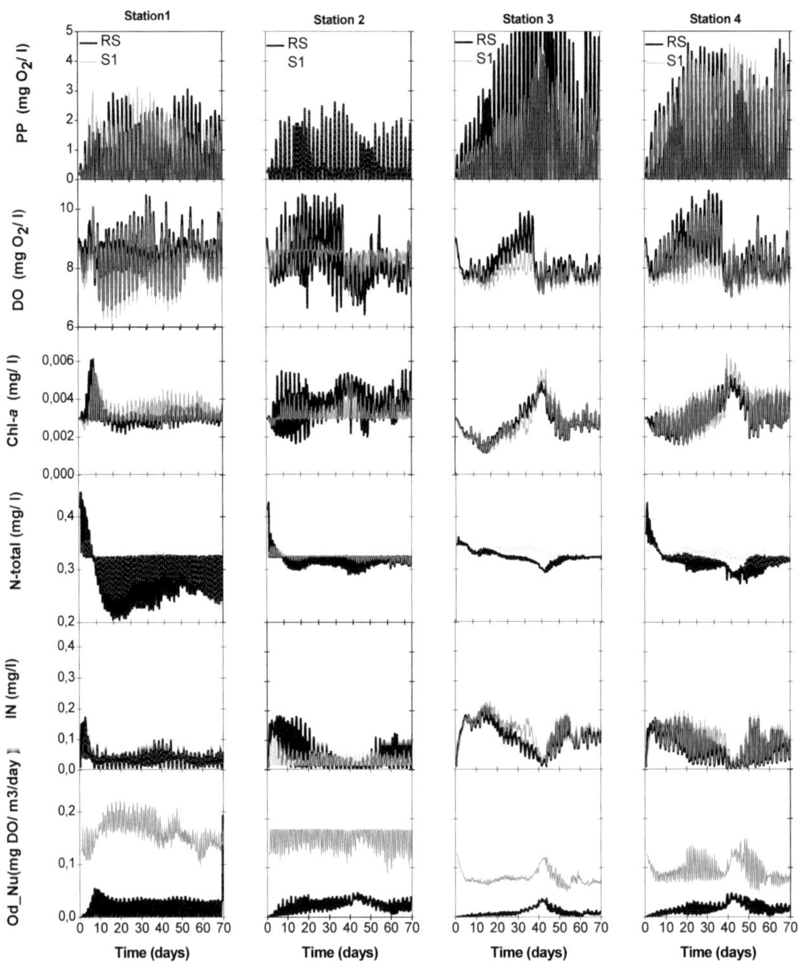

Fig. 9. Monthly time series for the main ecological model variables, for two different values of the depth of the oxygen penetration into sediments: (a) 0.0012m for the reference simulation (*RS*); (b) 0.012 m for the simulation (*SI3*).

In order to complement the above simulations, two other experiments have been performed: (1) a simulation corresponding to the proportionality factor for the sediment respiration, which value was chosen equal to 2 (named *SI4*) compared to the reference simulation (*RS*) for which the reference value for the parameter is 1; (2) a simulation corresponding to a high value of the nitrification rate, which was chosen equal to 9 (named *SI5*) compared to the reference simulation (*RS*) for which the reference value for the parameter is 4.8. Since the results are not very different from those presented in Fig. 9, no figure are presented concerning those cases. Concerning *SI4*, *DO* is slightly sensitive to the sediment respiration, with the concentration values ranging within the interval, 7-10 mg O_2 l^{-1}, and stations 1 and 3 showing the lowest values, reflecting an increase of the oxygen consumption and the influence of shallow areas. Concerning the other variables, unexpected concentration changes, even though small were observed. On the other hand, O*d_Det*, shows extreme high values for SI4, of the same range of that of Fig. 9, which reflects again, an increase of the detritus within the water column. These results may not be directly attributable to the specific process and location, but rather as direct consequence of the advective transport.
Concerning *SI5* it was observed that *DO* shows some significant amplitude fluctuations for the shallow stations, 1 and 2, which varies within the range of 6-10 mg O_2 l^{-1}, but remains quite unchanged for the deeper stations, 3 and 4. As expected no significant variations were observed for the other variables, even tough, small but significant *Chl-a* increase are observed for all the stations. This last result may be attributed to the increase of nutrients availability within the water column.

5. Discussion and conclusions

It is generally accepted that one of the main criteria for the assessment of the water quality of a given aquatic system is its oxygen content, as hypoxia or anoxia conditions reflect potential conditions chemical or biological contaminations. Dissolved oxygen in natural systems, as coastal and estuarine waters, reflects the state of these ecosystems and the influence of several physical, chemical and biological processes occurring in it, and is one of the most important indicators of their water quality state. Oxygen is involved in the main processes occurring in the water column, namely, photosynthesis, organic matter degradation and bacterial nitrification.
Sediments play, as well, an important role in the biogeochemical processes occurring in natural systems, namely coastal and estuarine waters, and affect the dissolved oxygen budget within the water column. During the transport and deposition, sediments may undergo many changes resulting from several chemical and biological processes. The biology of a sediment system might mediate the response of the bed to physical forcing. Accretion of sediment in intertidal areas can be influenced by vegetation: the role of vegetation is complex and is site specific, affected by a variety of physical factors as local topography, wave climate, sediment supply and vegetation patch size. The contribution of living root volume occupied by the different plant species may be important by the addition of the volume itself, as well as by binding sediments and minimising re-suspension (Wijsman, 1999; 2002; Brown et al, 1998).
Oxygen is involved in the main processes occurring in the upper layer of the bottom sediments, namely, oxygen penetration into the bottom sediments, sediment respiration, nitrification processes (Henriksen et al., 1981; Henriksen and Kemp, 1988) and denitrication processes (Kaspar, 1982) in sediments. In the presence of the oxygen, the ammonium

nitrogen can be converted into nitrate nitrogen (called nitrification). This process is affected by *pH*, flow condition, salinity, and turbidity. Under low dissolved oxygen condition, nitrate nitrogen can be converted into nitrogen (called denitrification). Its particulate fraction will settle and deposit on the bottom sediment when the water body is calm. Also the organic nitrogen can be converted into ammonium nitrogen by bacterial decomposition or mineralization before it can be taken up by phytoplankton (Zheng et al., 2004). Sediment respiration, which is directly linked to the flux of degradable organic matter to the sea floor, is the key factor for many early diagenesis processes in the sediment. Situations of low oxygen concentration of the bottom water and high rates of organic matter sedimentation increase the importance of the anaerobic degradation (Wijsman, 2002).

Ria de Aveiro is a temperate lagoon and a well oxygenated system due to its hydrodynamic feature of a very shallow and tidal dominated system (Dias et al., 2003; Lopes & Dias, 2007). Although some episodic hypoxia situations, have been observed during dry seasons, in situations of high oxygen demand waters at the far end areas (Alcântara et al., 1992; 1994; Almeida et al., 2005), the lagoon state, in overall, is characterised by a high content of dissolved oxygen, even reaching saturation levels for high phytoplankton productivity season. This feature evidences the influence of the physical and the biogeochemical processes. Previous applications of the water quality and ecological models have provided some understanding about the influence of these processes in the *DO* distribution of the lagoon (Lopes & Silva 2006; Lopes et al., 2008; 2010). It was shown that the lagoon may be divided into two areas: the main area where the dissolved oxygen concentrations are relatively high, due to the influence of the local production and of the dynamic processes induced by tides, which allows a constant interaction between the ocean and the lagoon and a renewal of the dissolved oxygen; the far end areas, under the influence of weak tidal currents, where the bio-geochemical processes may become dominant and the dissolved oxygen concentrations may reach extreme values: (1) very low ones, corresponding to hypoxia conditions, as t may occurs in situations of episodic organic contamination; (2) extremely high values, even oversaturation conditions, if favourable situations of high phytoplankton productivity, light availability and high nutrient concentration are met (Lopes & Silva, 2006; Lopes et al., 2008 2010). Indeed, the far end areas are characterized by low flushing rates, low residual currents and high residence time (Dias et al., 2003; Lopes & Dias, 2007). Therefore, in a particular situation, when the oxygen budget is negative, that is, when its consumption or degradation exceeds the production by the phytoplankton or the advection by tidal flow, hypoxic situations can be set up. Nevertheless, even considering the worst scenario, this condition tends to remain confined to the far end areas. In this situation, tidal transport, mixing and dilution between high and low oxygenated water occurs, contributing, therefore to a *DO* drop to values of the order of 7 mg O_2 l^{-1} in the central area (Lopes & Silva 2006; Lopes et al., 2008; 2010). Otherwise, *DO* tends to remain quite high, as under the influence of the tidal transport, ocean waters bring oxygenated waters to the central areas of the lagoon. Therefore, the high *DO* values for the central areas of the lagoon reflect, in general, the influence of the ocean waters and the role of the lagoon hydrodynamic in the dissolved oxygen distribution. On the other hand, states of very high dissolved oxygen within the water column, occurring far from the lagoon mouth, reflect conditions of high phytoplankton growth and productivity as well as high oxygen production. Plus et al. (2003) have shown for the Thau lagoon (France) that during winter, the oxygen concentrations remain rather high in the central areas of the lagoon, due to the phytoplankton production and water mixing by the wind, but in summer the oxygen concentrations are lower due to several factors leading to oxygen depletion: faster mineralization processes, lower solubility in warm waters and accumulation of dead organic material, coupled with higher oxygen consumption due to high temperatures. They showed that the influence of the macrophyte populations and of the mineralization of organic matter change significantly the dissolved oxygen distribution in the Thau lagoon (Denis & Grez, 2003). They found that the far end areas of the lagoon have potential risks of oxygen depletion and anoxic crisis, whereas the central and the southern areas of the lagoon may exhibit higher oxygen concentrations due the influence of marine oxygenated waters. Their results confirm that the oxygen concentrations not only depend on the in-situ biogeochemical processes, but also on the hydrodynamics of the lagoon. Malmgren-Hansen et al. (1984) modelled the oxygen depletion in the Limfjord, Denmark. They showed that the bottom oxygen concentration was sensitive to two main factors: The vertical mixing and the biological processes producing and consuming oxygen. They also demonstrated that the number of periods with oxygen depletion will increase exponentially with the rates of the oxygen consuming processes. Zeng et al., (2004) studying the water quality model for the Satilla River Estuary, Georgia, showed that the intertidal salt marsh has high sediment oxygen demand and acts as a major consumer of *DO* and is, therefore, the major responsible for the low *DO* values. They emphasised, as well, the importance of the heterotrophic bacterial respiration process in the *DO* distribution inside the estuary. Lianyuan et al. (2004) used the same approach as Zeng et al., (2004) and showed a *DO* distribution with a high-spatial variability, with high concentration values (7 mgl−1) in the inner shelf and mouth of the estuary, and lower values (below 4 mgl−1) in the upstream end of the estuary. They pointed out that *DO* balance is dominantly controlled by the *SOD*, reflecting the nature of an estuarine-salt marsh ecosystem, with low anthropogenic influence.

Until now, no modeling studies for the Ria de Aveiro lagoon have focused on the relationship between the biogeochemical states of the water column and the biogeochemical processes occurring in the upper layers of the bottom sediments. Assessing the influence of these processes is crucial to improve the understanding of the influence of the near bottom processes in the water column biogeochemical state, and, therefore, the role of the intertidal shallow areas in the biogeochemical budget inside the lagoon.

The results show that increasing suspended sediment (*SPS*) concentration contributes to a significant decay of the dissolved oxygen concentration, as well as of the chlorophyll-a concentrations and the primary production. This can be explained by the fact that in a situation of high background sediment concentrations, the light attenuation within the water column increases, and as less light is available for the photosynthesis, the mortality outpaces the growth of phytoplankton, which generates less oxygen production. On the other, an increase of the phytoplankton mortality may generate additional inorganic matter, which can be mineralised into nitrogen. The excess of nitrogen for the simulation, should, therefore, be attributed to the regeneration process. Furthermore, the concentration of organic substrates for the nitrifier bacterias, increases with the *SPS* concentration, leading to the rising of the heterotrophic nitrification rate. Therefore, increasing of the *SPS* concentration within the water column may generate low phytoplankton production and high consumption of oxygen. The results show, as well, that increasing the mineralisation rate of the sediment, induces a rise of of the *N-total* and *IN* within the water column, which is not surprising, as this process is responsible for the nutrient regeneration from the sediments. In this situation, an increase of *Chl-a* reflects the condition associated to the nitrogen availability within the water column.

Xinghui et al., 2009 showed that increase of *SPS* concentration, caused by the erosion, resuspension and sediment transport, generate the nitrification in aquatic systems. They showed that the nitrification rate increases with *SPS* concentration as a power function of the concentration. For the Yellow River, when *SPS* concentration increases from 0.001 g l^{-1} to 1, 5, 50, and 100 g l^{-1}, the nitrification rate increases about 1.1, 1.4, 2.1, and 2.3 times, respectively. On the other hand, Plus et al. (2003) showed that while during wet season, nutrient inputs from the watershed of the Thau lagoon support the uptake of macrophytes and phytoplankton, during dry periods, mineralisation processes, through mortality and decomposition, become dominant, allowing a primary production based mostly on nutrient regeneration. Part of this nitrogen stock can be released in the water and, even, feed phytoplankton growth.

When quantifying the nitrogen budget within the water column, nitrification process is not the only process to be taken into account. Denitrification, which is an inverse process by which combined nitrogen (nitrate, ammonium or organic forms) is reduced anaerobically to gaseous end products form of nitrogen (NO, N_2O or N_2) by heterotrophic bacteria (Devol et al, 1997; Devol 2008, Fenel et al., 2009, Kaspar, 1982), should be, as well, considered in the nitrogen budget of a given water body, even though its effect in sediments containing ample labile organic matter is often limited by the availability of nitrate or nitrite (Fenel et al., 2009). Trimmer et al. (1998) discussed the role of the bottom sediments in the *N* budget of the upper estuary of the Great Ouse, and showed that, although a fractional loss of riverine nutrient loads via either denitrification, primary production or sediment burial, within an estuary, seems to be, in part, related to the estuary flushing time or residence time (Balls 1994, Nixon et al. 1996), estuarine sediments may also add fixed available *N* (primarily NH_4^+) to the overlying water via NO_3^- and organic ammonifications. They showed, as well, that more than 90% of the total sedimentary N flux removing NO_3^- from the water column was due to *N* gases and the sediments of the upper estuary. Billen & Lancelot (1988) showed that estuarine sediments may constitute substantial source of *N* for pelagic primary production. Examining the nitrogen balance during spring bloom, they showed that microheterotrophic activity could play an important role in food-web dynamics by partly satisfying the nitrogen needs of phytoplankton, which represents about twice the mineral nitrogen stock initially present in the water column.

The results presented in the previous section concerning the Ria de Aveiro lagoon have emphasized the influence of the biogeochemical processes in the *DO* distribution of the lagoon. They have shown that the hydrodynamic is the dominant process defining the *DO* distribution of the lagoon. The good connection between Ria de Aveiro lagoon and the sea and the circulation pattern inside the lagoon, the exchanges with the atmosphere, as well as the primary productivity are, therefore, responsible for the relatively high level of oxygen content within the water column. Furthermore, even, in situation of intensification of the residual circulation toward the lagoon mouth (Dias et al., 2001; Dias & Lopes, 2007), and the export of low oxygenated waters from the river flows toward the lagoon central areas, there is no severe impact in *DO* distribution. Biogeochemical conditions associated to near bottom sediment processes influence the dissolved oxygen mass balance through the nutrients

State variables	Symbol	Unit
Phytoplankton carbon	*PC*	gC/m^3
Phytoplankton nitrogen	*PN*	gN/m^3
Phytoplankton phosphorus	*PP*	gP/m^3
Chlorophyll-a	*CH*	g/m^3
Zooplankton	*ZC*	gC/m^3
Detritus carbon	*DC*	gC/m^3
Detritus nitrogen	*DN*	gN/m^3
Detritus phosphorus	*DP*	gP/m^3
Ammonium	*NH*	gN/m^3
Nitrate	*N3*	gN/m^3
Inorganic phosphorus	*IP*	gP/m^3
Dissolved oxygen	*DO*	g/m^3
Benthic vegetation carbon	*BC*	gC/m^3

Table 1. The ecological/eutrofication model state variable

Phytoplankton Parameters	**Values**
Phytoplankton C growth rate	4.2 day^{-1}
Settling rate	0.1 day^{-1}
Settling velocity	0.2 $m*day^{-1}$
Coagulation induced settling at high phytoplankton concentration	0.02 mg^{-2} day^{-1}
1st order death rate	0.07 day^{-1}
Temperature growth rate coefficient	1.07
Coefficient for min. chlorophyll-a production	0.4
Coefficient for max. chlorophyll-a production	1.1
Min. intracellular concentration of nitrogen	0.07 g N $(g\ C)^{-1}$
Max. intracellular concentration of nitrogen	0.17 g N/g C
Min. intracellular concentration of phosphorus	0.003 g P/g C
Max. intracellular concentration of phosphorus	0.02 g P/g C
Shape factor for sigmoide for nitrate uptake function	0.01
Half-saturation constant for NH4 uptake	0.05 mg N l^{-1}
Half-saturation constant for NO3 uptake	0.1 mg N l^{-1}
Half-saturation constant for P uptake	0.2 mg P l^{-1}
Maximum N uptake	0.3 g $N(g\ C)^{-1}$
P uptake under limiting conditions	0.05 mg P l^{-1}
Fraction of nutrients release a phyto. death	0.1
Light saturation intensity at 20° C	200 μmole $(m)^{-2}(s)^{-1}$
Temperature dependency for light saturation intensity	1.04
Algae respiration rate	0.06 day^{-1}
Coefficient for nitrate dark uptake by phytoplankton	0.6
N release coefficient for N mineralisation	1
Zooplankton Parameters	
Maximum grazing rate	1 day^{-1}
Death rate (1st order) (2st order)	0.05 day^{-1} 0.002 day^{-1}
Grazing threshold	0.02 mg l^{-1}
Halfsaturation concentration for grazing PC	0.5 mg l^{-1}

Temperature dependency for maximum grazing rate	1.1
Temperature dependency for C mineralisation	1.14
N:C ratio	0.17 gN/gC
P:C ratio	0.03 gP/gC
Respiration due to specific dynamic	0.2
Basal metabolism	0.04 day^{-1}
Assimilation factor	0.28
Detritus c mineralization rate	0.05 day^{-1}
Detritus c settling rate <depth 2m	0.05 day^{-1}
Detritus c settling velocity >depth 2m	0.1 m day^{-1}
N release coefficient for N mineralisation	1
N release coefficient for N mineralisation	1
Macroalgae Parameters	
Sloughing rate at 20°	0.01 day^{-1}
Production rate at 20°	0.25 day^{-1}
Temperature dependency sloughing rate	1.07
Temperature dependency production rate	1.05
N:C ratios	0.137 gN/gC
P:C ratios	0.016 gP/gC
Halfsaturation concentration for N uptake	0.5 mg l^{-1}
Halfsaturation concentration for P uptake	0.01 mg l^{-1}
Light saturation intensity at 20°C	180 µEinstein m^{-2}s^{-1}
Temperature dependency for light saturation	1.04
Macroalgae Parameters	
Proportional factor for sediment respiration	1
Proportional factor for N release from sediment	1
Proportional factor for P release from sediment	1
Temperature dependency for N release from sediment	1.1
Temperature dependency for P release from sediment	1.1
N-release from anoxic conditions	0.05 gNm^{-2}day^{-1})
P-release from anoxic conditions	0.01 gPm^{-2}day^{-1})
Temperature factor for sediment respiration	1.07
Diffusion between coef. for NO_3 and NH_4between sediment and water	0.002 m^{2}day^{-1}
Max. Denitrification in sediment an 20°C	0.2 mg N l^{-1} day^{-1}
First constant for DO penetration into sediment	0.00124 m
Second constant for DO penetration into sediment	0.0000403 m^4 /gO_2
Third constant for DO penetration into sediment	0.0000473 m^3 *day/gO_2
Half-saturation conc. DO, for NH_4 or NO_3 release from sediment	3
Specific nitrication rate pelagic water, 20°C	0.05 day^{-1}
Teta value, nitrification	1.04
Half-saturation conc. NO_3, for denitrification	0.2 mg N l^{-1}
Half-saturation conc. , for DC denitrification	0.2 mg C l^{-1}
Half-saturation conc. DO denitrification	0.5 mg O_2 l^{-1}
g O2 used to oxidize 1 g NH_4 to 1 g NO_3	4.3 g O_2/ g NH_4 -N

Table 2. Main parameter values for the ecological/eutrofication model.

processes within sediments, namely at the lagoon far end, but as the transport by tidal flow is a very effective process, the *DO* status seems not very affected by these processes. Therefore, in a scenario where the geomorphodynamical and the hydrodynamical conditions of the Ria de Aveiro remain unchanged, the setup of hypoxic situation inside the lagoon is unlikely to occur, even in the worst scenarios. Changes in the lagoon morphodynamics resulting from the anthropogenic activities inside the lagoon and near its border, the infilling of the lagoon mouth or channels, may induce changes in the hydrodynamic regime, altering the lagoon flushing and residence times. Therefore, it is expected in this situation a negative impact of the biogeochemical processes in the *DO* distribution of the water column.

6. Acknowledgment

This work was supported by CESAM (Centro de Estudos do Ambiente e do Mar).

7. References

Abbott, M.B.; McCowan, A.D. & Warren, I.R. (1981). Numerical Modeling of Free-Surface Flows that are Two-Dimensional in Plan. *Transport Models for Inland and Coastal Waters,* Fischer, H.B. (Ed.), Academic Press, New York, pp. 362-407.

Alcântara, F.; Pereira, M. G.; Almeida, M. A. & Cunha, M.A. (1992). Qualidade microbiológica da água da Ria de Aveiro, 1991. *Polaveiro Repport.* Departamento de Biologia, Universidade de Aveiro, Portugal, pp 90.

Alcântara, F.; Pereira, M. G.; Almeida, M. A. & Cunha, M.A. (1994). Qualidade Microbiológica da Água da Ria de Aveiro, 1989-1993. *Polaveiro Repport* . Departamento de Biologia, Universidade de Aveiro, Portugal, pp. 85.

Almeida, M.A.; Cunha M.A. &Alcântara F. (2005). Relationship of bacterioplankton production with primary production and respiration in a shallow estuarine system (Ria de Aveiro, NW Portugal). *Microbiological Research,* No.160, pp. 315-328.

Ambrose Jr.; R.B., Wool; T.A. & Martin, J.L. (1993). *The Water Quality Analysis Simulation Program, WASP5, Part A.* Model Documentation. U.S. Environmental Protection Agency, Athens, Georgia, pp 202.

Archer, D. E.; Morford, J. L. & Emerson, S. R. (2002). A model of suboxic sedimentary diagenesis suitable for automatic tuning and gridded global domains. *Global Biogeochemical Cycles,* No.16, pp. 17, 1-21.

Balls, P. W. (1994). Nutrient inputs to estuaries from nine Scottish east coast rivers; influence of estuarine processes on inputs to the North Sea. *Estuarine, Coastal and Shelf Sci*ence, No.39, pp. 329–352.

Billen, G. & Lancelot, C. (1988). Modeling benthic nitrogen cycling in temperate coastal ecosystems. In: *Nitrogen Cycling in Coastal Marine Environments,* T.H. Blackburn and Sorensen (Eds.),. Wiley and Sons, New York, pp. 341-378.

Berg P.; Rysgaard S. & Thamdrup B. (2003). General dynamic modeling of early diagenesis and nutrient cycling; Applied to an Arctic marine sediment. *Journal of American Science,* No. 303, pp. 905-955.

Berner, R. A. (1980). *Early Diagenesis: A Theoretical Approach,* Princeton Univ. Press, Princeton, N.J., pp. 241.

Black, K.S. (1997). Microbiological factors contributing the erosion resistance in natural cohesive sediments. In: *Cohesive Sediments.* (Eds. N. Burt, R. Parker and J. Watts.

John Wiley&Sons, New York). Proc. 4th Nearshore and Estuarine Cohesive Sediment Transport Conference, 1994, Wallingford, U.K., pp. 231-244.
Blackburn T.H. & Henriksen K. (1983). Nitrogen cycling in different types of sediments from Danish Waters. Limnology Oceanography, Vol. 28, 3, 477-493.
Booij, K; Sundby, B. & Helder, W. (1994). Measuring the flux of oxygen to a muddy sediment with a cylindrical microcosm. *Netherlands Journal of Sea Research,* Vol.32, No.1, pp. 1-11.
Boudreau, B.P. (1996). A method-of-lines code for carbon and nutrient diagenesis in aquatic sediments. Computers & Geosciences , Vol. 22, No.5, pp. 479-496.
Boudreau, B.P. (1997). Diagenesis Models and Their Implementation: Modeling Transport and Reactions in Aquatic Sediments. Springer-Verlag,Berlin, pp. 414.
Brown, S.L.; Warman E. A.; McGrorty S.; Yates M.; Pakeman R.J.; Boorman L.A.; Goss-Custard J.D. & Gray A.J. (1998). Sediment fluxes in intertidal biotopes, BIOTA II. *Marine Pollution Bulletin,* 37, 173-181.
Cai, W.J. & F.L. Sayles (1996). Oxygen penetration depths and fluxes in marine sediments. *Marine Chemistry,* No.52, pp. 123-131
Casuli, V. A. (1999) Semi-Implicit Finite Difference Method for Non-Hydrostatic, Free-Surface Flows, *International Journal for Numerical Methods in Fluids,* No.30, pp. 425-440.
Devol A.H.; Codispoti L.A. & Christensen J.P. (1997). Summer and winter denitrification rates in western Arctic shelf sediments. *Continental Shelf Research,* Vol. 17, No.9, 1029-1050.
Devol A.H. (2008). Denitrification including anammox. In: *Nitrogen in the marine environment,* Capone D., Carpenter E., Mullholland M., Bronk D. (Eds). Elsevier, Amsterdam, pp. 263–302.
Denis L., Grenz C. (2003). Spatial variability in oxygen and nutrient fluxes at the sediment-water interface on the continental shelf in the Gulf of Lions (NW Mediterranean Sea). *Oceanological Acta,* No.26, pp. 373-389.
Dhakar, S. P. & Burdige, D. J. (1996). A coupled, non-linear, steady state model for early diagenesis processes in pelagic sediments. *American Journal of Science,* Vol. 296, pp. 296-330.
Dias, J.M. (2000). *Contribution to the study of the Ria de Aveiro hydrodynamics,* PhD thesis, Universidade de Aveiro, Portugal, pp. 288.
Dias, J. M.; Lopes, J. F. & Dekeyser, I. (1999). Hydrological characterisation of Ria de Aveiro lagoon, Portugal, in early summer. *Oceanological Acta,* No.22, pp. 473-485.
Dias, J. M.; Lopes, J. F. & Dekeyser, I. (2000). Tidal propagation in Ria de Aveiro lagoon, Portugal. *Physics and Chemistry of the Earth, B,* No.25, pp. 369-374.
Dias, J. M.; Lopes, J. F. & Dekeyser, I. (2001). Lagrangian transport of particles in Ria de Aveiro lagoon, Portugal. *Physics and Chemistry of the Earth, B,* No.26, pp. 729-734.
Dias, J. M.; Lopes J. F. & Dekeyser I. (2003). A numerical system to study the transport properties in the Ria de Aveiro lagoon. *Ocean Dynamics,* No.53, pp. 220-231.
Dias J. M. & Lopes J. F. (2006). Implementation and evaluation of hydrodynamic, salt and heat transport models: The case of Ria de Aveiro Lagoon (Portugal). *Environmental Modeling and Software,* No.21, pp. 1-15.
Droop, M.R (1973). Some thoughts on nutrient limitation in algae. *Journal of Phycology,* No.9, pp. 264-272.
Droop, M.R. (1975) The nutrient status of algal cells in batch cultures. *Journal of Marine Biology, Ass. U.K.,* No. 55, pp. 541-555.

Dyer, K.R. (1986). *Coastal and Estuarine Sediment Dynamics.* Wiley-Interscience, New York, p. 342.

Dyer, K.R. (1995). Sediment transport processes in Estuaries. In: *Geomorphology and Sedimentology of Estuaries*, Perillo, G.M.E. (Ed.), Elsevier, New York, pp. 423–449.

Erichsen A. C. & Rasch P. S. (2001). Two- and Three-Dimensional Model System Predicting the Water Quality of Tomorrow. In: *Proc. of the Seventh International Conference on Estuarine and Coastal Modeling*, Spaulding, M.L. (Eds.), American Society of Civil Engineers, pp. 165-184.

Fennel, K.; Brady D.; DiToro, D.; Fulweiler, R.; Gardner, W.; McCarthy, M.; Rao, A.; Seitzinger, S.; Thouvenot-Korppoo, M. & Tobias, C. (2009). Modeling denitrification in aquatic sediments. *Biogeochemistry*, No. 93, pp. 159-178.

Gill, A.E. (1982). Atmosphere-Ocean Dynamics. *International Geophysics Series*, 30, pp. 663.

Heip, C.; Goose N.; Herman P. M. J.; Krokamp J.; Middelburg J. & Soetaert K. (1995). Production and consumption of biological particles in temperate tidal estuaries. *Oceanography* and *Marine Biology* - An *Annual Review.* No.33, pp. 1–149.

Henriksen, K.; J. I. Hansen & T. H. Blackburn (1981). Rates of nitrification, distribution of nitrifying bacteria, and nitrate fluxes in different types of sediments from Danish waters. *Marine Biology*, No.61, pp. 299-304.

Henriksen, K. & Kemp W. (1988). Nitrification in estuarine and coastal marine sediments, In: *Nitrogen cycling in coastal marine environments,* T. H. Blackburn and J. Sorensen (Ed.), John Wiley & Sons, Inc., New York, pp. 201-249.

Herman, P. M. J., Soetaert, K., Middelburg J. J., Heip, C., Lohse, L., Epping, E., Helder, W., Antia, A. N. & Peinert, R. (2001). The seafloor as the ultimate sediment trap – using sediment properties to constrain benthic-pelagic exchange processes at the Goban Spur. *Deep-Sea Research II*, Vol. 48, pp. 3245–3264.

Jassby, A.D. & C.R. Goldman, 1974. Loss rates from a lake phytoplankton community. *Limnology and Oceanography*, No.21, pp. 540-547.

Jassby, A.D. & Platt, T. (1976). Mathematical formulation of the relationship between photosynthesisand light for phytoplankton. *Limnology and Oceanography*, 21,540-547.

Kaspar, H. F. (1982). Denitrification in marine sediment: measurement of capacity and estimate of *in situ* rate. *Applied and Environmental Microbiology*, No.43, pp. 522-527.

Kiørboe, T. & Nielsen T.G. (1994). Regulation of zooplankton biomass and production in a temperate, coastal ecosystem. I. Copepods. *Limnology and Oceanography*, No.39, pp. 493-507.

Lancelot, C. & Rousseau V. (1987). ICES intercalibration exercise on the 14C 33 method for estimating phytoplankton primary production. *Phase 2: experiments conducted on board of RV DANA.* Preliminary report, 35 pp.

Leendertse, J.J. & E.C. Gritton (1971). A Water-Quality Simulation Model for well-fixed Estuaries and Coastal Seas: Vol 2: *Computation Procedures,* The Rand Corporation, R-708-NYC, pp. 48.

Leonard, A. (1979). A stable and accurate convective modeling procedure based on quadratic upstream interpolation. *Computer Methods in Applied Mechanics and Engineering,* No.19, pp. 59-98.

Lessin G. & Raudsepp U. (2006). Water quality assessment using integrated modeling and monitoring in Narva Bay, Gulf of Finland. *Environmental Modeling & Assessment,* No. 11, pp. 315–332.

Lianyuan, Z. ; Chen C. & Zhang Y. F. (2004). Development of water quality model in the Satilla River Estuary, Georgia. *Ecological Modeling*, No. 178, pp. 457-482.

Lopes J.F.; J. M. Dias & A. C. Cardoso, C. I. Silva (2005). The water quality of the Ria de Aveiro lagoon, Portugal: from the observations to the implementation of a numerical model. *Marine Environmental Research*, No.60, pp. 594-628.
Lopes J.F.; Dias J.M. & Dekeyser I. (2006). Numerical Modeling of Cohesive Sediments Transport in the Ria de Aveiro Lagoon, Portugal. *Journal of Hydrology.* No. 319, pp. 176-198.
Lopes J.F. & Silva C. (2006). Temporal and spatial distribution of dissolved, *Ecological Modeling*. No. 197, pp. 67-88.
Lopes J.F. & Dias J.M. (2007). Residual circulation and sediments transport in Ria de Aveiro lagoon, Portugal. Journal of Marine Systems. Vol.68, pp. 507-528.
Lopes J.F.; Silva C. & Cardoso A.C. (2008). Validation of a water quality model for the Ria de Aveiro lagoon, Portugal. *Environmental Modeling & Software*, No.23, pp. 479-494.
Lopes J.F.; M.A. Almeida & Cunha M.A. (2010). Modeling the ecological patterns of a temperate lagoon in a very wet spring season. *Ecological Modeling*, Vol.221, No.19, pp. 2302-2322.
Luff, R.; Wallmann, K.; Grandel, S. & Schlüter, M. (2000). Numerical modeling of benthic processes in the deep Arabian Sea. *Deep-Sea Research II*, Vol. 47, pp. 3039-3072.
Malmgren-Hansen, A.; Mortensen, P. & Møller, B. (1984). Modeling of oxygen depletion in coastal waters. *Water Science and Technology*, No.17, pp. 967-978.
Meysman, F. J. R., Middelburg, J. J., Herman, P. M. J. & Heip, C. H. R. (2003a). Reactive transport in surface sediments. I. Model complexity and software quality. *Computers and Geosciences*, Vol. 29, pp. 291-300.
Meysman, F. J. R.; Middelburg, J. J.; Herman, P. M. J. & Heip, C. H. R. (2003b). Reactive transport in surface sediments. II. Media: an object-oriented problem-solving environment for early diagenesis. *Computers and Geosciences*, Vol. 29, pp. 301–318.
Mike3, 2005a. Flow Model, *Hydrodynamic and Transport Module*, DHI Water & Environment Denmark, pp. 42.
Mike3, 2005b. *Eutrophication Model, A scientific Description*, DHI Water & Environment. Denmark, pp. 36.
Mitchener, H. & Torfs, H. (1996).Mommaerts, J.P. (1978). *Systeembenadering van en gesloten mariene milieu, met de nadruk op de rol van het fytoplankton*. Doctoral thesis. Vrije Universiteit Brussel, pp. 335.
Nixon, S.W.; Arnmerman, J.W.; Atkinson, L.P.; Berounsky, V.M.; Billen, G.; Boicourt, W.C.; Boynton, W.R.; Church, T.M.; DiToro, D.M.; Ehgren, R.; Garber J.H. Giblin, A.E.; Jahnke, R.A.; Owens, N.J.P.; Pilson, M.E.Q. & Seitzinger S.P. (1996). The fate of nitrogen and phosphorus at the land-sea margin of the North Atlantic Ocean. *Biogeochemistry*, No. 35, pp. 141-180.
Nyholm, N. (1976). A Mathematical Model for the Growth of Phytoplankton. *Int. Symp. on Experimental Use of Algal Cultures in Limnology*, Sandefjord, Norway, Oct. 26-28 1976.
Pearson T.H. & Rosenberg R. (1978). Macrobenthic succession in relation to organic enrichment and pollution of the marine environment. *Oceanography and Marine Biology: an Annual Review*, No. 16, pp. 229-311.
Plus, M.; Chapelle, A.; Lazure, P.; Auby, I.; Levasseur, G.; Verlaque, M.; Belsher, T.; Deslous-Paoli, J.-M.; Zaldivar, J.-M. & Murray, C. N. (2003). Modeling of oxygen and nitrogen cycling as a function of macrophyte community in the Thau lagoon. *Continental Shelf Research*, No. 23, pp. 1877-1898
Roman M.; Gauzerns, A.; Rhinehart, W. & J. White (1993). Effects of low oxygen waters on Chesapeake Bay zooplankton. *Limnololgy and Oceanography*, No. 38, pp. 1603-1614.

Revsbech, N.P. (1983). In situ measurement of oxygen profiles of sediments by use of oxygen microelectrodes. In: *Polarographic oxygen sensors: Aquatic and physiological applications*, E. Gnaiger and H. Forstner (Eds.), pp. 265-273, Springer.

Revsbech, N.P. & Jorgensen, B.B. (1986). Microelectrodes: their use in microbial ecology. In: *Advances in Microbial Ecology*, KC. Marshall (Ed.), No 9, pp. 293-352, Plenum, NY.

Rysgaard, S. & Berg, P. (1996). Mineralization in a northeastern Greenland sediment: mathematical modeling, measured sediment pore water profiles and actual activities. *Aquatic Microbial Ecology*, No. 11, pp. 297-305.

Ruadij P. & Van Raaphorst W. (1995). Benthic nutrient regeneration in the ERSEM ecosystem model of the North Sea, Netherlands. *Journal of Sea Research*, Vol. 33, No. 3/4, pp. 453-483.

Rysgaard, S.; Finster, K. & Dahlgaard, H. (1996). Primary production, nutrient dynamics and mineralization in a northeastern Greenland fjord during the summer thaw. *Polar Biology*, No. 16, pp. 497-506.

Rysgaard, S.; Thamdrup, B.; Risgaard-Petersen, N.; Fossing, H.; Berg, P.; Christensen, P. B. & Dalsgaard,T. (1998). Seasonal carbon and nitrogen mineralization in a high-Arctic coastal marine sediment, Young Sound, Northeast Greenland. *Marine Ecology Progress Series*, Vol. 175, pp. 261-276.

Rysgaard, S.; Nielsen, T. G. & Hansen, B. (1999). Seasonal variation in nutrients, pelagic primary production and grazing in a high-Arctic coastal marine ecosystem, Young Sound, Northeast Greenland. *Marine Ecology Progress Series*, No. 179, pp. 13-25.

Silva, J. F., Duck, R.W., Hopkins, T.S. & Rodrigues, M., 2002. Evaluation of the nutrient inputs to a coastal lagoon: the case of the Ria de Aveiro, Portugal. *Hydrobiologia*, 475/476, pp. 379-385.

Scavia, D. (1979). Examination of Phosphorus Cycling and Control of Phytoplankton Dynamics in Lake Ontario with an Ecological Model. *Journal* of the *Fisheries Research Board* of Canada., No. 36, pp. 1336-1346.

Sorensen, J.; Hydes, D. J. & Wilson T. R. S. (1984). Denitrification in a deep-sea sediment core from the eastern equatorial Atlantic. *Limnology and Oceanography*, Vol. 29, pp. 653-657.

Soetaert, K.; Herman, P. M. J. & Middelburg, J. J. (1996). A model of early diagenesis processes from the shelf to abyssal depths. *Geochimica et Cosmochimica Acta*, Vol. 60, pp. 1019–1040.

Sundby B., Gobeil C. & Silverberg N. (1992). The Phosphorus cycle in coastal marine sediments. *Limnology and Oceanography*. Vol. 37, No. 6, pp. 1129-1145.

Sweerts, J-P.R.A. (1990). *Oxygen consumption, mineralization and nitrogen cycling at the sediment–water interface of North Temperate lakes*. PhD thesis, Groningen, p. 136.

Sweerts, J-P.R.A.; St. Louis, V. & Cappenberg, T.E. (1989). Oxygen concentration profiles and exchange in sediment cores with circulated overlying water. *Freshwater Biology* No. 21, pp. 401–409.

Sweerts, J-P.R.A.; Kelly, C.A.; Rudd, J.W.M.; Hesslein, R. & Cappenberg, T.E. (1991). Similarity of whole-sediment molecular diffusion coefficients in fresh water sediments of low and high porosity. *Limnology and Oceanography*, Vol. 36, No. 2, pp. 336-341, 1991.

Tett, P.; A. Edwards & K. Jones, 1986. A model for the growth of shelf-sea phytoplankton in summer. *Estuarine and Coast Shelf Science*, No. 23, pp. 641-672.

Torfs, H. (1995). *Erosion of mud/sand mixtures*. PhD thesis, Civil Eng. Dept., K.U. Leuven, p. 223.

Trimmer, M.; Nedwell, D.B.; Sivyer, D.B. & Malcolm S.J. (1999). Nitrogen fluxes through the lower estuary of the river Great Ouse, England:the role of the bottom sediments. *Marine Ecology Progress Series*, No. 163, pp. 109-124.

Van Cappellen, P. & Wang, Y. (1996). Cycling of iron and manganese in surface sediments: A general theory for the coupled transport and reaction of carbon, oxygen, nitrogen, sulfur, iron, and manganese: American Journal of Science, Vol. 296, pp. 197–243.

Van Cappellen, P.; Gaillard, J. F & Rabouille, C. (1993). Biogeochemical transformations in sediments: Kinetic models of early diagenesis, in Wollast, R., Mackenzie, F. T., and Chou, L., editors, *Interactions of C, N, P, and S Biogeochemical Cycles and Global Change*, Berlin, Heidelberg, New York, Springer-Verlag, p. 401–445.

Vested H.J.; Justesen P. & Ekebjærg L. (1992). Advection-dispersion modeling in three dimensions, *Applied Mathematical Modeling*, Vol. 16, pp 506-519.

Xinghui X., Zhifeng Y. & Xueqing Z. (2009). Effect of Suspended-sediment Concentration on Nitrification in River Water: Importance of Suspended Sediment-water *Interface, Environmental Science & Technology*, Vol.43, pp. 3681–3687.

Wijsman J.W.M.; Herman P.M.J. & Gomoiu M.T. (1999). Spatial distribution in sediment characteristics and benthic activity on the northwestern Black Sea shelf. *Marine Ecology,* No.181 , pp. 25–39.

Wijsman, J.W.M.; Herman, P.M.J.; Middelburg, J.J. & Soetaert, K. (2002). A model for early diagenesis processes in sediments of the continental shelf of the Black Sea. *Estuarine and Coast Shelf Science*, No. 54, pp. 403-421.

Windolf, J.; Jeppesen, E.; Jensen, J.P. & Kristensen, P. (1996). Modeling of seasonal variation in nitrogen retention and in-lake concentration: A four-year mass balance study in 16 shallow Danish Lakes. *Biogeochemistery*, No.33, 25-44.

Williams, P.J. LeB. (1975). Aspects of dissolved organic material in sea water. 1 In: *Chemical Oceanography* J.P. Riley & G. Skirrow, (Ed.), pp. 301-363, Academic Press, New York.

Zheng L. C. Chen, & F. Zhang, (2004). Development of water quality model in the Satilla River estuary, Georgia. *Ecological Modeling* , No.178, pp. 457-482.

10

Environmental Observations on the Kam Tin River, Hong Kong

Mervyn R. Peart, Lincoln Fok and Ji Chen
The University of Hong Kong
Hong Kong SAR
China

1. Introduction

Hong Kong, from being described as a barren rock with scarcely a house upon it in 1841, has undergone a remarkable transformation to a metropolis with, in 2008, a population of seven million. Whilst Hong Kong's economy has become increasingly service oriented since the 1980s, the industrial development of Hong Kong over the past few decades has been impressive (Poon and Tsang, 2010). Moreover, as Poon and Tsang (2010) note "despite the decline in manufacturing's share of GDP and employment, it would be a mistake to conclude that industries are of little relevance to the economy." They suggest that "In fact, the manufacturing and trading sectors have remained the most important sectors of the Hong Kong economy". However, the service sector has flourished in recent decades and has diversified to match the structural transformation of the Hong Kong economy (HKSAR Government, 2009).

Expansion of population, along with the growth and transformation of the economy, has been associated with considerable environmental change, both spatial and temporal, in Hong Kong. For example, urban areas have expanded and new towns have been developed. Moreover, as Wang (2010) observes "The process of urban development of Hong Kong in the past century has actually been a process of interaction between its urban transport system and its urban spatial expansion." This expansion of urban area has occurred into previously rural lowlands resulting in the loss of agricultural land and coastal wetlands (Corlett, 2010). Port and airport development has also accompanied the growth of Hong Kong. For example, Kai Tak Airport in Kowloon was built in 1925 but in the 1980s had reached saturation. A new airport was planned at Chek Lap Kok on Lantau Island and became operational in 1998. Moreover, expansion of port back-up facilities, such as container storage, has also contributed to land-use change. Electricity consumption has also increased in Hong Kong. Tso and Yau (2003) describe a substantial increase in domestic energy consumption in the 1990s. However, reflecting the structural changes in the economy Lam et al. (2010) report that the commercial sector was the largest electricity end user and that cooling-dominated office buildings accounted for most of the sector energy consumption.

The growth and development of Hong Kong has had consequences for the environment. This includes climate, ecology, land-use and upon streams and rivers for example. This paper will report some examples of the impact of this development upon the rivers and streams of Hong Kong by using the Kam Tin North River as a case study.

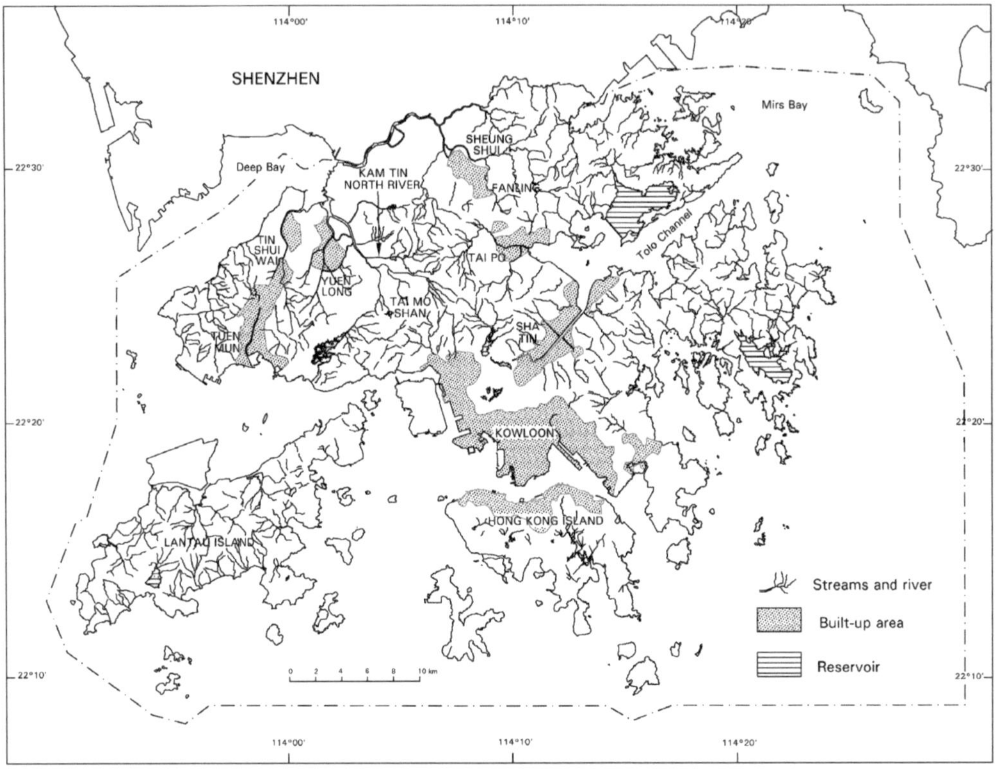

Fig. 1. Location of Hong Kong and the Kam Tin North River.

2. Study area

Hong Kong is a Special Administrative Region (SAR) of the People's Republic of China. It has a coastal location and structurally it is part of the South China coastal massif. Two large marine embayments exist on either side of the SAR, namely Deep Bay to the west and Mirs Bay to the east (Fig. 1). The SAR has an area of around 1,100 km^2 of which 72% is part of the Kowloon Peninsula and New Territories and which is part of the mainland. Some 18% of the SAR consists of around 230 offshore islands of which the two largest are Lantau Island (13% of the SAR) and Hong Kong Island. Despite such a small area, Hong Kong hasa variety of landscapes. Much of the land-area is mountainous with very little flat land: only 20% of the SAR has slope gradients of 0 to 5°, whilst 30% of terrain has slope angles of 15° or more (Styles & Hansen, 1989). The northwest New Territories has much of the lowland, but as Styles and Hansen (1989) note these often lie adjacent to steep slopes that form the uplands. Owen and Shaw (2007) reported that around 5% of Hong Kong is occupied by rivers and their floodplains, mostly in the northwest. It is in one of the drainage basins in the northwest of Hong Kong that this study is located: the Kam Tin River.

The Kam Tin drainage basin is located in the Northwest New Territories (Fig. 1) with an area of 44 km^2 (ERM, 1995; HKEPD, 2004), one of the largest in Hong Kong, and it discharges into Deep Bay. The drainage system, partly developed on one of the largest floodplains in Hong Kong, is composed of two main tributaries - the Kam Tin North and

South Rivers. This study is based on the Kam Tin North River, and it reflects the tension between development and environment.

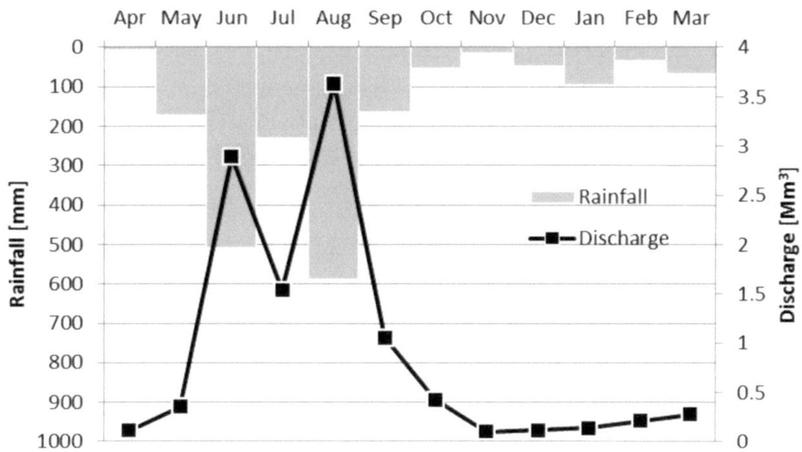

Fig. 2. Monthly rainfall and discharge of the Kam Tin North River, water year 1968.

In terms of climate, the location of Hong Kong puts it under the influence of the Asian monsoon system: accordingly a distinct seasonality exists. This is evidenced in Table 1 which presents climate data for two stations in the Kam Tin basin. These are located at 950 m near the summit of Tai Mo Shan (TMS), the highest peak in Hong Kong at 957 m, and the other sited on the lowland plain at Shek Kong (SK), at a height of around 5 m. In terms of runoff, and rainfall, the seasonality is illustrated in Fig. 2 which shows monthly rainfall and runoff at a Water Supplies Department gauging station for the 1968 water year. Based on data for the water years 1978 to 1988, on average, 68% of the annual runoff occurred during the summer wet season months of April to September. In the 1968 water year at Kam Tin the mean wet and dry season daily discharges were 52,254 m^3/d and 6,952 m^3/d respectively.

The major bedrock geology of the Kam Tin basin includes the volcanic rocks, largely tuffs, of the Shing Mun and Tao Mo Shan Formations and fine-grained Mesozoic granodiorites which form much of the uplands. The lowland areas of the basin are dominated by alluvial deposits, whilst extensive colluvial deposits occur along the foot-slopes of the uplands and extend up some of the stream systems into the hills.

In terms of land-use data for the lowland area of the Kam Tin River, information contained in the Outline Zoning Plans for the Kam Tin North, Shek Kong and Pat Heung districts (HKTPB, 2005a, b, c) show that residential development, village type development, industry, open storage, and major roads accounted for 6.3%, 18.6%, 4.5%, 5.6% and 1.9% respectively. Agriculture accounted for 26.8% of land-use with conservation areas occupying a further 25.9% of the area. There is also the Shek Kong Camp, a military base. However, as Jim (1997) indicates the lowland areas of Hong Kong (including the Kam Tin area) evidenced quite dramatic changes from the 1970s under the pressures of rapid urbanization, container-port expansion, village housing and increased cross-border linkages with China: rural areas, as Jim (1997) notes, had to accommodate multiple demands. The most significant physical consequence of these "demands" was the rapid decline of agriculture from the 1970s and the expansion of non-agricultural land-use, especially open storage in addition to the expansion village type housing under the small house policy implemented since 1972 (c.f. Tang, et al. 2005). More recently drainage improvement schemes for the Kam Tin River have introduced dramatic changes in the form of channelisation/training of the river. Work began in downstream areas in the mid-1990s and on the Kam Tin North main channel works were scheduled from Kam Tin San Tsuen upstream to Wang Toi Shan from mid-1999 to early 2004. In contrast to the lowlands, little 'formal' use is made of the steep uplands and they are covered in grassland, woodland and secondary forest. Hillfires are common on the upland slopes in Hong Kong and influence vegetation, soil development and erosion (c.f. Peart et al., 2009; Dudgeon and Corlett, 2004; Marafa and Chau, 1999).

Month	Mean Air T° [°C]		Relative Humidity [%]		Total Rainfall [mm]	
	SK	TMS	SK	TMS	SK	TMS
January	15.4	10.6	74	82	27.2	34.3
February	17.0	12.0	78	89	40.6	42.5
March	19.6	14.2	79	91	47.7	72.1
April	23.3	17.3	81	93	130.1	145.2
May	26.0	19.8	82	94	252.6	301.5
June	27.6	21.5	85	95	471.4	579.6
July	28.6	22.1	83	95	320.3	397.0
August	27.9	22.0	85	93	407.0	403.5
September	27.0	20.9	81	90	303.0	329.1
October	25.0	18.4	75	88	35.4	56.4
November	20.9	15.0	72	82	27.7	28.4
December	16.9	11.7	71	78	26.2	24.1

Table 1. Climate of the Kam Tin basin, Hong Kong.

3. Channel bank change

River channel bank erosion is a geomorphic process that occurs in many streams and rivers as an adjustment to accommodate the discharge and sediment supplied from the catchment area (Watson and Basher, 2005). It may occur during or shortly after stormflow events and, as Knighton (1998) observed, only limited erosion will occur in the absence of high discharges. Two main groups of processes may result in bank erosion, namely, hydraulic action at or below the water surface and mass failure (Knighton, 1998), although as noted by Watson and Basher (2005) these may be linked. Traditionally, river bank erosion has been regarded as having negative impacts such as loss of land along with the associated resources, damage to property and infrastructure (Piégay et. al. 1997). In addition river bank erosion delivers sediment from the channel banks to the river thereby contributing to the sediment load which may impact upon water quality and also contribute to the sedimentation problem. Kevin et al. (2008) have suggested that "the erosion of channel bank material is a major source of sediment for most rivers" and they also note that because eroded bank material is delivered directly to the river it may have an immediate impact. The importance of channel bank erosion as a source of sediment has been confirmed in a number studies evaluating sediment sources such as Juracek and Ziegler

(2009), Collins and Anthony (2008) and Mukundar et al. (2010). Bank erosion may now be regarded as a desirable attribute of rivers (Floursheim et at. 2008). This is because bank erosion may provide sediment that creates riparian habitat and active channel banks that may lead to the creation and maintenance of a variety of habitats of diverse structure (Florsheim et al. 2008). Pieguay et al. (2005) note that bank erosion provides ecosystem services and other benefits.

There is an array of factors that may influence bank erosion including flow properties, channel bank materials, climate, subsurface conditions, channel geometry, biotic factors including human impact (c.f. Knighton, 1998). Consequently, the amount, along with spatial and temporal distribution, of channel bank erosion are highly variable, as noted by Knighton (1998). The data from this study evidence both spatial and temporal variability of channel bank erosion on sections of the Kam Tin North River. These will be discussed in the following.

Information upon the change in channel bank position has been obtained from two sources, namely graphical comparison of plan-form changes and direct observation. A graphical comparison of plan-form changes was undertaken based upon 1:1200 and 1:1000 maps for the period 1959 to 1983 and for which channel plan-forms were superimposed using Arc/Info. Maps at a scale of 1:1000 showing plan-form change were developed for a study reach identified in Fig. 3 (Top panel). Direct observations by field measurement of channel banks has also been undertaken on two active meanders near Kam Tin which are located in the area identified as C in Fig. 3. The two active meander bends associated with section C in Fig. 3 have been monitored using a fixed baseline against which to measure channel bank retreat by means of repeat survey. Baseline lengths were over 30 m and the channel bank height for the sections were 1 to 3 m above low-flow water levels and channel width was around 14 m. Channel bank materials were sand, silt and clays with a sand content ranging from 38 to 96% and clay content from 0.5 to 6% by weight. A third meander scar (see erosion pin site in Fig. 3 top panel) had erosion pins emplaced around an 8 meter arc and at this location channel bank height was over 3 m. This meander arc was developed in a resistant cobble, sand and silt mix. Actively eroding sections around the three meanders were devoid of vegetation, and often vertical in profile, although at one site in particular gravitational failure frequently resulted in a slump deposit below an upper cliff-section. A further set of channel bank observations were carried out using over 70 erosion pins located in a relatively straight section of channel some 400 m upstream of the meander bends identified as C in Fig. 3. At this location the deepest channel bank height was around 3 m and channel bank materials were a mix of sand, silt and clay; with clay content typically being less than 5%, with sand content ranging from 38 to 80% by weight. Erosion pins were emplaced over a 200 m reach and arranged to take account of bank height.

Evidence of river bank erosion obtained from plan-form changes along the Kam Tin North River is presented in Fig. 3 for three active sections of the channel. Superimposition of plan-form maps for the period 1959 to 1983 for three reaches of the river reveal lateral migration of the channel which may provide sediment for the river. Meander section A has a maximum lateral migration of 12.5 m while the two bends in section B both exhibited a maximum migration of 16 m. Section C contains three bends which exhibited maximum lateral migrations of 12, 11 and 10 m for the period 1959 to 1983. These maxima convert into average annual migration rates that range from 0.66 m/year for both bends in B; 0.5 m/year for A and 0.48, 0.44 and 0.4 m/year respectively for the three bends in section C of Fig. 3. There is, therefore, spatial variability between active meanders and Peart and Wong (2002) provide further details.

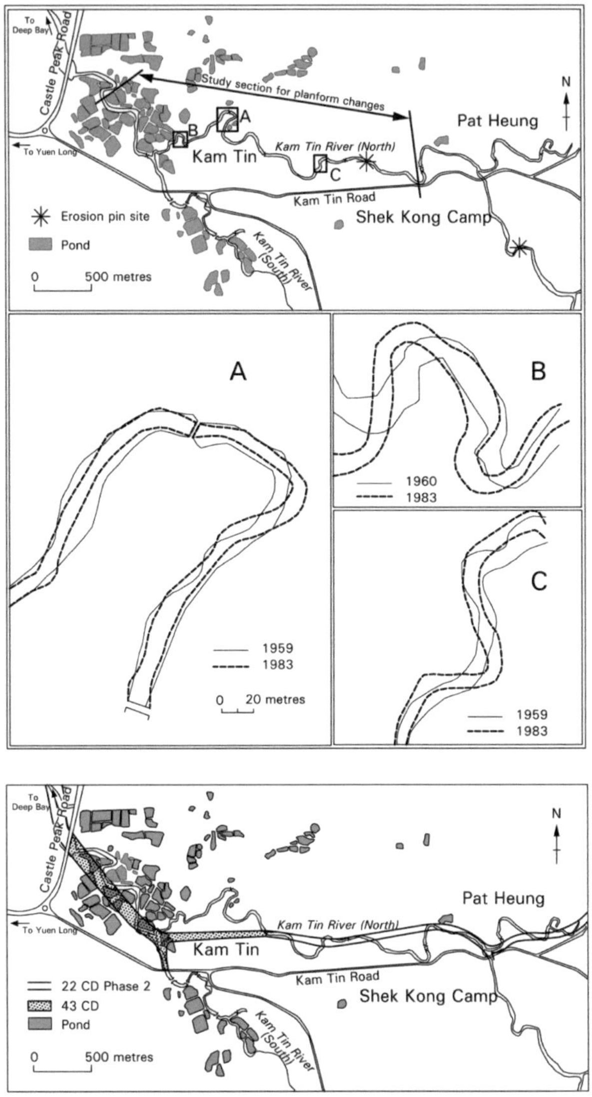

Fig. 3. Plan-form changes of the Kam Tin River. Top: Locations of the studied sections; Mid: Three active meander sections; Bottom: channalisation of the Kam Tin North River.

Fig. 4 presents the bank retreat data for the two meander scars derived from re-survey of channel bank position relative to the baseline. Meander scars one and two evidence spatial and temporal variation in the rate of retreat around the meander arcs. For meander scars one and two it is possible to compare erosion rates during the summer wet season of 1999 and median bank retreat values of 0.6 and 1.81 m, are obtained. Meander scar two permits observations of channel bank retreat for three summer wet seasons, namely 1997, 1998 and 1999, to be made. Respective median rates of bank retreat of 0.96, 1.23 and 1.81 m are obtained. Temporal variability is also evidenced by meander scar one in that for the year 1999, as noted above, a median bank retreat of 0.6 m was recorded; however by mid-April 2000 a median rate of bank retreat of 1.18 m had already occurred over the same length of baseline. Earlier observations from a previous baseline on meander scar one, over a 21 m baseline, confirms the active nature of this meander: over the period from 25th February 1997 to 6th June 1998 a median bank retreat of 1.97 m was observed over the 15 months of observations. Spatial variation around the meander arcs are also evidenced in Fig. 4. For example, in meander one most change occurs from around 20 m whilst on meander two the most active section is from 10 - 20 m. A third meander scar, located in more resistant materials, has been monitored using erosion pins. During 1997 and 1998 median bank retreats of 0.02 m and 0.018 m respectively were recorded, far lower than recorded on the other two meander scars. This may reflect the more resistant nature of the channel bank materials at this site.

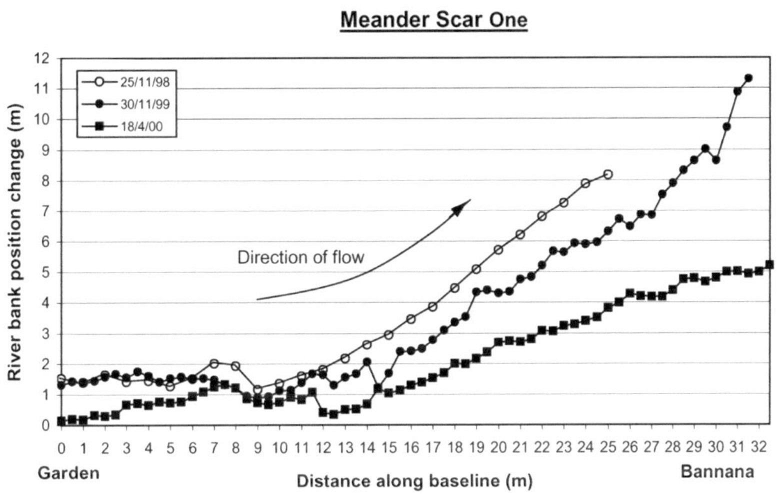

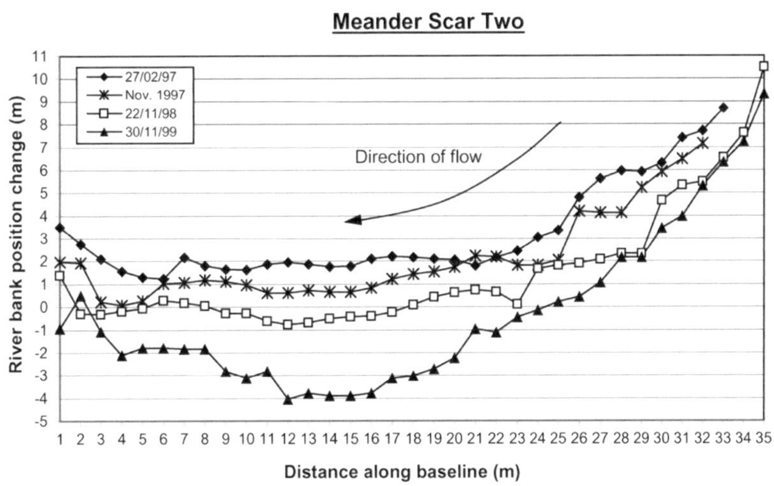

Fig. 4. River bank migration at two meander scars of the Kam Tin River.

Erosion pins have also been emplaced on a relatively straight section of the Kam Tin North River located around 400 m upstream of the two meander scars measured by baseline re-survey. During the year of 1996, for 70 pins on which measurements were obtained, a median loss of 0.052 m was recorded: it should also be noted that of the 70 pins, 16 (23%) recorded a gain of material, evidencing deposition. Compared to the two "active" meanders downstream this evidences much less active river bank position change.

It is necessary to place the observations of channel bank erosion on the Kam Tin North River in perspective for Hong Kong. It should be noted that currently the Kam Tin North River has been channelised (following the bottom panel given in Fig. 3) which has drastically altered channel plan-form and effectively de-coupled the river from the floodplain in order to reduce the flood hazard: it is now for much of its lowland course an artificial channel. Channelisation is one of the human impacts pertinent to change in river channels identified by Gregory (2006). It should also be noted that the Kam Tin North River is not the only river to be channelised or trained in order to mitigate the flood hazard in Hong Kong and the Beas and Indus Rivers afford further examples (c.f. HKDSD, 2008).

The data obtained on channel bank change on the Kam Tin North River in Hong Kong can be placed in perspective with reference to the literature. Odgaard and Abdad (2008) cite Hooke (1980) who plotted bank erosion rate versus drainage area for 11 streams in Devon, UK, along with data for 43 other streams and rivers from the literature. Rates of change cited by Odgaard and Abdad (2008) from Hooke (1980) ranged from 0.05 m/yr for a 3 km^2 drainage basin to 800 m/yr for a 1 million km^2 basin. From the plot of erosion rate against drainage basin area Odgaard and Abdad (2008), based upon Hooke (1980), report an approximate relationship as mean erosion equals 0.05 times the square root of drainage basin area in square kilometers. Substituting a drainage basin area of 11.72 km^2 for the Kam Tin North River at the old Water Supplies Department gauging station (located adjacent to the erosion pin site and just 400 m upstream of the two meanders shown in Fig. 3 and 4, gives a mean erosion rate of 0.171 m/year. This value is within the range of observations reported in this study. Odgaard and Adbad (2008) also cite the work of Brice (1982) who found an approximate relationship of mean erosion rate in meters per year being 0.01 times channel width in meters based upon a study of 36 US rivers. Using a mapped width of 13.75 m for the meander bends at section C shown in Fig. 3 gives a mean rate of channel bank change of 0.1375 m/yr: rather lower than the baseline data but not too dissimilar to the maximum lateral migration of 0.66 m/yr derived for this location from map data for 1959/60 to 1983 (Peart and Wong, 2002). The straight reach above the meanders had a measured channel width of 8.7 m and using the formula from Brice (1982) a mean rate of bank erosion of 0.087 m/yr is obtained: similar to those recorded by the erosion pins for this site.

4. Sediment quality

A sampling programme has been undertaken to monitor the C and N content of suspended matter on the Kam Tin North River on the main channel near Kam Tin. At the sampling site under stable-flow conditions large volume samples of around 5 L were obtained whilst samples obtained under stormflow conditions varied in volume from 500 ml to around 5 L. The suspended sediment was separated by filtration using GF/C filter papers. These were subsequently air dried and the sediment removed and disaggregated before analysis. Suspended matter was not pre-treated to remove carbonates prior to analysis. Carbon and N were measured using a Perkin Elmer model 2400 elemental analyser. Particulate C and N in water can be expressed as either a percentage (by weight) of the suspended matter or as a mass per unit volume of water: both have been used to characterise C and N in the Kam Tin North River, and the results are presented in Table 2 for the year 2004.

Type of sample		C [%]	C [mg/L]	N [%]	N [mg/L]	C/N	SSC [mg/L]
Stable-flow*	Mean	28.20	24.82	3.88	3.18	7.44	90.8
	Median	29.93	18.45	4.06	2.39	7.62	61.97
Stormflow **	Mean	12.39	48.99	1.50	6.23	8.81	357.0
	Median	11.12	27.42	1.40	3.31	8.52	235.1

* n = 96 ; ** n = 33.

Table 2. C and N of suspended sediment of the Kam Tin River.

It can be seen from Table 2 that there is a contrast in C and N in the Kam Tin North River between samples collected under stable-flow and stormflow conditions: the median C and N percent by weight for samples collected under stable-flow being 2.69 and 2.90 times higher respectively. However, in terms of concentration under stormflow conditions, the median concentrations of C and N are 1.49 and 1.38 times higher than the stable-flow samples. These results can be placed in temporal perspective by comparison to Peart (2000) who reports percent by weight observations for samples collected on the Kam Tin North River in 1998 - 1999. He reports median percent by weight C and N for stormflow samples of 12.29 and 1.7 respectively which are similar to the values recorded in 2004 for this study. Moreover a median C/N ratio value of 8.52 is given in Table 2 which is broadly comparable with the value of 7.56 reported by Peart (2000) for stormflow samples. Regarding samples collected under stable-flow conditions, Peart (2000) reports median percent C and N and their ratio of 33.02, 4.27 and 7.77 respectively, which are very similar to those given in Table 2 for samples collected in 2004. It would appear that in the 5 year period from 1999 there has been no great change in C and N. This suggests stable sources exist in the Kam Tin North River.

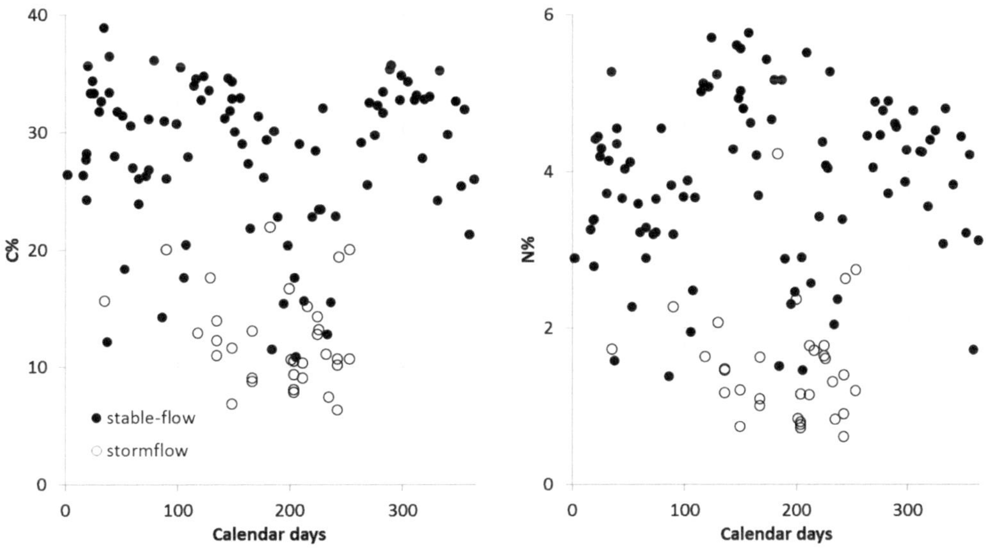

Fig. 5. C% and N% variation through the year 2004.

Fig. 5 plots the percent by weight C and N values for both stormflow and stable-flow samples collected during 2004. The plots reveal the contrast between the two types of sample (stormflow vs stable-flow). However, there is some evidence of seasonality in Fig. 5, especially for C which is comparatively lower in the summer wet season: this is reflected both in the stormflow samples, but can also be discerned in the stable-flow data. Peart (2000) also detected seasonality in percent by weight C and N in suspended matter of the Kam Tin North River.

It is also of interest to examine the sediment concentration data in Table 2 in terms of the water quality objectives utilised by the Hong Kong Government. These state that the median suspended sediment concentrations on an annual basis should not exceed 20 mg/L: Table 2 reveals that median suspended sediment concentrations in the Kam Tin North River are much higher. Indeed, some 99.6 and 90.4 percent respectively of stormflow and stable-flow samples exceed 20 mg/L in the Kam Tin North River in Hong Kong. Collins and Anthony (2008) adopted an annual average SSC of 25 mg/L to indicate achievement of 'good ecological status' for rivers in England and Wales. Utilising this value for the data on suspended sediment concentrations presented in Table 2 for the Kam Tin North River reveals that only 1 of the 33 stormflow samples is below this value while only 6 (6.3%) of the stable-flow samples were below 25 mg/L. In terms of stable-flow samples Peart (1999) reports mean and median suspended sediment concentrations of 88.9 and 54.4 mg/L for 77 samples collected in 1997-1998. Based upon the use of 25 mg/L, the Kam Tin North River would not meet 'good ecological status' as defined by Collins and Anthony (2008) for England and Wales.

A comparison can be made between the C and N data from the sampling site located in the lowlands near Kam Tin to those in an upland headwater stream in the same basin. Based upon samples collected in the period 1998 - 2001, Peart (2003) reports median percent by weight values of 12.8 and 1.0 respectively under stable-flow conditions and 12.6 and 1.0 respectively under stormflow for C and N in an upland headwater stream. Kong (2005), for the same upland stream and based upon samples collected in 2003, provides further information on C and N in the upland stream. Under stormflow conditions he reported median C and N content by weight of 15.2% and 1.15%. The equivalent values for samples collected under stable-flow conditions are 14.37% and 1.07%. Kong (2005) also gives median C/N ratio values of 13.26 and 13.46 for stormflow and stable-flow samples respectively. A number of contrasts are immediately apparent to samples collected at the Kam Tin North River. Firstly, C and N content in suspended sediment in the upland basin are much lower than observed in the main channel. Secondly, there is no great contrast between stormflow and stable-flow samples in the upland basin, in contrast to the Kam Tin North River and, finally, the C/N ratio of the upland stream at 12.5 to 13.5 is much higher than observed in the Kam Tin North River

It is interesting to compare the C and N data of the Kam Tin North River to that of the channel bank material and other potential sources. Peart and Wong (2002) report low C and N by weight values for channel bank materials. This observation is supported by data on C and N obtained from two vertical sections in the meander scars at location C in Fig. 3. Some 23 and 25 samples were analysed in vertical bank profiles 2.2 and 2.5 m deep respectively. The 2.2 m vertical profile had median C, N and C/N ratio values of 0.44%, 0.04% and 16.60 respectively whilst the second 2.5 m vertical profile had equivalent values of 0.41%, 0.05% and 9.47 respectively. A number of algae samples have been collected from the Kam Tin North River basin and their analysis gives C, N and C/N ratios ranging from 28.2 to 48.7; 5.22 to 9.78 and 4.46 to 6.17 respectively. The algal C and N characteristics are similar to those of the suspended matter in the Kam Tin North River especially samples collected under stable-flow conditions. Li et al. (2009) present N and C/N ratio data for 7 common plant species in Hong Kong and the mean N% and C/N ratios for these 7 species ranged from 1.0 to. 2.2% N and 17.4 to 42.3 for the C/N ratio, with these end members being defined by *Ficus fistulosa* and *Pandanus furcatus*: these C/N ratio values are much higher than observed in the Kam Tin North suspended matter, suggesting that fresh leaf litter is not a major source.

In addition data on C and N in the suspended matter of the Kam Tin North River information has also been collected, under stormflow conditions, of major elements along with trace metals and rare earth elements in the particulate matter. These data are presented in Table 3. The dominance of SiO2 is clear, it being 3 times higher than the next most dominant element, Al2O3. Both Langford et al. (1989) and Davis (1952) report the dominance of SiO2 and A12O3 in the bedrock geology of the Kam Tin North basin: SiO2 content ranged from 56.7 to 68.8% and Al2O3 from 13.4 to 23.8%, which are similar to the values observed in the suspended sediment. Langford et al. (1989) also reports that MgO is higher than MnO for both granodiorite and rocks of the Tai Mo Shan Formation which is also their rank order in the suspended matter of the Kam Tin North River. Concentrations of the major elements are broadly comparable to those obtained from the East River (Fok and Peart, 2010); Hong Kong background mean (Sewell, 1999) and the average river particulate matter (McLennan and Murray, 1999), with CaO being the exception. CaO is lower in the Kam Tin North River compared to average river particulate matter, probably reflecting the lack of calcareous bedrock in the basin. Compared with average crustal values for China (Li, 1994) enrichment occurs in the suspended sediment of the Kam Tin North River for aluminum (Al), manganese (Mn), potassium (K) and phosphorus (P) oxides whilst those of magnesium (Mg) and calcium (Ca) are depleted by comparison.

Enrichment of Zn (9.6 times), W (8.7 x), Pb (7.1 x), Cu (5.7 x), Sb (3.6 x), S (2.1 x), As (1.8 x), Ta (1.7 x), Be (1.2 x) and Rb (1.2 x) against the crustal averages are also observed for Kam Tin samples whilst, U (4 x), Th (2.1 x) along with rare earth elements (Y and Lanthanum series), which were known to be associated with granites, are also found enriched. The study of Cheung et al. (2003) allows further contextualization of the data. They sampled riverbed sediments in the Pearl River Delta region including Hong Kong. For the six sites in Hong Kong they report that Cd, Cr, Cu, Ni, Pb and Zn had ranges of 1.2 - 2.7, 1.9 - 46.1, 8.7 – 140.0, 6.3 – 31.9, 48.1 – 139.3 and 46.4 – 533.3 mg/kg respectively. One of their sites is also comparable to that utilised in this study on the Kam Tin North River and mean concentrations of around 1.6, 15, 72, 14, 128 and 330 mg/kg respectively for Cd, Cr, Cu, Ni, Pb and Zn were reported. For the range of concentrations observed at six Hong Kong sites by Cheung et al. (2003) the values detected in suspended sediment of the Kam Tin North River for Cr, Ni, Pb and Zn fall within the limits. However, Cd and Cu are respectively lower and higher than the range of values reported by Cheung et al. (2003) for Hong Kong. In direct comparison of the data, the suspended sediment of the Kam Tin North River, Cd and Pb were 2.3 and 1.4 times higher respectively than in the bottom sediments analysed by Cheung et al. (2003). In contrast Cr (2.5 x), Cu (2.6 x), Ni (1.4 x) and Zn (1.5 x) were enriched in the suspended sediment samples analysed in this study. These contrasts may, at least, in part reflect the differing total C amounts in the samples: Cheung et al. (2003) report that C is around 2% whilst the average C content of the 19 suspended sediment samples is 4.4%, around 2 times higher. With reference to sediment quality guidelines developed for Hong Kong (Chapman et al. 1999), As, Cu and Pb concentrations for the suspended sediment of the Kam Tin North River exceeded the threshold level (ISQV-1) guideline. Zn in particular, slightly exceeds the effect level (ISQV-h) guideline. These elements are known to be related to anthropogenic activities.

Analyte	Unit	This study		ER	Sewell	Csl	Susp.	ISQV-l	ISQV-h
		Median	MAD						
SiO2	%	54	6	55		58	62		
Al2O3	%	17	3	20		9.6	18		
Fe2O3	%	5	0.65	6.9	4.1	4.7	6.9		
MnO	%	0.082	0.016	0.18	0.13	0.03	0.14		
MgO	%	0.81	0.13	0.76		2.3	2.0		
CaO	%	0.8	0.58	0.36	0.67	10	3.1		
Na2O	%	0.36	0.1	0.23		0.30	0.96		
K2O	%	3	0.19	2.8		2	2.4		
TiO2	%	0.6	0.055	0.88	0.6	0.44	0.93		
P2O5	%	0.42	0.28	0.19	0.11	0.16	0.26		
As	ppm	14	3	29	27	5		8	70
Ba	ppm	480	34	471	317	260	600		
Be	ppm	4	0	4		1.8			
Br	ppm	3.9	0.3	5.7		3.6			
Co	ppm	10	0.74	20	6	33	20		
Cr	ppm	38	7.7	94	49	52	100	80	370
Cs	ppm	12	2.2	14	12	6.9	6		
Cu	ppm	188	142	36	102	28	100	65	270
Hf	ppm	6.8	1.3	13		3.9	6		
Ni	ppm	20	1.5	35	11	25	90	40	
Pb	ppm	89	13	76	104	11	150	75	218
Rb	ppm	210	30	180	216	95	100		
Sb	ppm	1.8	0.15	2	2.8	0.39			
S	%	0.13	0.11	0.029		0.042			
Sc	ppm	10	1.6	16	9.4	10	18		
Sr	ppm	48	8.9	45	35	330	150		
Ta	ppm	2.4	0.3	2.7		0.9	1.2		
Th	ppm	27	3.9	27	44	8.7	14		
U	ppm	10	0.89	8.3	8.7	2	3		
V	ppm	68	12	110	43	54	170		
W	ppm	11	1.5	13	11	1.2			
Y	ppm	47	7.4	65	51	20	28		
Zn	ppm	477	305	136	141	45	350	200	410
Zr	ppm	187	73	438	802	130			
La	ppm	52	8.7	90	52	34	46		
Ce	ppm	98	13	150	127	64	88		
Nd	ppm	51	4.4	70		28	33		
Sm	ppm	9	1.3	12		5.5	7		
Eu	ppm	1.8	0.24	2.2		1.1	1.5		
Tb	ppm	1.5	0.3	1.6		0.76	0.89		
Yb	ppm	5	0.61	5.8		2	3.2		
Lu	ppm	0.7	0.089	0.86		0.3	0.52		

MAD: median absolute deviation; ER: East River bed sediments (<63 μm) median value (Fok and Peart, 2010); Sewell (1999): Geochemical Atlas of Hong Kong, fluvial bed sediments (<150 μm); Csl: China's sedimentary layer abundance (Li, 1994); Susp.: Average river particulate (McLennan and Murray, 1999); ISQV: Interim sediment quality threshold (-l) and effect (-h) level value of Hong Kong (Chapman et al., 1999).

Table 3. Median and median absolute deviation for geochemistry of Kam Tin River sediment samples, along with reference values and quality guidelines.

The information presented on suspended sediment geochemistry for the Kam Tin North River should be viewed in the context of the observation of Walling (2009) that whilst measurements of sediment dynamics in rivers have traditionally revolved around magnitude, timing and concentrations there is increasing recognition of the importance of the quality dimension of sediment fluxes. This is because, Walling (2009) suggests, that sediment associated nutrients and contaminants may generate environmental problems. Other commentators who identify the importance of knowledge on sediment geochemistry in rivers include Miller and Orbock-Miller (2007) whilst Owens et al. (2005) also outline the importance of quality in sediment fluxes and they observe that many rivers are showing evidence of increasing concentrations of contaminants and nutrients which are often approaching or above available sediment quality guidelines. The National Sediment Quality Survey of the USA (USEPA, 1997) further attests to the importance of sediment geochemistry as does the work reported by Chapman et al. (1999) in Hong Kong.

5. Floatable debris

In 2009, 6,014 tonnes per day of domestic waste were collected in Hong Kong; 10 years earlier in 1999 this value was 7,426 tonnes per day. Regarding municipal solid waste (consisting of domestic, commercial and industrial waste) in 2009 and 1999 some 8,963 and 9,269 tonnes per day were produced. For Yuen Long District, which is adjacent to the Kam Tin basin, some 496 and 472 tonnes per day of domestic waste were produced in 2009 and 1999 respectively. In terms of municipal waste in Yuen Long district in 2009 and 1999 respectively 721 and 646 tonnes per day were generated. These are significant volumes of materials and Hong Kong is facing a challenge in terms of waste disposal (HKEPD, 2005). In terms of Hong Kong's environment waste can pose a problem. As Yung (2001) notes, with respect to marine waters and beaches, floating refuse has been a problem for many years. In fact Morton (1976) identifies rubbish as being a problem in Hong Kong coastal waters and beaches. In terms of origins of this floating rubbish Yung (2001) states that some of the debris may be derived from open waters, bought in by wave and tides, whilst some materials are derived from land sources consequent upon illegal dumping in storm water drains and open nullahs. It is this latter source that is of interest herein. Whilst floating rubbish and debris in beach and coastal waters has received considerable attention in Hong Kong it has been largely ignored in local streams and rivers, with only Peart (1999) offering any data.

Regarding the composition of domestic waste disposed of at waste facilities in 1999, the HKEPD report that glass bottles accounted for 2.7%, paper, including cardboard 25%, and plastics 18.9% of which 1.7% was beverage bottles, 11.6% plastic bags, and 10% polyfoams (HKEPD, 1999). In 2009 the composition of domestic waste delivered to waste facilities was similar to that of 1999 (HKEPD, 2009).

Observations have been made of floating debris and rubbish on the Kam Tin North River prior to channelisation/training. During May 1998, in excess of 8 m^3 of floating rubbish and debris accumulated under and against the bridge at Kam Hing Wai, part of the settlement of Kam Tin. This material was washed down during storms of the previous week. Some 897 pieces of debris/rubbish were counted and classified with plastic distilled water bottles of less than 2 liters in volume forming the largest category (163 pieces or 18.2%). If other categories of drinks and miscellaneous contents are included, this volume of plastic bottles accounted for 284 pieces. The next most common type of debris were polystyrene food containers (107 pieces or 11.9%) followed by drink boxes (84 pieces or 9.4%). Detergent, shampoo and other such containers accounted for 68 pieces or 7.6% of the sample. A wide range of other rubbish was recorded including tin cans, motor oil containers, glass bottles and jars, plastic pots and jars, plastic bags, flip-flops, light bulbs, wood, aerosol cans and basket-balls. Large debris not included in the count were car tyres and wheels, a 100 gallon oil drum, several 20 litre drums, wood pallets, 14 cupboard draws, and a small polystyrene surf board. Surveys of counts of debris moving down the river during subsequent storm events revealed the above findings to be not unusual. For example, in the storm of 9 June 1998 during a 15 minute counting period, 42 pieces of rubbish/debris were counted. Polystyrene accounted for 21.4%, plastic bottles of less than 2 L, 16.7% of the sample. Plastic materials accounted for a further 14.3% while tin cans were 4.8% of the sample whilst Peart (1999) provides further details.

Additional monitoring in 1999 and 2000 of floating debris and rubbish involved counting a further 460 pieces of debris and a major survey was undertaken after the storm of 21/6/99 when a further 601 pieces were counted. For the 460 pieces of debris (year 2000), plastic bags accounted for 20%, drink PETs 7.8% and polystyrene food trays/lunch boxes 10.6%. Wood and glass were 5% and 4.8% of the debris respectively. For year June 1999 storm, the debris mainly comprised of plastic bags (20%), drink PETs (9.2%), polystyrene food trays (16%), wood (1%) and glass (4%).

The observations on rubbish and floatable debris can be placed in perspective with reference to the composition of domestic waste in 1999 reported previously (HKEPD, 1999). To recap briefly, they report that glass bottles, paper and plastics (including bags, EPS food/drink, PET bottles, polyfoams etc.) accounted for 2.7%, 25% and 18.9% of domestic refuse by weight. Moreover, Yung (2001) cites the Hong Kong Marine Conservation Society territory wide Marine Debris Survey in 1996 - 1997 as finding that 46.8% of debris was plastic with polystyrene foam accounting for 16.1%, glass 13%, paper 8.2% and wood 6.2%. Both the data on domestic waste composition and the data from the Marine Debris Survey demonstrate a degree of similarity to the observations on domestic rubbish in the Kam Tin North River. This finding does suggest that some marine floatable debris were derived from rivers, as previously indicated by Yung (2001).

6. Conclusion

Channel plan-form change and bank erosion have been documented on the Kam Tin North River whilst observations on floating debris and rubbish have been presented. Information on sediment quantity and quality in the Kam Tin North River has also been made. These observations on channel bank erosion are important for they illustrate an "active" river which has implications for ecosystem function contributing to habitat and biodiversity maintenance. However, channelisation/training has halted this process and affords an example of the human impact upon the river. The observations on floating debris also confirm the human impact upon the river. It may also be a damming verdict of society values --- it would appear that the river was being used as a waste disposal system. The data on sediment quality confirms the need to consider this aspect of sediment fluxes in rivers: there is evidence that some elements in the suspended matter of the Kam Tin North River may exceed sediment quality guidelines and these elements are associated with human activities.

7. References

Brice, J.C. (1982). *Stream Channel Stability Assessment*. Report No. FHWA/RD/82/021, Washington D.C.: Department of Transportation Office of Research and Development, 42 pp.

Chapman, P.M., Allard, P.J. & Vigers, G.A. 1999. Development of Sediment Quality Values for Hong Kong Special Administrative Region: A Possible Model for Other Jurisdictions. *Marine Pollution Bulletin*, 38, 161-169.

Cheung, K.C., Poon, B.H.T., Lon, C.Y. and Wong M.H. (2003) Assessment of Metal and Nutrient Concentrations in River Water and Sediment Collected from the Cities in the Pearl River Delta, South China, *Chempospher*e, 52, 1431-1440.

Collins, A.L. and Anthony, S.G. (2008) Assessing the Likelihood of Catchments Across England and Wales Meeting 'Good Ecological' Status due to Sediment Contributions from Agricultural Sources. *Environmental Science and Policy*, 11, 163-170.

Corlett, R.T. (2010) Ecology and Biodiversity. In: *A New Geography of Hong Kong*, Vol. II, pp. 114-169.

Davis, G.G. (1952) *The Geology of Hong Kong*. Hong Kong Government Printer.

Dudgeon, D. and Corlett, R. (2004) *The Ecology and Biodiversity of Hong Kong*. Joint Publishing, Hong Kong, 336 pp.

ERM (1995) *Main Drainage Channels for Ngau Tam Mei, Yuen Long and Kam Tin: Environmental Impact Assessment*. Hong Kong: ERM - Hong Kong Ltd, Territory Department Office, MTN Development office.

Florsheim, J.L., Mount, J.F. and Chin, A. (2008). Bank Erosion as a Desirable Attribute of Rivers. *BioScience*, 58(6), 519-529.

Fok, L. & Peart, M.R. 2010. Determination of the Geochemical Baseline of the East River Basin, China. *In:* Kazimierz, B. (ed.) *Sediment Dynamics for a Changing Future.* Wallingford: IAHS Publ.

Gregory, K.J. (2006) The Human Role in Changing River Channels. *Geomorphology*, 79, 172-191.

HKDSD (2008) *Sewerage and Flood Protection: Drainage Services 1841-2008*. Hong Kong: Drainage Services Department, 204 pp.

HKEPD (2009). *Monitoring of Solid Waste in Hong Kong: Waste Statistics for 2009*. Hong Kong: Environmental Protection Department.

HKEPD (2005). *A Policy Framework for the Management of Municipial Solid Waste (2005-2014).* Hong Kong: Environmental Protection Department.

HKEPD (2004). *River Water Quality in Hong Kong in 2003*. Hong Kong: Environmental Protection Department.

HKEPD (1999). *Monitoring of Solid Waste in Hong Kong 1999*. Hong Kong: Environmental Protection Department.

HKSAR Government (2009). *Hong Kong Yearbook*. Hong Kong: HKSAR Government.

HKTPB (2005a). *Approved Kam Tin North Outline Zoning Plan No. S/YL/KTN/6*. Hong Kong: Town Planning Board.

HKTPB (2005b). *Approved Pat Heung Outline Zoning Plan No. S/YL/PH/6*. Hong Kong: Town Planning Board.

HKTPB (2005c). *Approved Shek Kong Outline Zoning Plan No. S/YL/SK/4*. Hong Kong: Town Planning Board.

Hooke, J.M. (1980). Magnitude and distribution of rates of river bank erosion. *Earth Surface Processes & Landforms*, 5, 143-157.

Jim, C.Y. (1997). Rural Blight and Land Use Planning in Hong Kong. *The Environmentalist*, 17, 269-281.

Juracek, K.E. and Ziegler, A.C. (2009). Estimation of Sediment Sources Using Selected Chemical Tracers in the Perry Lake Basin, Kansas, USA. *International Journal of Sediment Research*, 24, 108-125.

Kevin, G.T., Philip, N.O., Ramon, J.B. & Celso, G. (2008). Sediment and contaminant sources and transfers in river basins. *Sustainable Management of Sediment Resources*, 4, 83-135.

Kong, S.-p. (2005). Carbon and nitrogen content of suspended matter in a headwater catchment in Hong Kong. Hong Kong: University of Hong Kong, Unpublished Thesis.

Knighton, D. (1998). Fluvial Forms & Processes: a new perspective. London: Arnold.

Langford, R.L., Lai, K.W., Arthurton, R.S. & Shaw, R. (1989) *Geology of the Western* New *Territories*, 1:2000 sheets 2, 5 and 6. Hong Kong Geological Survey Memoir no. 3, Hong Kong Government Printer.

Li, T. (1994). Element Abundances of China's Continental Crust and its Sedimentary Layer and Upper Continental Crust (in Chinese). *Geochimica*, 23, 140-145.

Li, A., Ng, L. & Dudgeon, D. (2009). Effects of leaf toughness and nitrogen content on litter breakdown and macroinvertebrates in a tropical stream. *Aquatic Sciences - Research Across Boundaries* 71(1): 80-93.

Marafa, L.M. & Chau, K.C. (1999) *Effect of Hill Fire on Upland Soil in Hong Kong*. Forest Ecology and Management, 120, 97-104.

McLennan, S.M. & Murray, R.W. 1999. Geochemistry of Sediments. *In:* Marshall, C.P. & Fairbridge, R.W. (eds.) *Encyclopedia of geochemistry*, Dordrecht: Kluwer Academic.

Miller, J.R. and Orbock-Miller, S.M. (2007). *Contaminated Rivers: A Geomorphological-Geochemical Approach to Site Assessment and Remediation*. Springer, 418 pp.

Morton, B. (1976). The Hong Kong sea-shore: an environment in crisis. *Environmental Conservation*, 3(4), 243-254.

Mukundan, R., Radcliffe, D.E., Ritchie, J.C., Risse, L.M. and McKinley, R.A. (2010). Sediment Fingerprinting to Determine the Source of Suspended Sediment in a Southern Piedmont Stream. *J. Environmental Quality*, 39, 1328-1337.

Odgaard, A.J. and Abdad, J.D. (2008). River Menadering and Channel Stability. P.439-460. In: *Sedimentation Engineering: Processes, Measurements, Modeling and Practice*. M.H. Garcia (ed.). American Society of Civil Engineers.

Owen, B. and Shaw, R. (2007). *Hong Kong Landscapes: Shaping the Barren Rock*, HKU Press, 253 p.

Owens, P.N., Batalla, R.J., Collins, A.J., Gomez, B., Hicks, D.M., Horowitz, A.J., Kondolf, G.M., Marden, M., Page, M.J., Peacock, D.H., Petticrew, E.L., Salomons, W. and

Trustrum, N.A. (2005) Fine-Grained Sediment in River Systems: Environmental Significance and Management Issues, *River Research Applications* 21: 693-717.

Peart, M.R. (1999) *Development and Water Quality in the Kam Tin Basin, Hong Kong. In: Impacts of Urban Growth on Surface Water and Groundwater Quality.* Proceedings Birmingham, U.K., Symposium. IAHS Publ. No. 259, 315-320.

Peart, M.R. (2000) Some Observations on Total Carbon and Nitrogen in Suspended Matter in the Kam Tin River, Hong Kong. In: *The Role of Erosion and Sediment Transport in Nutrient and Contaminant Transfer*, Proc. Waterloo Symposium, IAHS Publ. 363, p. 149-157.

Peart, M.R. (2003) Carbon and Nitrogen Content of Suspended Matter in a Hong Kong Drainage Basin. *Hydrobiologia*, 494, 215-222.

Peart, M.R., Hill, R.D.H., Fok, L. (2009) Environmental Change, Hillslope Erosion and Suspended Sediment: Some Observations from Hong Kong. *Catena*, 79(3), 198-204.

Peart, M.R. and Wong, H. (2002) Channel Floodplain Linkages on the Kam Tin River, Hong Kong, p. 255-261. In: *The Structure, Function and Management Implications of Fluvial Sedimentary Systems*, F.J. Dyer, M.C. Thomas and J.M. Olley (Eds.). IAHS Publication No. 276.

Piégay, H., Cuaz, M., Javelle. E. & Mandier, P. (1997). Bank erosion management based on geomorphoogical, ecological and economic criteria on the Galuaure River, France. *Regulated Rivers: Research and Management*, 13, 433-448.

Piégay, H., Darby, S.E., Mosselman, E. and Surian, N. (2005) A Review of Techniques Available for Delimiting the Erodible River Corridor: A Sustainable Approach to Managing Bank Erosion. *River Research and Applications*, 21, 773-789.

Poon, Y.W. and Tsang, C.C. (2010) Industry. In: *A New Geography of Hong Kong, Vol. II*, pp. 113-133.

Sewell, R.J. 1999. *Geochemical Atlas of Hong Kong,* Hong Kong, Geotechnical Engineering Office.

Styles, K.A. and Hansen, A. (1989) *Geotechnical Area Studies Programme: Territory of Hong Kong*. GASP Report XII, 346 p. Geotechnical Control Office, Civil Engineering Services Department, Hong Kong.

Tang, Bo-sin, Wong Siu-wai and Lee, A.K.W. (2005) Green Belt, Countryside Conservation and Local Politics: A Hong Kong Case Study. *RURDS*. 17, 230-247.

Tso, G.K.F. and Yau, K.K.W. (2003) A study of domestic energy usage patterns in Hong Kong, *Energy*, 28, pp. 1671–1682.

USEPA (1997) The Incidence and Severity of Sediment Contamination in Surface Waters of the United States. Vol. 1, *The National Sediment Quality Survey*. EPA-823-R-97-006.

Walling, D.E. (2009) *The Impact of Global Change on Erosion and Sediment Transport by Rivers: Current Progress and Future Challenges*. The United Nations World Water Assessment Programme: Scientific Paper, 26 pp.

Wang, J. (2010). Transport Geography. In: *A New Geography of Hong Kong, Vol. II*, pp. 135-158. Cosmos Books, Hong Kong.

Watson, A.J. and Basher, L.R. (2005). *Stream Bank Erosion: A Review of Processes of Bank Failure, Measurement and Assessment Techniques, and Modelling Approaches*. Motueka Integrated Catchment Management (Motueka ICM) Programme Report Series. Landcare Research, New Zealand. 32 pp.

Yung, P. (2001). Marine and beach refuse: Background paper. *Civic Exchange*, 12 p.

11

Unraveling Sediment Transport Along Glaciated Margins (the Northwestern Nordic Seas) Using Quantitative X-Ray Diffraction of Bulk (< 2mm) Sediment

J.T. Andrews
Institute of Arctic and Alpine Research
and Department of Geological Sciences,
University of Colorado, Boulder
USA

1. Introduction

In most environments, sediment transport is controlled by the density and velocity of the transporting medium, and the grain-size of the sediment. Thus in a very simplistic sense, transport by wind and/or water results in a settling of grains and invariably sediments become more fine-grained along the transport path. Major exceptions to this rule are those areas where sediment is entrained and transported by sea ice or icebergs. In these polar and sub-polar areas the normal rules of sediment transport do not apply. Hence, coarse-grained sediments can be, and have been, transported 100's to 1000's of kilometers beyond their point of entrainment in sea ice or a glacier/ice stream. However, a critical but frequently overlooked aspect of sediment transport is that of provenance. On-land, the sources of sediment delivered to a stream channel are essentially known as they are delimited and restricted mineralogically to the bedrock that outcrops within the drainage basin (Eberl, 2004). The marine environment, and especially the glacial marine environment, is much more open, so that abrupt changes in sediment transport from different bedrock outcrops are not uncommon. An obvious example is that of the major changes in sediment provenance associated with the North Atlantic Heinrich (H-) events of the last glacial cycle, which involved massive discharges of melt-water and icebergs transported from the Hudson Strait ice stream of the Laurentide Ice Sheet (Heinrich, 1988; Andrews and Tedesco, 1992; Bond et al., 1992; MacAyeal, 1993; Dowdeswell et al., 1995; Hesse, 1995; Andrews, 1998; Hemming, 2004). These large-scale glaciological events resulted in the transport of massive amounts of glacially derived detrital carbonate into the North Atlantic, which were then transported as far east as the margins of Portugal and the British Isles (Lebreiro et al., 1996; Peck et al., 2007).

A variety of methods have been used to distinguish changes in the transport and provenance of glacial marine sediments. These have included studies of the mineralogy and characteristics of the sand-size fraction (Bond et al., 1997), radiogenic isotopic signatures that allow identification of the probable bedrock outcrops (Grousset et al., 1993, 2001; Farmer et al., 2003; Verplanck et al., 2009), argon ages which allow a source identification (Hemming et al., 2002a; 2002b), magnetic properties (Pirrung et al., 2002; Andrews and Hardardottir, 2009), sediment reflectance (Ortiz, 2009), and quantitative X-ray diffraction (qXRD) of the bulk sediment (Moros et al., 2004; Andrews and Eberl, 2007). It is the purpose of this paper to show how qXRD is a relatively cheap, simple, but powerful approach to determine changes in sediment provenance, hence sediment transport, along the glaciated margin of E/NE Greenland and across the Denmark Strait to Iceland.

2. Background

The area of Denmark Strait is a contact zone between southward flowing Polar/Arctic water masses sourced from the Arctic Ocean (East Greenland Current and East Iceland Current) and warmer, more saline northward flowing Atlantic Water (Irminger and North Iceland Irminger Current) (Stefansson, 1962; Malmberg, 1985; Hopkins, 1991) (Fig. 1). Sea ice is pervasive on the East Greenland shelf and in many years also extends onto the NW and N Iceland shelf (Fig. 1), but only during extreme years does the ice wrap around Iceland and impact the E/SW coasts (Gray, 1881a; Ogilvie, 1996). This SW-NE pattern of sea ice extent in the Nordic Seas (Fig. 1) also appears as a feature of the LGM (Li et al., 2010). Ocean fronts form at the contact between the Polar and Atlantic water masses and define the North Iceland Front and the Polar Front (Fig. 2). These fronts are often areas of high marine productivity (Jennings et al., 2011) and they also mark boundaries between low to higher rates of drift ice melt, hence sites of enhanced sediment deposition.

Fig. 1. Location of research area showing the major surface currents---TransPolar Drift, East Greenlany.

Icebergs calved from tidewater margins of E/NE Greenland ice streams (Fig. 3) are frequently retained within the fjords for months to years because of the presence of sikkussuaq and land-fast sea ice (Dwyer, 1995; Syvitski et al., 1996; Reeh et al., 2001). Thus during summers, when there is little or no removal of the land-fast sea ice, iceberg sediment transport is severely curtailed and the bulk of the deposition will take place within the fjord

(Reeh et al., 1999; Mugford and Dowdeswell, 2010). Conversely, during warmer periods when the barriers to transport are removed, icebergs can exit the fjords onto the shelf and still retain a sediment load.
Mafic-rich Tertiary and Quaternary flood basalts crop-out on either side of Denmark Strait (Larsen, 1983). On the Greenland side the outcrop extends from the south shore of Scoresby Sund southward to Kangerlussuaq Fjord; it also extends offshore (Larsen, 1983; Brooks, 1990). To the north of Scoresby Sund, the geology is complex (Henriksen, 2008) but the bedrock is dominated by igneous and metamorphic rocks, with some Paleozoic sandstones (red beds) (Pirrung et al., 2002) and carbonates. Offshore sediments are rich in quartz and k-feldspars, thus they present a clear mineralogical contrast with similar processes affecting the basalts of East Greenland and Iceland (Andrews et al., 2010).

3. Sediment loads and transport

A key question, and one that is difficult to answer in absolute terms, is the magnitude of the sediment load that is transported in the icebergs and sea ice, and which is thus available to melt-out and be deposited on the seafloor (Hebbein, 2000; Dethleff, 2005; Dethleff and Kuhlmann, 2009, 2010). In glacial marine environments there is another reworking and transport mechanism, which is the impact of large icebergs on the seafloor, and which causes sediment reworking and resuspension (Dowdeswell et al., 1994, 2010; Syvitski et al., 2001). The flux of water in the East Greenland Current through Fram Strait (EGC) is of the order of 3000-5000 km^3/yr (Foldvik et al., 1988), compared with an annual iceberg flux north of Kangerlussuaq Trough of ~100 km^3 (Bigg, 1999) and a sea ice flux in the range of ~700 km^3/yr (Kwok, 2009). The magnitude of the iceberg sediment load along the East Greenland margin is essentially unknown, however, there are some critical observations that need to be considered. In fast-flowing tidewater ice streams the sediment is usually held within the lowermost 1-10 m (Dowdeswell, 1986), although, if there is a pronounced sill, considerable thickness of sediment can be added through the freezing on of super-cooled basal melt-water (Alley et al., 1997, 1998; Lawson et al., 1998). A key consideration is that along the E/NE Greenland margin, icebergs are restricted in moving out of the fjords and onto the shelf by the presence of land-fast sea ice and/or the sikkussuaq (a mélange of sea ice, bergy bits, and icebergs) (Syvitski et al., 1996; Reeh et al., 1999, 2001). Thus the icebergs suffer significant mass loss during their "enforced captivity, " hence most probably lose a considerable fraction of their sediment. An order of magnitude estimate of the sediment load in icebergs (100 km^3 of ice calved (Bigg, 1999), average iceberg thickness of 200 m (Dowdeswell et al., 1992), hence 500 km^2 area coverage, a 2 m sediment thickness) is 2600 x 10^9 kg, or a mass accumulation rate (MAR) ~0.5$mg/cm^2/yr$ if distributed evenly over the 500,000 km^2 shelf of NE Greenland, north of 68°N. Melting of the margin of the Greenland Ice Sheet below the Equilibrium Line produces ca 273 Gt (km^3) of melt-water per year (van den Broeke et al., 2009), probably < ¼ of this from NE Greenland. However, from a sediment transport viewpoint, the flux of sediment entrained in the melt-water plumes decreases exponentially away from the ice front (Andrews and Syvistki, 1994; Syvitski et al., 1996; Mugford and Dowdeswell, in press) with a half-distance transport length of 10's of km, thus resulting in massive deposition within the fjords (Smith and Andrews, 2000) but with relatively little impact on the mid- and outer shelf.

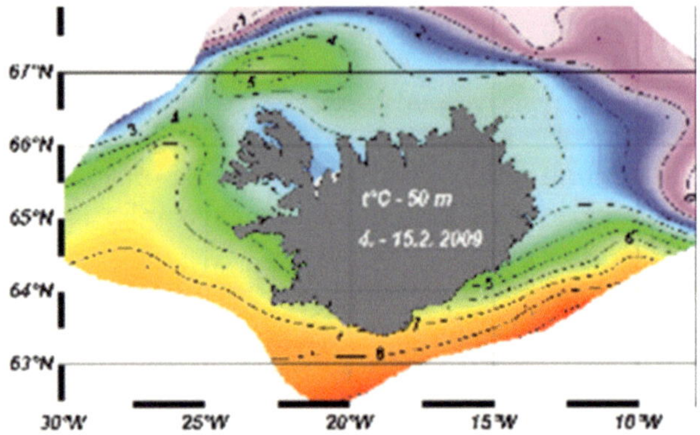

Fig. 2. Temperatures at 50 m water depth across the North Iceland Front (Belkin et al., 2009) (www.hafro.is)

The sediment load in sea ice exiting through Fram Strait (Fig. 1) has been estimated at ca 158 x 10^9 kg (Dethleff and Kuhlmann, 2010), thus a potentially smaller MAR (~0.03 mg/ cm^2/yr) than the crude estimate (above) for the iceberg sediment load. These estimates will vary spatially and temporally, thus a sediment trap in Fram Strait recorded a MAR of 0.29 mg/cm^2/yr from the melting of sea ice (Hebbein, 2000).

Sea ice, sediment, and driftwood are exported from the Arctic Ocean via the TransPolar Drift (Fig. 1) (Eggertsson, 1993; Dyke et al., 1997; Tremblay et al., 1997). Sediments in the sea ice "…contain on average more than 94% silt and clay" (Dethleff and Kuhlmann, 2010), and the process of sediment entrainment in sea ice (Reimnitz et al., 1987; Darby, 2003; Darby et al., 2009; Dethleff and Kuhlmann, 2009) results in an enhancement of the fine-fractions relative to sediments on the sea floor. Dethleff and Kuhlmann (2010) were able to distinguish between eastern (Kara Sea) and and western (Laptev Sea) sources based on clay mineralogy. Similarily, Darby and Bischoff (1996) were able to distinguish sources based on the Fe-oxide compositon of grains in the 250-40 µm fraction; Darby also identified grains from the Laptev and Kara seas in cores off N Iceland (Andrews et al., 2009b).

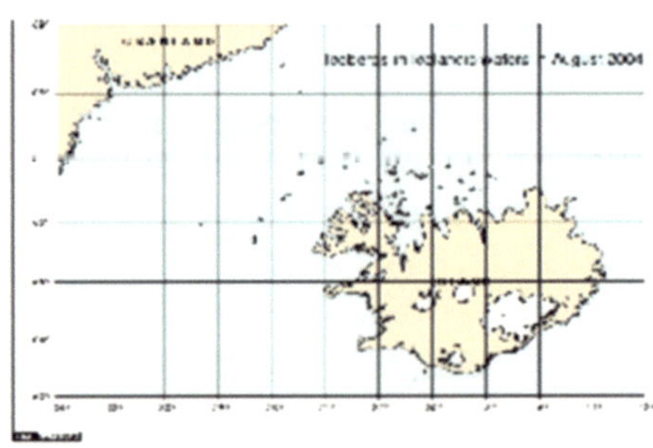

Fig. 3. Distribution of icebergs off NW and N Iceland in August 2004 (www.harfo.is)

Sediment transport via sea ice will of course end at the limit of sea ice. This is a spatially and temporally variable limit (Parkinson, 2000; Parkinson et al., 2001) with significantly expanded aerial extent during the Little Ice Age (Gray, 1881b; Koch, 1945; Wallevik and

Sigurjonsson, 1998; Ogilvie and Jonsdottir, 2000; Divine and Dick, 2006) (Fig. 1). SW Iceland is predicted to be essentially "ice free" whereas NW/N Iceland have intermittent sea ice and icebergs (Figs. 1 and 3). However, there is, unfortunately, no record of iceberg numbers along the East Greenland margin, such as there is for the eastern Canadian margin off Newfoundland (Miller and Hotzel, 1984; Marko et al., 1994; Bigg, 1999). Icebergs are noted in recent observations on the Iceland shelf (Sigurdsson, 1969) (Fig. 3). The driftwood on Iceland beaches (Eggertsson, 1993) is brought on and in sea ice, and originates from the Russian mainland west of the Ural Mountains, thus firmly indicating long distant transport.

4. Methods

This chapter is based on published core data from areas of the Kara Sea (Stein et al., 2004) southward to 65°N off East Greenland (Fig. 1) (Andrews et al., 2010), and sites around Iceland (Andrews et al., 2009b) (Table 1). Estimates of the sediment accumulation rates (SAR) and mass accumulation rates (MAR) are based on published data and on the sediment properties of the two shelf areas (Andrews et al., 2002).

There are a variety of different approaches to obtaining qXRD weight % estimates (wt%) of non-clay and clay minerals (Kolla et al., 1979; Ward et al., 1999; Eberl, 2003; Moros et al., 2004). The method used in this chapter is described by Eberl (2003); it has been very successful (2nd and 3rd place) in the International Reynolds Cup, in which "unknown" mixtures of minerals are sent to laboratories around the world for identification (McCarty, 2002). Briefly, 1 g of dry bulk sediment in the < 2 mm size range (i.e. sand, silt, and clay) is mixed with 0.111 g of ZnO and ground for 5 minutes in a McCrone mill. The samples are then dried, side packed in aluminum holders and loaded into a 40-sample carousel on a D5000 Siemens X-ray diffraction instrument. The samples are run between 5° and 65° 2-theta with a scan of 2 sec every 0.02° two-theta (100 minutes per sample). This results in 3000 counts per sample. The digital output is loaded into a 50 MB Excel macro program called RockJock v6 (Eberl, 2003), which has 125 mineral standards. The XRD sample pattern is matched to an estimated XRD pattern based on the selection of minerals that experience indicates may be present. A graph is generated that shows major residual XRD peaks so that the mineral selection can be altered. In practice, for the down core analyses, a single selection is made of around 23 non-clay and 10 clay mineral species (Andrews and Eberl, 2007). A degree-of-fit (DOF) is iteratively calculated between the observed and expected patterns and a selection is finally arrived at after 5 iterations with no change in the 4th decimal place. This usually takes 20-30 minutes on a regular PC. In glacial marine environments the focus has been on the non-clay mineralogy as this usually includes anywhere from 70-90 wt% of the bulk sediment (Andrews and Eberl, 2007; Andrews et al., 2010).

An additional research step is undertaken using an unmixing algorithm "MinUnMix" (Eberl, 2004; Andrews et al., 2010) and the latest version, which is termed "SedUnMix" (Andrews and Eberl, subm.). This program uses linear algebra programming (using the "Solver" in Excel). The steps are illustrated in Table 2 with an abbreviated list of minerals (8) that are summed to 100%. The objective is to consider how well the mineral composition from NW Iceland (Fig. 4) can be explained by contributions from 4 sources (Table 2A), Iceland, Kara Sea, Scoresby Sund, and Nansen Fjord (Geikie Plateau) (Figs. 1, 4, 5 and 6). Core JR51GC35 (Bendle and Rosell-Mele, 2007) (Fig. 5) is used. The program iteratively seeks to minimize the absolute differences between the target compositions (samples from.

Core Id	Long	Lat	#1	#2	#3	#4	#5	#6
MD99-2258	-24.44	63.96	0.16	0.075	68	16.40		0.30
HU93030-003I	-24.23	64.30	0.22	0.100	140			NA
MD99-2256	-24.21	64.32		0.030	24	7.90		0.10
B997-343	-24.49	64.78		0.030	237	NA		NA
B997-330	-21.08	65.87		0.060	47	1.40		0.50
JM96-1216	-30.63	65.96		0.026	21	5.50	4.00	9.40
BS1191--K18B	-30.63	65.96		0.006	5	9.90	9.80	5.60
MD99-2266	-23.27	66.23		0.135	106	17.70		0.00
MD99-2269	-20.86	66.64		0.250	197	9.20		3.60
MD99-2263	-24.20	66.68	0.08	0.040	28	28.60		1.20
MD99-2264	-24.20	66.68		0.021	16	28.60		0.90
B997-315	-24.33	66.74		0.008		26.90		1.00
JR51GC-35	-17.96	67.00		0.040	32	3.00		1.90
MD99-2322	-30.83	67.14		0.117	95	3.20		9.20
HU93030-019B	-30.82	67.15	0.20		90	6.10	0.80	6.60
JM96-1207	-29.35	68.10		0.015	12	0.80		7.70
MD99-2317	-27.86	68.10		0.054	44	0.00		3.70
HU93030-023	-31.88	68.14	0.42		430	NA		NA
JM96-1212	-31.36	68.16	0.31		250	NA		NA
JM96-1210	-29.60	68.18	0.22	0.125	170	0.00		6.00
BS1191-K8	-30.09	68.26		0.372	305	0.30	NA	10.50
JM96-1209	-29.80	68.30	1.36		1010	NA		NA

1 = SAR cm/yr 210PB

2=SAR cm/yr 14C

3=mg/cm2/yr

4=wt% calcite

5=wt% bio silica

6=wt% qrtz

Table 1. Core data and locations (Fig. 5). Not all core identifications are shown.

JR51GC35) and calculated compositions (Table 2B) based on fractions of contributions from the four sources (Table 2C). Errors on source contributions (± one standard deviation) are computed from randonly generated iterations (Table 2A)---in this example I show the results for 2 samples dated ~620 andf 1550 yr BP based on 10 iterations (Table 2C). The data are not constrained to sum to 100% but usually range within ± 20% of that total. The results (Table 2C) indicate a dominance of local Icelandic-sourced sediments but with contributions from East Greenland

Sources
1 = Iceland
2 = Kara Sea
3 = Scoresby Sund
4 = Nansen Fjord

A	**Random sample 1 to 5**	1.3	2.1	3.4	4.1
	Quartz	0.0	28.8	10.6	8.6
	intermediate Microcline feldspar	0.0	9.0	3.2	7.4
	Anorthoclase feldspar	37.4	6.1	28.4	14.1
	Bytownite feldspar	10.0	0.3	6.6	13.8
	Pyroxene	12.2	1.9	15.3	19.8
	Saponite	28.5	18.8	19.5	20.6
	Ferruginous smectite	0.0	13.8	9.8	6.6
	Fe-Chlorite	11.8	21.1	6.7	9.1
	Random sample 1 to 5	1.2	2.5	3.4	4.5
	Quartz	1.1	34.1	10.0	11.6
	intermediate Microcline feldspar	0.2	7.4	7.0	3.0
	Anorthoclase feldspar	36.7	14.4	29.0	17.6
	Bytownite feldspar	8.1	8.7	6.4	10.5
	Pyroxene	13.9	7.9	12.9	16.6
	Saponite	27.5	8.4	15.2	24.4
	Ferruginous smectite	0.0	9.1	9.9	5.7
	Fe-Chlorite	12.6	10.0	9.7	10.5
	k iterations 1 to 200				

B	**Minerals present**	Target 620 yr	Calculated	Difference
	Quartz	1.9	2.1	-0.2
	intermediate Microcline feldspar	1.5	0.9	0.5
	Anorthoclase feldspar	37.1	34.5	2.6
	Bytownite feldspar	10.3	10.2	0.0
	Pyroxene (diopside)	14.7	13.7	1.0
	Saponite	19.1	25.9	-6.8
	Ferruginous smectite	0.7	1.5	-0.8
	Fe-Chlorite (Tusc)	14.7	11.8	2.9
	Mean absolute difference			1.8
	Minerals present	Target 1550 yr	Calculated	Meas-Calc
	Quartz	2.3	2.0	0.3
	intermediate Microcline feldspar	0.3	0.7	-0.4
	Anorthoclase feldspar	37.9	35.0	2.8
	Bytownite feldspar	10.6	10.6	0.0
	Pyroxene (diopside)	17.5	14.4	3.2
	Saponite	18.1	26.4	-8.3

Ferruginous smectite		0.8	1.1	-0.2	
Fe-Chlorite (Tusc)		12.5	12.4	0.1	
Mean absolute difference				1.9	

C

Mixed sample JR51GC35		1	±	2	±
	620	0.98	0.03	0.00	0.00
	1550	1.00	0.01	0.00	0.01
Mixed sample JR51GC35		3	±	4	±
	620	0.07	0.09	0.08	0.09
	1550	0.05	0.08	0.13	0.08

All ages in this chapter are calibrated yr BP (Stuiver et al., 1998) but are simply noted as yr BP.

Table 2. Explanation of the steps in the mineral unmixing program SedUnMix

5. Research design

A glaciated margin, such as E/NE Greenland (Fig. 1) consists of a series of point sources (ice streams and tidewater glaciers), which contribute sediment along fjords and into the main north-→south flow of the East Greenland Current (Fig. 4). Another potential sediment source to the E Greenland and Iceland margins are sediments entrained in sea ice and exported through Fram Strait (Figs. 1 and 4). The water, sea ice, and iceberg transport along the NE/E Greenland shelf are variable, with the former two being primarily forced by changes in sea-level pressure associated with the Arctic Oscillation (Thompson and Wallace, 1998; Schmith and Hanssen, 2003; Rogers et al., 2005).

Based on our Experimental design (Fig. 4) the chapter will examine the following hypotheses:

Hypothesis 1: The sediments deposited on the NW Iceland shelf and central East Greenland shelf are largely derived from sea ice entrained sediments originating on the shallow shelves of the Arctic Ocean.

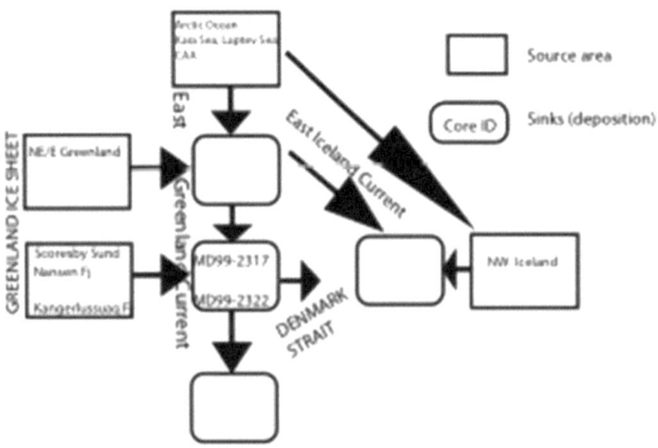

Fig. 4. Research desing in terms of sediment transport from the Artic Ocean (sea ice) and East Greenland (icebergs) to depositional stes off East Greenland and N Iceland

Alternate Hypothesis: The sediments deposited on the NW Iceland shelf and central East Greenland shelf are largely derived from NE/E Greenland icebergs and melt-water.
In order to investigate these hypotheses there is a need to determine: 1) the mass accumulation rates in the areas of interest—the deposition areas, 2) the mineralogy of the bulk sediment in both the source areas and the areas of deposition (sinks), and 3) estimate the relative sediment source contributions in the areas of deposition. In order to undertake the project I will use results from seafloor surface sediments from the Kara Sea, NE/E Greenland, and Iceland, and down-core data from cores MD99-2269, MD99-2317, and JR51GC35 (Fig. 5) (Andrews et al., 2009b, 2010). I used the qXRD method (Eberl, 2003) to identify 22 non-clay and 9 clay minerals. However, several species occurred infrequently or with low (<1%) wt% and the number of minerals was reduced to 20 for analysis with "SedUnMix".

6. Mass accumulation rates

Sediment accumulation rates (SAR) and mass accumulation rates (MAR) provide initial guidelines on sediment transport. Mass accumulation rates (mg/cm^2/yr) have been estimated from a combination of sources, including ^{210}Pb measurements, and radiocarbon dated SAR (Smith et al., 2002; Andrews et al., 2009a; Axford et al., in press). In the latter case the measurements have been converted to mass accumulation rates using average dry densities (Andrews et al., 2002). The values for MAR range from 1000 to 5 mg/cm^2/yr (Table 1). The estimates across Denmark Strait (Fig. 5) show the expected pattern of high inshore values and decreasing toward the shelf break. An exception is the high MAR (~200) from MD99-2269 off N Iceland (Stoner et al., 2007). MAR decreases non-linearly from the East Greenland coast to the shelf break, such that the MAR has decreased from 1000 to ~100 mg/cm^2/yr in 150 km (Andrews et al., 1997; Smith et al., 2002). At the moment there are no reliable MARs for sites in and north of Scoresby Sund.

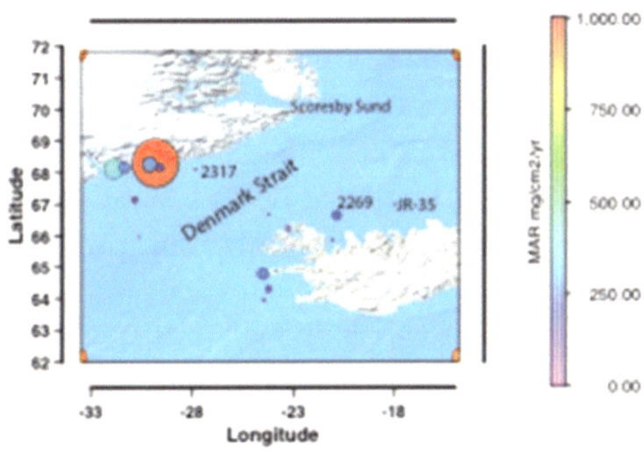

Fig. 5. Late Holocene mass accumulation rates across Denmark Strait (mg/cm^2/yr) based on ^{210}Pb and ^{14}C estimates. Core locations are shown for sites (Table 1)illustrated further in Figures 8 and 10.

The MAR estimates for Iceland are biased toward larger values because of the export by wind and water of volcanic glass; roughly ½ of the sediment is estimated to be derived from Holocene volcanic activity (Andrews, 2007). In marine situations, MARs consist of a transported mineral component and an *in situ* biological component formed by the production of biogenic calcite (foraminifera and coccoliths), aragonite, and silica (diatoms). As part of the qXRD analyses all three of these components are determined (Table 1). Apart from sites off NW Iceland, which are enriched in calcite due to reworking of shelf sediments, the biogenic wt% estimates usually total < 6-wt%. The impact of the intermediate Irminger Current waters in the Kangerlussuaq Trough (Jennings et al., 2011) are readily apparent from the values of calcite wt% versus very low values at sites to the north (Table 1).

7. Mineralogy of sources and sinks

Figure 6 illustrates the mineral compositions of source areas (Fig. 4), which provide sediments for transport to distal sites either in sea ice or icebergs. The 11 of the 20 minerals used are summed to 100% for the pie diagrams.

Kara Sea to N Iceland: A potential source for sea ice sediments and driftwood off Iceland is the area west of the Ural Mountains and especially the shallow (< 50 m wd) shelf of the Kara Sea (Eggertsson, 1993). Sand, silt, and clay separates were obtained from surface samples from ther Kara Sea (Stein et al., 2004) and were processed for qXRD. I focus on the combined silt and clay mineralogy as these fractions are enhanced in sea ice sediments due to the entrainment process (Dethleff, 2005). The silt fraction has high values in quartz, various feldspars and about 15-wt% of smectites. Pyroxene is present but with a low wt%. However, the clay fraction consists of about 10% quartz and nearly 70-wt% are smectites (saponite and ferruginous smectite) due to weathering of the 2 x 10km^2 outcrop of the 250 Ma old Siberian Traps. Kaolinite (not included) was ≤ 5-wt% in the clay fraction.

Given the sea ice limits around Iceland, the trajectory of the TransPolar Drift and East Iceland Current (Figs. 1 & 2), and the location of the marine North Iceland Front (Fig. 5), then a sea ice sediment transport from the Kara Sea, and deposition to the N Iceland margin is probable (Moros et al., 2006). In terms of mineral tracers for a Kara Sea sediment source (Fig. 6), it has higher wt% of quartz, saponite, ferruginous smectite, and Fe-chlorite than samples from East Greenland or Iceland. Quartz and k-feldspars do not occur in any significant amounts within the bedrock of Iceland (Eiriksson et al., 2000) and they have, therefore, been used as an index of ice-rafting debris (IRD) (Eiriksson et al., 2000; Moros et al., 2006; Andrews, 2009; Andrews et al., 2009b) (e.g. Fig. 6, and Table 2A Iceland samples).

Core MD99-2269 is a well-dated 25-m long core off N Iceland (Stoner et al., 2007) (Fig. 7A), and which has been studied with a large number of proxies (Andrews et al., 2003; Giraudeau et al., 2004; Moros et al., 2006; Kristjansdottir et al., 2007). Figure 7A shows a plot of the quartz wt% (Moros et al., 2006) versus the silt and clay fractions in this core---notice silt plus clay total between 86 and 98%, with a persistent increase in this fraction over the last 8000 yr BP. Over the last 4000 yr BP the < 63 μm fraction and quartz wt% track each other, but there are significant discrepancies in the association between 7000 and 12,000 yr BP.

Because the last 4000-6000 yr BP has seen the regrowth of glaciers and re-expansion of the Greenland Ice Sheet and Iceland glaciers (Geirsdottir et al., 2009; Kelly and Lowell, 2009) it is surprising that N Iceland and E Greenland sediment (Andrews et al., 2010) show a pervasive fining-upward signal. Given the grain-size selectivity of sea ice sediment entrainment (Dethleff and Kuhlmann, 2009, 2010) the fining-upward sediments may reflect an increase in sediment transport from the Arctic Ocean.

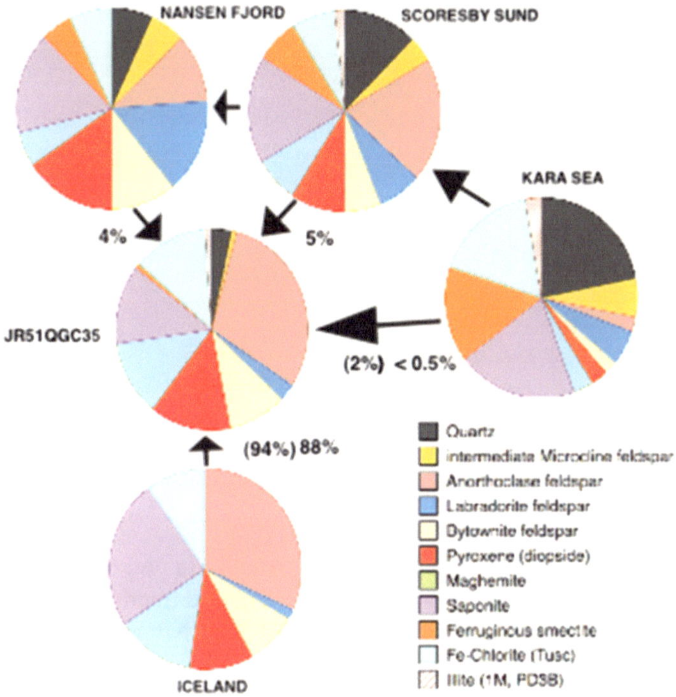

Fig. 6. Potential transport diagram for sediments at core site JR51GC35 (Fig. 5), N Iceland. For clarity only the most plentiful species out of 20 total are shown in the pie diagrams. Figures refer to the average contributoin from the sources to JR51GC35, those in brackets relate to the two end member solution.

To test the hypothesis that Iceland offshore sediments retain an imprint of sediment transport from the Arctic Ocdean via sea ice, I examine the mixing of sediments on the N Iceland outer shelf on core JR51- GC35 (Bendle and Rosell-Mele, 2007) (Tables 1 and 2, Fig. 5) between two sources; the first being local inner shelf Iceland sediments (represented by MD99-2258, (Axford et al., in press), Table 1) and the other being Kara Sea surface sediments (Fig. 6). The average sampling interval for the qXRD data is 85 yr. I use data from JR51-GC35 because the qXRD results on MD99-2269 only reported quartz, plagioclase, and calcite wt%s (Moros et al., 2006). However, the quartz wt% data from JR51-GC35 closely parallels the MD99-2269 data (Andrews, 2009) (Fig. 7B). Twenty non-clay and clay mineral species were identified by qXRD and were processed using the linear progamming software in the Excel "Solver," and using the program SedUnMix (Table 2) (Eberl, 2004; Andrews and Eberl, subm.). The two sources were each represented by 5 samples. The average absolute difference in mineralogy was 3.1 wt% on 20 species (e.g. Table 2B). The estimated contribution from the Kara Sea only averaged 2% (Fig. 6) but shows a systematic increase over the last 6000 yr BP (Fig. 7B), which roughly parallels the change in quartz wt% (note that quartz is only 1/20 of the minerals in the SedUnMix calculation). However, it is known that icebergs drift into NW/N Iceland waters (Fig. 3) thus E/N Greenland could also be a source of the felsic minerals in JR51GC35 (see below).

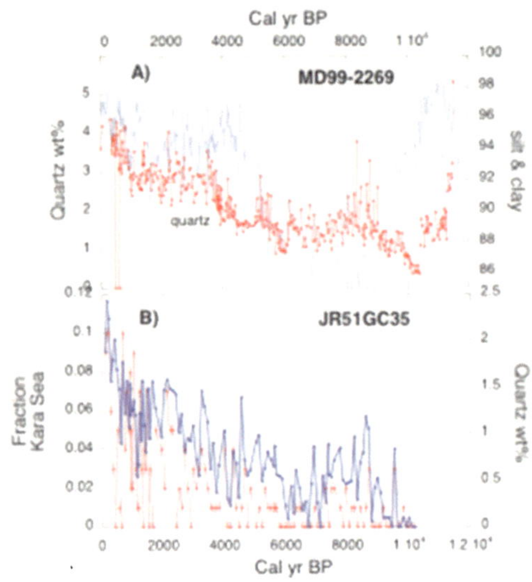

Fig. 7. A) Quartz wt% and the fraction of sediment <63μm (slit and clay) in MD99-2269 off N Iceland. B) Quarz wt% (blue) at JR51-GC35 and the estimated fraction of sediment contributed from the Kara Sea (red) comparated with a local N Iceland composition (see Fig. 6.)with ± standard error of the estimate.

NE to E Greenland: The TransPolar Drift exports sea ice sediments from the Arctic Ocean along the NW/E Greenland margin, with sediment also being contributed from local glaciers and ice streams (Fig. 4) (Evans et al., 2002). If sea ice were the pervasive sediment transport medium then the influence of the local glacier sources should be small. In particular, as sea ice traverses along the Tertiary basalt outcrop from the south shore of Scoresby Sund to Kangerlussuaq Fjord (Fig. 1) we would expect only small changes in sediment composition. However, Andrews et al. (2010) showed a major shift in mineral composition across this northern boundary. Earlier, Pirrung et al. (2002) showed a boundary in mass magnetic susceptibility extending seaward from Scoresby Sund with high values along the Tertiary basalt coastline. U-channel rock magnetic measurements on cores south of Scoresby Sund showed a gradual southward decrease in values, which was associated with a decrease in maghemite and in increase in felsic minerals (Andrews and Hardardottir, 2009). A similar abrupt break is present in the qXRD data (Fig. 8A & B), where an abrupt change in mineralogy occurs north and south of Scoresby Sund from felsic-rich to mafic-rich sediments. These results strongly suggest that the seaward transport of glacially derived, mafic-rich sediments from the early Tertiary volcanic outcrop overwhelms the southward transport of felsic minerals from NE Greenland and the Arctic Ocean. However, single grains in the 250-40 μm range can still be used for identification (Fe-oxide composition) of source areas around the margin of the Arctic Ocean (Darby, 2003; Darby and Bischof, 2004; Andrews et al., 2009b).

2.3 Denmark Strait: Figure 9a & B shows scaled weight% values of quartz and pyroxene for surface samples from Scoresby Sund south and across to Iceland. These data extend the analysis of surface samples from NE Greenland (Fig. 8) and confirm the sharp boundary in sediment mineralogy associated with the Tertiary outcrop of the Geikie Plateau and the younger basalts of Iceland. Very low quartz weight% values are found in Grivel Basin (GB, Fig. 9A) and values ≤ 1% occur off NW and SW Iceland (Axford et al., in press). The quartz wt% values suggest two distinct transport paths for quartz (Fig. 9A) with low values characterizing the intervening area of the Denmark Strait area. The variations in quartz wt% in MD99-2269 and JR51-GC35 (Fig. 7) are associated with the transport of IRD in sea ice (possibly some icebergs, Fig. 3) in the East Greenland/East Iceland Current and deposition via melting as the polar waters reach the North Iceland Front (Fig. 2). The western quartz transport path (Fig. 9A) is directed from Kangerlussuaq Fjord and the large ice stream that calves in that fjord (Dwyer, 1993, 1995; Syvitski et al., 1996; Howat et al., 2008; Mugford and Dowdeswell, 2010). Use of an earlier version of SedUnMix on MD99-2322 from Kangerlussuaq Trough suggested that the contributions of sediment from the fjord (more felsic in nature) had steadily increased over the last 5000 cal yr BP (Andrews et al., 2010). Direct evidence for west to east transport across Denmark Strait during earlier times is seen in the early Tertiary ages of basaltic clasts in sediments off NW Iceland (Principato, 2003).

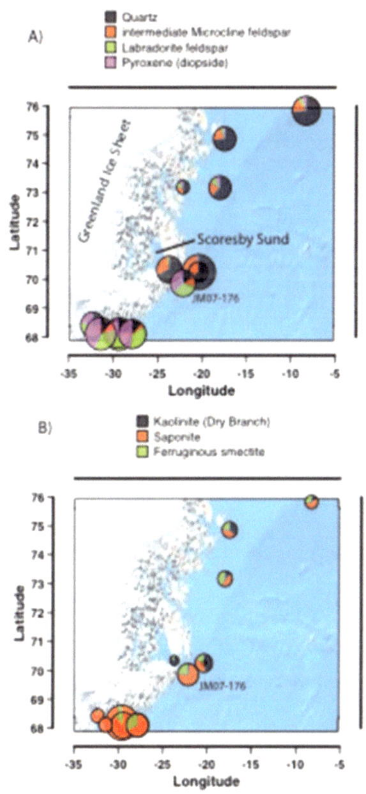

Fig. 8. A) Scaled pie diagrams to the sum of the four non-clay minerals on surface samples along the NE/E Greenland margin between 68° and 76°N. JM07-176 at 22.131° W and 69.859° N, just south of Scoresby Sund marks the break from felsic-rich to mafic-rich sediments. B) As for 8A. but showing wt% of kaolinite, saponite and Fe-smectite

In another run of SedUnMix on the JR51GC35 data, two additional source areas were added, namely Scoresby Sund and inner shelf/fjord sediments off the Geikie Plateau (Figs. 2 and 6). The results indicated that the dominant source is still the Iceland mainland (Fig. 5) and ~88% of the sediment on average is associated with this transport path (Fig. 6). In this 2nd run, SedUnMix assigns some 8% of the sediment in JR51GC35 to transport from across Denmark Strait and only < 0.5% is associated with the Kara Sea. The Kara Sea contributions only commenced during the last 4000 yr BP. Sediments attributed to the two East Greenland sources both show a steady increase over the Holocene. The comparison between the 2 or 4 source areas (Fig. 6) highlights a common problem in attempts to quantify sediment transport paths if the tracer(s) are not unique and have attributes in common (Grousset et al., 2000; Farmer et al., 2003).

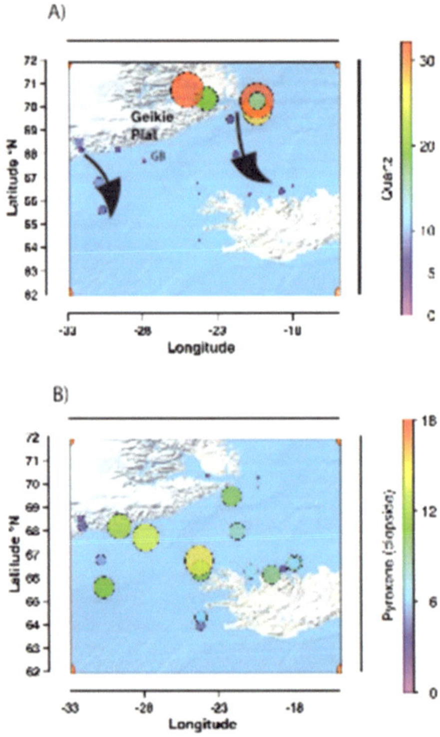

Fig. 9. Bubble plot of the weight% of A) Quartz and B) Pyroxene wt% from sites across Denmark Strait. Notice the sharp change in values along the southern edge of Scoresby Sund. See text for discussion of the arrows on A.

SedUnMix was then used to estimate the contribution from various sources to MD99-2317, a well-dated core in the Grivel Basin (GB, Fig. 9) with a sampling interval averaging 90 yr/sample---the core does not include the last 2000 yr BP or so (Jennings et al., 2002; Andrews et al., 2010). Potential sources of sediment include the Kara Sea (and other shallow Arctic Ocean shelves (Fig. 6)), Scoresby Sund, and local ice caps and glaciers on the Geikie Plateau—the latter is represented by sediments from core JM96-1210 in Nansen Fjord (Jennings and Weiner, 1996). The sediments from Nansen Fjord contain small amounts of quartz derived from erosion of the underlying Precambrian basement rocks (Brooks, 1990). I thus include as a "pure" basaltic end member sediments from SW Iceland (Axford et al., in press). The average absolute difference on target minus the calculated mineral weight %s averages 2.1 ± 0.43 wt% (e.g. Table 2B). The down core results (Fig. 10) indicate a dominance of transport seaward from the adjacent Geikie Plateau, as represented by the sediments from Nansen Fjord (91.9% of the total). The percentage of sediment estimated to have been transported from the Kara Sea in sea ice is only 0.27%, and the more felsic-rich contribution from Scoresby Sund (Fig. 9A) is 1.3%. A pure basaltic contribution is estimated at 6.5%. The contributions from the 4 sources are not randomly distributed during the Holocene (Fig. 10). Of particular note are the sharp peaks in sediments attributed to a Kara Sea source (Fig. 10A). Overall the analysis of the sediment composition indicates a steady increase in transport and deposition from the adjacent glaciated plateau whereas the contributions from felsic sources in Scoresby Sund have declined to zero by 4000 yr ago (Fig. 10B). The apparent periodicity in the sediment input from Nansen Fjord (Fig. 10B) was tested using the multi-taper method (MTM) (Mann and Lees, 1996; Ghil et al., 2002) on residual detrended (linear) data, integrated at 100 yr intervals (Paillard et al., 1996). The results indicated two dominant periodicities at 910 (90% CI) and 350 yr (95% CI) (Fig. 10C). The 350 yr periodicity has been identified in several paleoclimate records and associated with solar variability (Clemens, 2005).

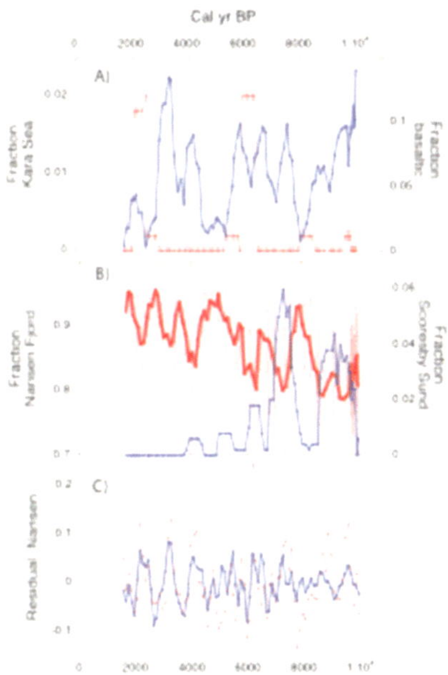

Fig. 10. Estimates of the sediment composition of MD99-2317 from Grivel Basin (GB, Fig. 9). Estimates based on the SedUnMix program (Eberl, 2004: Andrews and Eberl, subm.) and are 5-piont moving averages. A) Fractions of sediment from Kara Sea (red) or from basalt (blue)...error bars are ± one standard error. B) Fraction of sediment from Nansen Fjord (red) and Scoresby Sund (blue). The error bars of the Nansen Fjord fractions are ± one standard deviation based on 10 iterations of 4 end members with 5 samples each. C) Residuals from a linear trend of the fraction of Nansen Fjord sediments in MD99-2317 (red), versus the reconstructed time-series using the 350 and 910 yr periodicities.

A similar analysis was performed on the linear detrended residuals from the combined E Greenland contributions to JR51GC35. MTM analysis of those data indicated two weak periodicities (90% CI) of 445 and 263 yr, the latter often linked to solar variability (Clemens, 2005). A comparison of the residual East Greenland fractions from JR51GC35 and MD99-2317 (Fig. 11) suggests several common intervals of enhanced sediment transport with some of the largest residuals occurring between 6000 and 7000 yr BP.

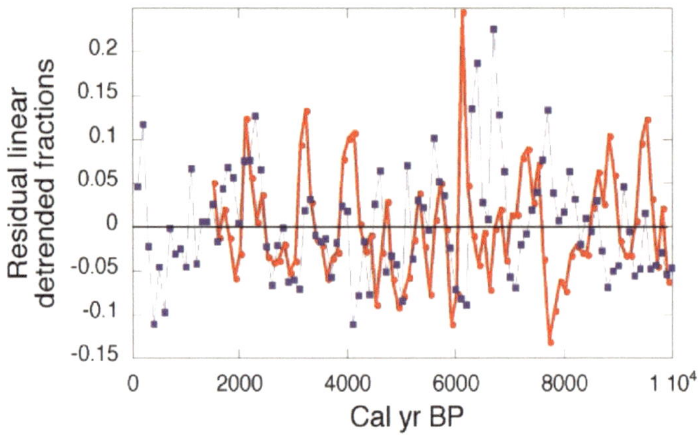

Fig. 11. Residuals from linear detrended data of the fraction of sediment contributed from Nansen Fjord to MD99-2317 (red) and from East Greenland sources (Fig. 6) to JR51GC35, N Iceland (blue).

8. Conclusions

Sediment transported by either sea ice or icebergs, i.e. ice-rafted, has the ability to be moved considerable distances from source areas. Thus potentially they provide key indicators of large-scale changes in either the distribution of sea ice or icebergs. The most dramatic examples are the so-called Heinrich events, which are a dramatic feature of the North Atlantic marine sediment archives during the last glaciation, and are recognized largely by their detrital carbonate content (Heinrich, 1988; Andrews and Tedesco, 1992; Bond et al., 1992). During the last 12,000 yr BP the changes in sediment transport from the Arctic Ocean and NE/E Greenland margin have been subtler, but this chapter demonstrates that quantitative X-ray diffraction analysis of marine sediments can be used to untangle changes in sediment sources and transport pathways (Figs. 4 to 10). Thus, as noted by Moros et al. (2004), quantitative mineral identification is an extremely powerful tool in establishing the connections between sediment sources→transportation→deposition.

Initially two hypotheses were proposed; the first stressed the importance of the Arctic Ocean sea ice as a sediment source, whereas the alternative thesis was that the bulk of the sediment was supplied from NE/E Greenland. The latter is more strongly supported by the data. Analysis of both the surface sediments (Figs. 8 and 9) and the down-core sediment composition (Fig. 10) indicate that the mineralogy immediately south of Scoresby Sund shifts abruptly from being felsic- to mafic-rich with < 2% of the sediment being attributed to sources from the Kara Sea or Scoresby Sund. These data demonstrate that glacial erosion of the basalt is overwhelming any contribution from sediments entrained in the East

Greenland Current from sources north of 70°N. On the east side of Denmark Strait, the north coast of Iceland (Fig. 5, JR51GC35) may record sediment input from the Kara Sea (Fig. 7B) while the relationship between the silt % and quartz% over the last few thousand years (Fig. 7A) may reflect the export of silt-rich sediments from the shallow Arctic Ocean shelves via sea ice. However, further analysis suggested that the major external sediment sources to JR51GC35 were more closely affiliated mineralogically to sources across Denmark Strait than the Kara Sea.

9. Acknowledgements

I wish to thank the National Science Foundation for funding much of the research that led to this chapter, in particular grants ATM-0502515, OPP-070713755, and OCE-0823535. I also wish to thank INSTAAR and the Department of Geological Sciencese of Colorado at Boulder, for their support over 4 decades. I wish to thank Dennis Eberl from the USGS in Boulder who has worked with me (see references) on several projects and shared his experience and knowledge. Work on sample preparation has been assisted by assistance from Casey Coleman and the maintenance of the INSTAAR XRD unit has been in the capable hands of Wendy Roth. Drs R. Stein, H. Grobe, and J. Evans kindly provided samples from the Kara Sea and NE Greenland.

10. References

Alley, R. B., Cuffey, K. M., Evenson, E. B., Lawson, D. E., and Strasser, J. C., 1997: How glaciers entrain and trasnport basal sediment: Physical constraints. *Quaternary Science Review,* 16: 1017-1038.

Alley, R. B., Lawson, D. E., Evenson, E. B., Strasser, J. C., and Larson, G. J., 1998: Glaciohydraulic supercooling: a freeze-on mechanism to create stratified, debris-rich basal ice: II. Theory. *Journal of Glaciology,* 144: 563-569.

Andrews, J. T. and Tedesco, K., 1992: Detrital carbonate-rich sediments, northwestern Labrador Sea: Implications for ice-sheet dynamics and iceberg rafting (Heinrich) events in the North Atlantic. *Geology,* 20: 1087-1090.

Andrews, J. T. and Syvistki, J. P. M., 1994: Sediment fluxes along high latitude continental margins (NE Canada and E. Greenland). *In* Hay, W. (ed.), *Material fluxes on the surface of the Earth.* Washington, D.C.: National Academy Press, 99-115.

Andrews, J. T., Smith, L. M., Preston, R., Cooper, T., and Jennings, A. E., 1997: Spatial and temporal patterns of iceberg rafting (IRD) along the East Greenland margin, ca. 68 N, over the last 14 cal.ka. *Journal of Quaternary Science,* 12: 1-13.

Andrews, J. T., 1998: Abrupt changes (Heinrich events) in late Quaternary North Atlantic marine envrionments: a history and review of data and concepts. *Journal of Quaternary Science,* 13: 3-16.

Andrews, J. T., Kihl, R., Kristjánsdóttir, G. B., Smith, L. M., Helgadóttir, G., Geirsdóttir, Á., and Jennings, A. E., 2002: Holocene sediment properties of the East Greenland and Iceland continental shelves bordering Denmark Strait (64°-68°N), North Atlantic. *Sedimentology,* 49: 5-24.

Andrews, J. T., Hardardottir, J., Stoner, J. S., Mann, M. E., Kristjansdottir, G. B., and Koc, N., 2003: Decadal to millennial-scale periodicities in North Iceland shelf sediments over the last 12,000 cal yrs: long-term North Atlantic oceanographic variability and Solar forcing. *Earth and Planetary Science Letters,* 210: 453-465.

Andrews, J. T. and Eberl, D. D., 2007: Quantitative mineralogy of surface sediments on the Iceland shelf, and application to down-core studies of Holocene ice-rafted sediments. Journal of Sedimentary Research, 77: 469-479.

Andrews, J. T., 2007: Holocene denudation of Iceland as determined from accumulation of sediments on the continental margin. *Boreas,* 36: 240-252.

Andrews, J. T., 2009: Seeking a Holocene drift ice proxy: non-clay mineral variations from the SW to N-central Iceland shelf: trends, regime shifts, and periodicities. *Journal of Quaternary Science,* 24: 664-676.

Andrews, J. T., Belt, S. T., Olafsdottir, S., Masse, G., and Vare, L., 2009a: Sea ice and marine climate variability for NW Iceland/Denmark Strait over the last 2000 cal. yr BP. *The Holocene,* 19: 775-784.

Andrews, J. T., Darby, D. A., Eberl, D. D., Jennings, A. E., Moros, M., and Ogilvie, A., 2009b: A robust multi-site Holocene history of drift ice off northern Iceland: Implications for North Atlantic climate. *The Holocene,* 19: 71-78.

Andrews, J. T. and Hardardottir, J., 2009: A comparison of Holocene sediment- and paleo-magnetic characteristics from the margins of Iceland and East Greenland. *Jokull,* 59: 51-66.

Andrews, J. T., Jennings, A. E., Coleman, C. G., and Eberl, D., 2010: Holocene variations in mineral and grain-size composition along the East Greenland glaciated margin (ca 67-70°N): local versus long-distant sediment transport. *Quaternary Science Reviews,* 29: 2619-2632.

Andrews, J. T. and Eberl, D. D., subm.: Determination of sediment provenance by the unmixing the mineralogies of < 2 mm source-area sediments

Axford, Y., Andresen, C., Andrews, J. T., Belt, S. T., Geirsdottir, A., Masse, G., Miller, G. H., Olafsdottir, S., and Vare, L. L., in press: Do paleoclimate proxies agree? Statistical comparison of climate and sea-ice reconstructions from Icelandic marine and lake sediments, 300-1900 AD. *Journal Quaternary Research.*

Bendle, J. A. P. and Rosell-Mele, A., 2007: High-resolution alkenone sea surface temperature variability on the North Icelandic Shelf: implications for Nordic Seas paleoclimatic development during the Holocene. *The Holocene,* 17: 9-24.

Bigg, G. R., 1999: An estimate of the flux of iceberg calving from Greenland. *Arctic, Antarctic, and Alpine Research,* 31: 174-178.

Bond, G., Heinrich, H., Broecker, W. S., Labeyrie, L., McManus, J., Andrews, J. T., Huon, S., Jantschik, R., Clasen, S., Simet, C., Tedesco, K., KLas, M., Bonani, G., and Ivy, S., 1992: Evidence for massive discharges of icebergs into the glacial Northern Atlantic. *Nature,* 360: 245-249.

Bond, G., Showers, W., Cheseby, M., Lotti, R., Almasi, P., deMenocal, P., Priore, P., Cullen, H., Hajdas, I., and Bonani, G., 1997: A Pervasive Millennial-Scale Cycle in North Atlantic Holocene and Glacial Climates. *Science,* 278: 1257-1266.

Brooks, C. K., 1990: Kangerdlugssuaq Studies: Processes at a Rifted Continental Margin: *Geological Institute, Univ. of Copenhagen,* 100 pp.

Clemens, S. C., 2005: Millennial-band climate spectrum resolved and linked to centennial-scale solar cycles. *Quaternary Science Reviews,* 24: 521-531.

Darby, D. A. and Bischoff, J. F., 1996: A statistical approach to source determination of lithic and Fe oxide grains: An example from the Alpha Ridge, Arctic Ocean. *Journal of Sedimentology Research,* 66: 599-607.

Darby, D. A., 2003: Sources of sediment found in the sea ice from the western Arctic Ocean, new insights into processes of entrainment and drift patterns. *Journal of Geophysical Research,* 108: 13-11 to 13-10. doi: 10, 1111029/1112002JC1001350, 1112003.

Darby, D. A. and Bischof, J., 2004: A Holocene record of changing Arctic Ocean ice drift analogous to the effects of the Arctic Oscilation. *Palaeoceanography,* 19: 1 of 9. doi: 10.1029/2003PA000961, 002004.

Darby, D. A., Ortiz, J., Polyak, L., Lund, S., Jakobsson, M., and Woodgate, R. A., 2009: The role of currents and sea ice in both slowly deposited central Arctic and rapidly deposited Chukchi-Alaskan margin sediments. *Global Planetary Change.,* 68: 58-72, doi:10.1016/j.gloplacha.2009.1002.1007.

Dethleff, D., 2005: Entrainment and export of Laptev Sea ice sediments, Siberian Arctic (vol 110, art no C07009, 2005). *Journal of Geophysical Research-Oceans,* 110.

Dethleff, D. and Kuhlmann, G., 2009: Entrainment of fine-grained surface deposits into new ice in the southwestern Kara Sea, Siberian Arctic. *Continental Shelf Research,* 29: 691-701.

Dethleff, D. and Kuhlmann, G., 2010: Fram Strait sea-ice sediment provenance based on silt and clay compositions identify Siberian and Kara and Laptev sxeas as main source reegions. *Polar Research,* 29: 265-282.

Divine, D. V. and Dick, C., 2006: Historical variability of the sea ice edge position in the Nordic Seas. *Journal of Geophysical Research,* 111: 1 of 14, doi:10.1029/2004JC002851.

Dowdeswell, J. A., 1986: The Distribution and Character of Sediments in a Tidewater Glacier, Southern Baffin Island, N.W.T., Canada. *Arctic and Alpine Research,* 18: 45-46.

Dowdeswell, J. A., Whittington, R. J., and Hodgkins, R., 1992: The sizes, frequencies, and freeboards of East Greenland icebergs observed using ship radar and sextant. *Journal Geophysical Research,* 97: 3515-3528.

Dowdeswell, J. A., Whittington, R. J., and Marienfeld, P., 1994: The origin of massive diamicton facies by iceberg rafting and scouring, Scorsby Sund, East Greenland. *Sedimentology,* 41: 21-35.

Dowdeswell, J. A., Maslin, M. A., Andrews, J. T., and McCave, I. N., 1995: Iceberg production, debris rafting, and the extent and thickness of Heinrich layers (H-1, H-2) in North Atlantic sediments. *Geology,* 23: 301-304.

Dowdeswell, J. A., Evans, J., and Cofaigh, C. O., 2010: Submarine landforms and shallow acoustic stratigraphy of a 400 km-long fjord-shelf-slope transect, Kangerlussuaq margin, East Greenland. *Quaternary Science Reviews,* 29: 3359-3369.

Dwyer, J. L., 1993: Monitoring Characteristics of Glaciation in the Kangerdlugssuaq Fjord Region, East Greenland, Using Digital LANDSAT MSS and TM Data. MSc, University of Colorado, Boulder. 238 pp.

Dwyer, J. L., 1995: Mapping tide-water glacier dynamics in East Greenland using Landsat data. *Journal of Glaciology,* 41: 584-596.

Dyke, A. S., England, J., Reimnitz, E., and Jette, H., 1997: Changes in Driftwood Delivery to the Canadian Arctic Archipelago: The Hypothesis of Postglacial Oscillations of the Transpolar Drift. *Arctic,* 50: 1-16.

Eberl, D. D., 2003: *User guide to RockJock: A program for determining quantitative mineralogy from X-ray diffraction data.* United States Geological Survey, Open File Report 03-78, 40 pp, Washington, DC.

Eberl, D. D., 2004: Quantitative mineralogy of the Yukon River system: Variations with reach and season, and determining sediment provenance. *American Mineralogist,* 89: 1784-1794.

Eggertsson, O., 1993: Origin of the driftwood on the coasts of Iceland: A dendrochronological study. *Jokull,* 43: 15-32.

Eiriksson, J., Knudsen, K. L., Haflidason, H., and Henriksen, P., 2000: Late-glacial and Holocene paleoceanography of the North Iceland Shelf. *Journal of Quaternary Science,* 15: 23-42.

Evans, J., Dowdeswell, J. A., Grobe, H., Niessen, F., Stein, R., Hubberten, H.-W., and Whittington, R. J., 2002: Late Quaternary sedimentation in Kejser Joseph Fjord and the contiental margin of East Greenland. *In* Dowdeswell, J. A. and O'Cofaigh (eds.), *Glacier-influenced sedimentation on High-Latitude conintental margins.* London: Geological Society 149-179.

Farmer, G. L., Barber, D. C., and Andrews, J. T., 2003: Provenance of Late Quaternary ice-proximal sediments in the North Atlantic: Nd, Sr and Pd isotopic evidence. *Earth and Planetary Science Letters,* 209: 227-243.

Foldvik, A., Aagaard, K., and Torresen, T., 1988: On the velocity field of the East Greenland Current. *Deep Sea Research,* 35: 1335-1354.

Geirsdottir, A., Miller, G. H., Axford, Y., and Olafsdottir, S., 2009: Holocene and latest Pleistocene climate and glacier fluctuations in Iceland. *Quaternary Science Reviews,* 28: 2107-2118.

Ghil, M., Allen, M. R., Dettinger, M. D., Ide, K., Kondrashov, D., Mann, M. E., Roberston, A. W., Saunders, A., Tian, Y., Varadi, F., and Yiou, P., 2002: Advanced spectral methods for climatic time series. *Reviews of Geophysics,* 40: 3-1 to 3-41 1003, doi:1029/2000RG000092.

Giraudeau, J., Jennings, A. E., and Andrews, J. T., 2004: Timing and mechanisms of surface and intermediate water circulation changes in the Nordic Sea over the last 10,000 cal years: a view from the North Iceland shelf. *Quaternary Science Reviews,* 23: 2127-2139.

Gray, D., 1881a: Ice chart of the Arctic Ocean between Greenland and Spitsbergen. From observations by Capt. David Gray: Royal Geographical Society, Control # 503747. Scale 503741:503746,503750,503000.

Gray, D., 1881b: The recent advance of the Polar Ice in the Greenland and Spitzbergen Sea. *Proceedings of the Royal Geographiocal Society and monthly record of Geography,* 3: 740-741.

Grousset, F. E., Labeyrie, L., Sinko, J. A., Cremer, M., Bond, G., Duprat, J., Cortijo, E., and Huon, S., 1993: Patterns of ice-rafted detritus in the glacial North Atlantic (40-55°N). *Paleoceanography,* 8: 175-192.

Grousset, F. E., Pujol, C., Labeyrie, L., Auffret, G., and Boelaert, A., 2000: Were the North Atlantic Heinrich events triggered by the behavior of the European ice sheets. *Geology,* 28: 123-126.

Grousset, F. E., Cortijo, E., Huon, S., Herve, L., Richter, T., Burdloff, D., Duprat, J., and Weber, O., 2001: Zooming in on Heinrich layers. *Paleoceanography,* 16: 240-259.

Hebbein, D., 2000: Flux of ice-rafted detritus from sea ice in the Fram Strait. *Deep-Sea Research II,* 47: 1773-1790.

Heinrich, H., 1988: Origin and consequences of cyclic ice rafting in the Northeast Atlantic Ocean during the past 130,000 years. *Quat. Res.,* 29: 143-152.

Hemming, S., Vorren, T., and Kleman, J., 2002a: Provinciality of ice rafting in the North Atlantic: application of 40Ar/39Ar dating of individual ice rafted hornblende grains. *Quaternary International,* 95-96: 75-85.

Hemming, S. R., Hall, C. M., Biscaye, P. E., Higgins, S. M., Bond, G. C., McManus, J. F., Barber, D. C., Andrews, J. T., and Broecker, W. S., 2002b: 40Ar/39Ar ages and 40Ar concentrations of Fine-Grained Sediment Fractions from North Atlantic Heinrich Layers. *Chemical Geology,* 182: 583-603.

Hemming, S. R., 2004: Heinrich Events: Massive late Pleistocene detritus layers of the North Atlantic and their global climate imprint. *Reviews of Geophysics,* 42: RG1005/2004.

Henriksen, H., 2008: *Geological history of Greenland.* Copenhagen: Geological Survey of Denmark and Greenland, 272 pp.

Hesse, R., 1995: Continental slope and basin sedimentation adjacent to an ice-margin: a continous sleeve-gun profile across the Labrador Slope, Rise and Basin. *In* Pickering, K. T. e. a. (ed.), *Atlas of Deep Water Environments Architectural style in turbidite sytems,* 14-17.

Hopkins, T. S., 1991: The GIN Sea- A synthesis of its physical oceanography and literature review 1972-1985. *Earth Science Reviews,* 30: 175-318.

Howat, I. M., Joughin, I., Fahnestock, M., Smith, B. E., and Scambos, T. A., 2008: Synchronous retreat and acceleration of southeast Greenland outlet glaciers 2000-06: ice dynamics and coupling to climate. *Journal of Glaciology,* 54: 646-660.

Jennings, A. E. and Weiner, N. J., 1996: Environmental change on eastern Greenland during the last 1300 years: Evidence from Foraminifera and Lithofacies in Nansen Fjord, 68°N. *The Holocene,* 6: 179-191.

Jennings, A. E., Gronvold, K., Hilberman, R., Smith, M., and Hald, M., 2002: High resolution study of Icelandic tephras in the Kangerlussuaq Trough, southeast Greenland, during the last deglaciation. *Journal of Quaternary Science,* 17: 747-757.

Jennings, A. E., Andrews, J. T., and Wilson, L., 2011: Holocene environmental evolution of th SE Greenland Shelt north and south of the Denmark Strait: Irminger and East Greenland current interactions. *Quaternary Science Reviews.*

Kelly, M. A. and Lowell, T. V., 2009: Fluctuations of local glaciers in Greenland during the latest Pleistocene and Holocene time. *Quaternary Science Reviews,* 28: 2088-2106.

Koch, L., 1945: The East Greenland Ice. *Meddelelser om Gronland,* 130 (3): 346.

Kolla, V., Biscaye, P. E., and Hanley, A. F., 1979: Distribution of quartz in Late Quaternary sediments in relation to climate. *Quaternary Research,* 11: 261-277.

Kristjansdottir, G. B., Stoner, J. S., Gronvold, K., Andrews, J. T., and Jennings, A. E., 2007: Geochemistry of Holocene cryptotephras from the North Iceland Shelf (MD99-2269): Intercalibration with radiocarbon and paleomagnetic chronostratigraphies. *The Holocene,* 17: 155-176.

Kwok, R., 2009: Outflow of Arctic Ocean Sea Ice into the Greenland and Barents Seas: 1979-2007. *Journal of Climate,* 22: 2438-2457.

Larsen, B., 1983: Geology of the Greenland-Iceland Ridge in the Denmark Strait. *In* Bott, M. H. P., Saxov, S., Talwani, M., and Thiede, J. (eds.), *Structure and Development of the Greenland-Scotland Ridge*. London: Plenum Publishing Corp., 425-444.

Lawson, D. E., Strasser, J. C., Evenson, E. B., Alley, R. B., Larson, G. J., and Arcone, S. A., 1998: Glaciohydraulic supercooling: a freeze-on mechanism to create stratified, debris-rich basal ice: I. Field Evidence. *Journal of Glaciology,* 44: 547-562.

Lebreiro, S. M., Moreno, J. C., McCave, I. N., and Weaver, P. P. E., 1996: Evidence for Heinrich layers off Portugal (Tore Seamount: 39N, 12W). *Marine Geology,* 131: 47-56.

Li, C., Battisti, D. S., and Bitz, C. M., 2010: Can North Atlantic Sea Ice Anomalies Account for Dansgaard-Oeschger Climate Signals? *Journal of Climate,* 23: 5457-5475.

MacAyeal, D. R., 1993: Binge/purge oscillations of the Laurentide Ice Sheet as a cause of North Atlantic's Heinrich events. *Paleoceanography,* 8: 775-784.

Malmberg, S.-A., 1985: The water masses between Iceland and Greenland. *Journal Marine Research Institute,* 9: 127-140.

Mann, M. E. and Lees, J. M., 1996: Robust estimation of background noise and signal detection in climatic time series. *Climatic Change,* 33: 409-445.

Marko, J. R., Fissel, D. B., Wadhams, P., Kelly, P. M., and Brown, R. D., 1994: Iceberg severity off Eastern North America---its relationship to sea-ice variability and climate change. *Journal of Climate,* 7: 1335-1351.

McCarty, D. K., 2002: Quantitative mineral analysis of clay-bearing mixtures: The "Reynolds Cup" contest. *Innternational Union of Crystallography, Newsletter,* No. 27: 12-16.

Miller, J. D. and Hotzel, I. S., 1984: Iceberg flux estimation in the Labrador Sea. International Offshore Mechanics and Arctic Engineering Symposium, 3rd, New Orleans, Louisiana, Feb. 12-17, 1984, 3: 298-304.

Moros, M., McManus, J., Rasmussen, T., Kuijpers, A., Dokken, T., Snowball, I., Nielsen, T., and Jansen, E., 2004: Quartz content and the quartz-to-plagioclase ratio determined by X-ray diffraction: a proxy for ice rafting in the northern North Atlantic? *Earth and Planetary Science Letters,* 218: 389-401.

Moros, M., Andrews, J. T., Eberl, D. D., and Jansen, E., 2006: The Holocene history of drift ice in the northern North Atlantic: Evidence for different spatial and temporal modes. *Palaeoceanography,* 21: 1 of 10. doi:10.1029/2005PA001214.

Mugford, R. I. and Dowdeswell, J. A., 2010: Modeling iceberg-rafted sedimentation in high-latitude fjord environments. *Journal of Geophysical Research-Earth Surface,* 115.

Mugford, R. I. and Dowdeswell, J. A., in press: Modeling glacial meltwater plume dynamics and sedimentation in high-latitude fjords. *Journal of Geophysical Research.*

Ogilvie, A. E. J., 1996: Sea-ice conditions off the coasts of Iceland A.D. 1601-1850 with special reference to part of the Maunder Minimum period (1675-1715). *North European climate data in the latter part of the Maunder Minimum period A.D. 1675-1715,* AmS-Varia 25: 9-12.

Ogilvie, A. E. J. and Jonsdottir, I., 2000: Sea ice, climate, amd Icelandic fisheries in the eighteenth and nineteenth centuries. *Arctic,* 53: 383-394.

Ortiz, J. D., Polyak, L., Grebmeier, J. M., Darby, D., Eberl, D. D., Naidu, S., and Nof, D., 2009: Provenance of Holocene sediment on the Chukchi-Alaskan margin based on combined diffuse spectral reflectance and quantitative X-Ray Diffraction analysis. *Global and Planetary Change,* 68: 71-84.

Paillard, D., Labeyrie, L., and Yiou, P., 1996: Macintosh Program Performs Time-Series Analysis. *EOS,* 77: 379.

Parkinson, C. L., 2000: recent trend reversals in Arctic sea ice extents: possible connection to the North Atlantic oscillation. *Polar Geography,* 24: 1-12.

Parkinson, C. L., Rind, D., Healy, R. J., and Martinson, D. G., 2001: The impact of sea ice concentration accuracies on climate model simulations with the GISS GCM. *Journal of Climate,* 14: 2606-2623.

Peck, V. L., Hall, I. R., Zahn, R., Grousset, F. E., Hemming, S. R., and Scourse, J. D., 2007: The relationship between Heinrich events and their European precursors over the past 60 ka BP: a mulit-proxy ice-rafted debris provenance study in the North East Atlantic. *Quaternary Science Reviews,* 26: 862-875.

Pirrung, M., Futtere, D., Grobe, H., Matthiessen, J., and Niessen, F., 2002: Magnetic susceptibility and ice-rafted debris in surface sediments of the Nordic Seas: Implications for Isotope Stage 3 oscillations. *Geo-Marine Letters,* 22: 1-11.

Principato, S. M., 2003: The late Quaternary history of eastern Vestfirdir, NW Iceland. PhD, Geological Sciences, University of Colorado, Boulder. 258 pp.

Reeh, N., Mayer, C., Miller, H., Thomsen, H. H., and Weidick, A., 1999: Present and past climate control on fjord glaciations in Greenland: Implications for IRD-deposition in the sea. *Geophysical Research Letters,* 26: 1039-1042.

Reeh, N., Thomsen, H. H., Higgins, A. K., and Weidick, A., 2001: Sea ice and the stability of north and northeast Greenland floating glaciers. *In* Jeffries, M. O. and Eicken, H. (eds.), *Annals of Glaciology,* 33, 474-480.

Reimnitz, E., Barnes, P. W., and Kempema, E. W., 1987: Anchor Ice, Seabed Freezing, and Sediment Dynamics in Shallow Arctic Seas. *Journal of Geophysical Research,* 92: 14, 671-614, 678.

Rogers, J. C., Yang, L., and Li, L., 2005: The role of Fram Strait winter cyclones on sea ice flux and on Spistbergen air temperatures. *Geophysical Research Letters,* 32: 1 of 4, doi:10.1029/2004GL022262.

Schmith, T. and Hanssen, C., 2003: Fram Strait ice export during the nineteenth and twentieth centuries reconstructed from a multiyear sea ice index from Southwestern Greenland. *Journal of Climate,* 16: 2782-2791.

Sigurdsson, F. H., 1969: Report on Sea Ice off the Icelandic Coasts October 1967 to September 1968. *Jokull,* 19: 77-93.

Smith, L. M. and Andrews, J. T., 2000: Sediment characteristics in iceberg dominated fjords, Kangerlussuaq region. East Greenland. *Sedimentary Geology,* 130: 11-25.

Smith, L. M., Alexander, C., and Jennings, A. E., 2002: Accumulation in East Greenland Fjords and on the continental shelves adjacent to the Denmark Strait over the last century based on 210Pb geochronology. *Arctic,* 55: 109-122.

Stefansson, U., 1962: North Icelandic Waters. *Rit Fiskideildar,* III. Bind, Vol 3, 269 pp.

Stein, R., Dittmers, K., Fahl, K., Kraus, M., Matthiessen, J., Niessen, F., Pirrung, M., Polyakova, Y., Schoster, F., Steinke, T., and Futterer, D. K., 2004: Arctic (palaeo) river discharge and environmental change: evidence from the Holocene Kara Sea sedimentary record. *Quaternary Science Reviews,* 23: 1485-1511.

Stoner, J. S., Jennings, A. E., Kristjansdottir, G. B., Andrews, J. T., Dunhill, G., and Hardardottir, J., 2007: A paleomagnetic approach toward refining Holocene radiocarbon based chronostratigraphies: Paleoceanographic records from North

Iceland (MD99-2269) and East Greenland (MD99-2322) margins. *Palaeoceanography,* 22: 1 0f 23. PA1209, doi:1210:1029/2006PA001285, 002007.

Stuiver, M., Reimer, P. J., Bard, E., Beck, J. W., Hughen, K. A., Kromer, B., McCormack, F. G., v.d. Plicht, J., and Spurk, M., 1998: INTCAL98 Radiocarbon age calibration 24,000-0 cal BP. *Radiocarbon,* 40: 1041-1083.

Syvitski, J. P. M., Andrews, J. T., and Dowdeswell, J. A., 1996: Sediment deposition in an iceberg-dominated Glacimarine Environment, East Greenland: Basin Fill Implications. *Global and Planetary Change,* 12: 251-270.

Syvitski, J. P. M., Stein, A., Andrews, J. T., and Milliman, J. D., 2001: Icebergs and seafloor of the East Greenland (Kangerlussuaq) continental margin. *Arctic, Antarctic and Alpine Research,* 33: 52-61.

Thompson, D. W. J. and Wallace, J. M., 1998: The Arctic oscillation signature in wintertime geopotential height and temperature fields. *Geophysical Research Letters,* 25: 1297-1300.

Tremblay, L. B., Mysak, L. A., and Dyke, A. S., 1997: Evidence from driftwood records for century-to-millennial scale variations of the high latitude atmospheric circulation during the Holocene. *Geophysical Research Letters,* 24: 2027-2030.

van den Broeke, M., Bamber, J., Ettema, J., Rignot, E., Schrama, E., van de Berg, W. J., van Meijgaard, E., Velicogna, I., and Wouters, B., 2009: Partitioning Recent Greenland Mass Loss. *Science,* 326: 984-986.

Verplanck, E. P., Farmer, G. L., Andrews, J., Dunhill, G., and Millo, C., 2009: Provenance of Quaternary glacial and glacimarine sediments along the southeast Greenland margin. *Earth and Planetary Science Letters,* 286: 52-62.

Wallevik, J. E. and Sigurjonsson, H., 1998: *The Koch index: Formulation, corrections and extension.* Icelandic Meteorological Office Report

Ward, C. R., Taylor, J. C., and Cohen, D. R., 1999: Quantitative mineraology of sandstones by X-ray diffractometry and normative analysis. *Journal Sedimentary Research,* 69: 1050-1062.

12

Reconstruction of the Kinematics of Landslide and Debris Flow Through Numerical Modeling Supported by Multidisciplinary Data: The 2009 Siaolin, Taiwan Landslide

Chien-chih Chen[1], Jia-Jyun Dong[2], Chih-Yu Kuo[3], Ruey-Der Hwang[4], Ming-Hsu Li[5] and Chyi-Tyi Lee[2]

[1]*Grad. Inst. Geophys. & Dept. Earth Sciences, Nat'l Central Univ., Jhongli, Taoyuan*
[2]*Grad. Inst. Applied Geology, Nat'l Central Univ., Jhongli, Taoyuan*
[3]*Research Center for Applied Sciences, Academia Sinica, Nankang, Taipei*
[4]*Dept. Geology, Chinese Culture Univ., Taipei*
[5]*Grad. Inst. Hydrological & Oceanic Sciences, Nat'l Central Univ., Jhongli, Taoyuan*
Taiwan

1. Introduction

Typhoon Morakot struck southern Taiwan in the summer of 2009, causing the region's most severe flooding since the 1950s. In the early morning of August 9 (local time), a rainfall-triggered landslide and debris-flow extinguished the township of Siaolin Village, Kaohsiung. The interviews of many survived villagers could shed light on some parts of the story about the landslide catastrophe (Lee et al., 2009). Scientifically, a simulation of the landslide/debris-flow can be used to examine its complex kinematic characteristics. A landslide/debris-flow simulation is conventionally performed using constrained digital terrain models, field geological mappings of channel cross-sections or laboratory measurements of slope material properties. Here, we employ seismological and near-surface magnetic data in a novel way to validate and reinforce our simulation of the catastrophic Siaolin debris avalanche.

Siaolin is identified from a high-resolution aerial photograph taken before Typhoon Morakot (Figure 1). Changes in elevation before and after Typhoon Morakot (Figure 1) were derived from two digital terrain models (DTMs) with a precision of five meters by the Agriculture and Forestry Aerial Survey Institute (AFASI) of Taiwan. Based on the DTMs before and after Typhoon Morakot, the major landslide body (cool colors in Figure 1) had an extent of 57 hectares, and the volume of the landslide was about 23.87 million m^3. According to extensive field investigation, the sliding block was mainly composed of Pliocene shale and Quaternary colluvium. The sliding surface may have been located along the interface between the fresh and weathered Pliocene shale. The volumes of the deposits (warm colors in Figure 1) on the unnamed creek and the west bank of the Cishan River were estimated to have loose-measure volumes of 10.91 and 4.53 million m^3, respectively. Most of the deposits laid down in the Cishan River, which had dammed the river for a short time after the debris avalanche, were washed away. The deposits that were washed away had an estimated volume of 10.83 million m^3, assuming that the loose-measure volume of the sliding body yielded an expansion of 10%. Consequently, the total bulk of the natural landslide dam (i.e., the sum of the washed-away materials and the deposits left on the west bank of the Cishan River) was 15.36 million m^3.

2. Frictional experiments of the sliding materials

Frictional properties of sliding surfaces at low to high slip rates are crucial for conducting numerical simulations of landsides/debris-flows. Recently, researchers have utilized the rotary-shear high-velocity friction apparatus to measure the frictional coefficients of the sliding surfaces of landside materials under common conditions present in landslide motion (Mizoguchi et al., 2007; Ferri et al., 2009; Miyamoto et al., 2009; Yano et al., 2009; Togo et al., 2009). The high-velocity friction apparatus and the experimental procedures we utilized in this study are described in papers by Mizoguchi et al. (2007) and Togo et al. (2009).

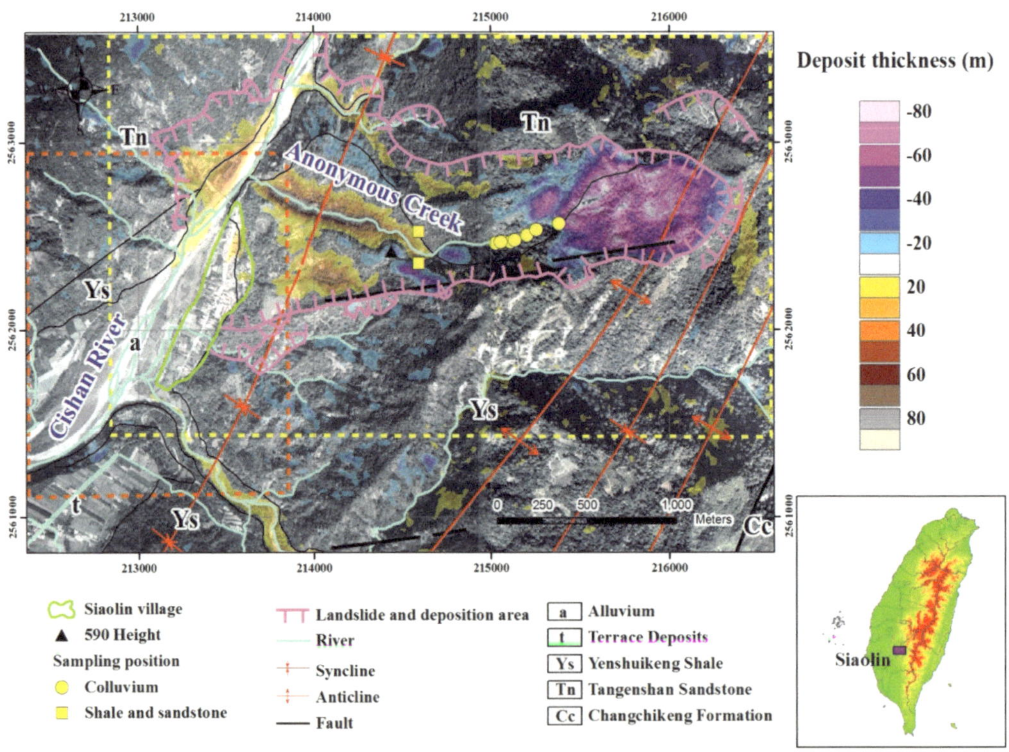

Fig. 1. Aerial photo before the catastrophic Siaolin landslide superposed by the elevation differences between two versions of DTM (before and after Typhoon Morakot) from AFASI. Cool colors represent the sliding mass while warm colors represent the deposits. The yellow and red dashed-line rectangles indicate the areas shown in Figure 3 and in Figure 6, respectively.

Shale and colluvium in the Siaolin area were carefully sampled. The sampling locations are indicated in Figure 1. The tested samples were sustained under a normal stress of around 1

MPa, which accurately represents the *in-situ* normal stress for these sliding surfaces. For the colluvium samples, the initial water content was kept equal to the natural water content of the colluvium collected in the field (~20%). For the wet shale samples, we mixed 0.8 g of shale powder with 0.2 g of pure water, corresponding again to an initial water content of 20%. A wide velocity spectrum from 0.01 through 1.3 m/s, covering the range of likely sliding velocities, was adopted.

Figure 2A shows the frictional behavior of the wet colluvium at different slip rates. Except at 0.01 m/s showing almost constant friction coefficient of 0.3, friction usually decreased with shear displacement after its peak was reached, and the residual friction fell in the range of 0.05 to 0.2. The 'V-shaped' friction curve, i.e., slip weakening then slip hardening, observed in some experiments at 0.7 and 1.3 m/s may be due to the fluid pressure buildup followed by the fluid escape from the slip zone.

Figure 2B displays the frictional behavior for the wet shale samples at three different slip rates. All showed that the friction decreased with shear displacement from the peak value of 0.2 - 0.3 to its residual value of 0.15. Again, the friction at 1.3 m/s showed the 'V-shaped' behavior at displacement of 30 - 40 m. This indicates the dried shale gouge, after the release of fluid pressure, may have quite low frictional strength at such a high slip rate of 1.3 m/s.

In both cases of colluvium and shale, the residual friction coefficients are mostly around 0.1 - 0.2. The experimental results thus suggest that a frictional coefficient of around 0.2 is a reasonable estimate for the relatively high-velocity sliding regime before the moving block collapsed and transformed itself into a granular flow. Although the frictional coefficient of a granular flow is difficult to determine experimentally, an equivalent frictional coefficient of ~0.2 also falls within the range between 0.1 and 0.2 inferred from theoretical calculations (Okura et al., 2000).

3. Simulation of Siaolin landslide

The apparent friction coefficients determined in the laboratory tests are helpful in the applications of continuum landslide models because they can be directly associated with the friction force. In the present study, we use a two-dimensional shallow-water model extended from the paper of Tai and Kuo (2008) to simulate the Siaolin landslide. The effects of the general topography on the landslide flow are brought into account. The flow is simplified as an incompressible and inviscid fluid with a basal Coulomb friction force. The sole rheological parameter, the friction coefficient, is assumed to be a constant. Though with such simplifications, this type of models has been successively validated to be able to reproduce experimental and natural granular flows (Gray et al. 2003; Kuo et al. 2009).

In addition to be constrained by the frictional experimental results, the Coulomb friction coefficient in the simulation is further referenced to a universal scaling law (Staron and Lajeunesse, 2009) and reconfirmed by an iterative optimization scheme. Inspecting extensive practical landslide data and discrete element simulations, the friction coefficient is found to follow a simple geometrical relation, the universal scaling law, which is inversely proportional to the one-third power of the landslide volume (Staron and Lajeunesse, 2009). Based on the friction angle 6° found in the Tsaoling landslide with a volume 0.126 km^3 (Kuo et al., 2009), we estimate the friction angle for the Siaolin landslide (about 1/6 in volume) is about 11° according to the geometrical scaling relation. This value is well constrained in the range suggested by the aforementioned frictional tests (assuming the friction coefficients asymptote to constant values). The initial volume is taken from the DTMs without a presumed volume dilation. The major slid mass on the east side of the Cishan River is released from the rest at t = 0 (Figure 1). The computation domain 3,710×2,220 m² is discretized into a 10×10 m² grid mesh and the simulation time is 180 seconds. With the

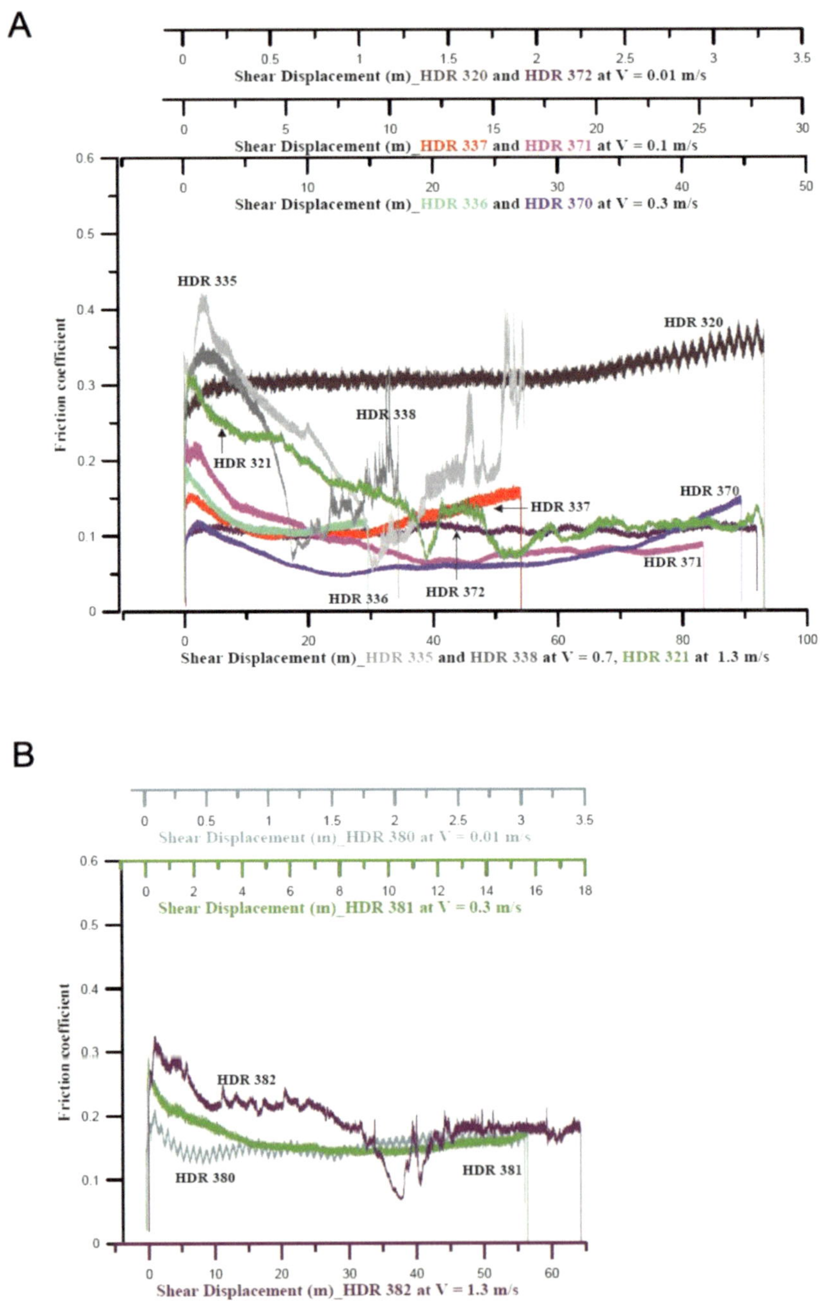

Fig. 2. Frictional coefficients of (A) colluvium and (B) shale powder sustained by accumulative shear displacement under different shearing velocities. Note that we have used different horizontal axes to indicate different conditions of shearing velocities.

best-fit algorithm to minimize the difference between the simulation and the field measurement, it is confirmed that the optimal value of the friction coefficient is 11.5°, with just a minute offset from the straightforward estimation. Fuller theoretical and numerical details of the simulation have been reported in a designated paper (Kuo et al., 2011).

Two snapshots at t = 32 and t = 60 seconds are shown in Figures 3A and 3B, respectively. These times correspond to when the landslide flow reached the 590 Height (cf. Figure 1) and the west bank of the Cishan River. Upon hitting the 590 Height, the flow split into two sliding courses (Figure 3A). The main stream was diverted to flow along the valley of the unnamed creek, and the second flow moved through the southern part of the 590 Height into the Siaolin Village. The simulation shows that when the flow hit the west bank of the Cishan River, a portion of the volume of debris overflowed from the guarding northeastern ridge into the Siaolin Village about 60 seconds after the initiation of the landslide (Figure 3B). The velocity of the landslide's front reached a magnitude of about 50 m/s, and the duration of the landslide was about 110 seconds. The simulated deposit volume (Figure 3C) in the Cishan River was about 21.1 million m^3; this formed a dam that was subsequently flushed away when it was breached about 30 minutes later. An animation of the whole process of the Siaolin landslide can be found in the online supplementary material.

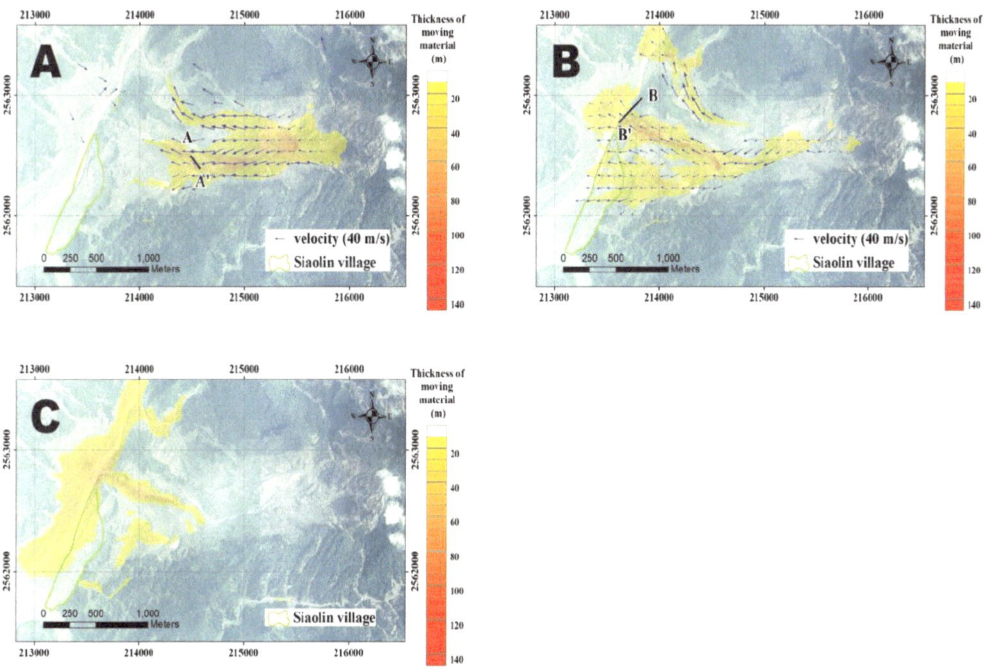

Fig. 3. Snapshots of the Siaolin debris avalanche. (A) Simulated sliding mass hitting the 590 Height (benchmarked with a black line AA') 32 seconds after landslide initiation and (B) entering the Cishan River channel (benchmarked with a black line BB') at 60 seconds, hitting the river bed and bank. Arrows indicate the flow velocities. (C) Final deposition depths due to the landslide. Note that a landslide dam formed along the Cishan River channel soon after the landslide.

4. Geophysical data validating the landslide simulation

The scenario replayed in the simulation agrees with two aspects of the data obtained from seismographs and a near-surface magnetic survey. First, several broadband TCWBSN (Taiwan Central Weather Bureau Seismic Network) seismographs recorded ground motions, especially Rayleigh waves, generated by the Siaolin landslide. We used the arrival times of the 25-s Rayleigh waves recorded at 8 stations to locate the original position of the landslide. A half-space model with 3.75 km/sec of S wave and an inversion algorithm (Stein and Wysession, 2003) were adopted for the determination of the landslide location. Estimated from the Rayleigh-wave calculation is the landslide location (Figure 4), together with its occurrence time of 6:16 in the moring.

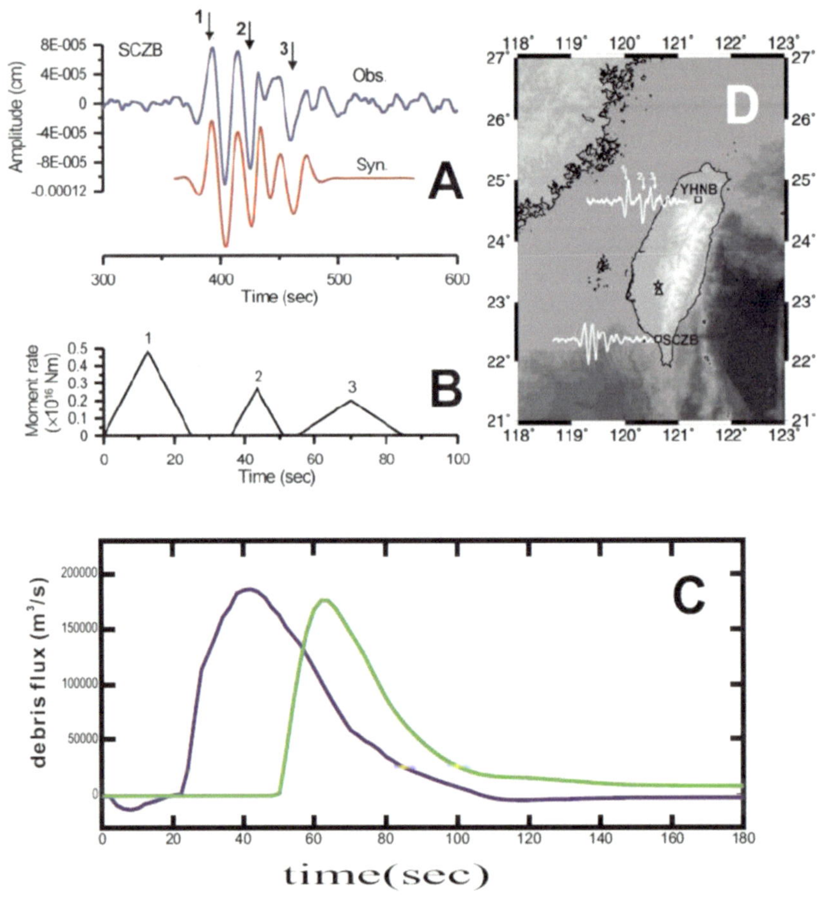

Fig. 4. (A) Observed (blue) and synthetic (red) Rayleigh waves recorded at TCWBSN Station SCZB. (B) Source time function of three sub-events used for modeling the Rayleigh waves in (A). (C) Simulated debris flux past the two benchmarks AA′ (blue) and BB′ (green) in the Siaolin area (Figure 3). (D) Geographic locations of landslide (23.2637°N, 120.6267°E; located by means of seismic recordings) and of Siaolin Village (23.1625°N, 120.6444°E) marked by a star and a triangle, respectively. Also shown are two seismograms observed at Stations SCZB and YHNB (squares).

We note that at least three impact sub-events (marked by the arrows in Figures 4A and 4D) during the landslide can be clearly identified by the Rayleigh waves recorded, for example, at the distant Station YHNB. The seismogram recorded at Station SCZB (blue line in Figure 4A) also shows the visible sub-events during the landslide. Having the waveforms as recorded by TCWBSN, we simulated the waveforms of Rayleigh waves at a frequency band of 0.033-0.05 Hz. The focal mechanism of a horizontal thrust faulting, i.e. an *equivalent* single-force representation of shallow landslide, was used in the waveform modeling (Dahlen, 1993). The velocity structure is again a simple half-space model with 6.5 km/sec of P wave, 3.75 km/sec of S wave and 2.8 g/cm^3 of density. The program of synthetic waveform of surface wave was developed by Wang (1981). Also, to avoid the interferences from the velocity structure and the rupture directivity of source, we simulated the waveforms at the closer Station SCZB where is located to the south of the landslide. Three parameters in the forward modeling of the occurrence time, duration and moment for each sub-event were fitted to ensure that the synthetic waveform (red line in Figure 4A) matches the observed one.

Three sub-events thus obtained in the modeling can fit well the waveform observed at Station SCZB indeed. Seismic moments for these three sub-events are 6.0, 1.4 and 3.0 $\times 10^{16}$ Nm, respectively (Figure 4B). This calculation shows that the second and third sub-events occurred 34 and 55 seconds, respectively, after the occurrence of the first sub-event. If we assume that the first sub-event represents the landslide's initiation, these two moments are quite coincident to the above-mentioned moments when the landslide flow reached the 590 Height and the channel of the Cishan River. After calculating the debris fluxes at two benchmarks near the 590 Height (AA′ in Figure 3A) and the Cishan River (BB′ in Figure 3B), we find that the large impacts to the 590 Height (blue line in Figure 4C) and the Cishan River (green line in Figure 4C) occurred around 40 and 60 seconds after the landslide's initiation, respectively. The waveforms recorded at Station SCZB are thus consistent with the kinematics resulting from the simulation of the Siaolin landslide event.

In addition, the results of the near-surface magnetic survey (Figure 5) provide another confirmation of the landslide simulation regarding the flow direction of the debris. High-resolution near-surface magnetic data covering the main township of Siaolin Village were obtained in October. The track lines of magnetic measurement were along the direction of NW to SE (with an interval of five meters). Portable proton precession magnetometers (Geometrics model G-856, with a sensitivity of 0.1 nT) were used in the magnetic surveys. The total-field magnetic data were collected at 1.22 meters and 1.82 meters above the ground. To ensure measurement reliability, the magnetic data were measured at least three times at each location within a few minutes. We refer readers to the paper of Doo et al. (2011) for the detailed information about magnetic survey and data analyses.

The magnetic anomaly map (colors in Figure 5A) shows several anomalous patterns of magnetic dipole fields, and it can be used to derive the amplitude distribution (contours in Figure 5A) of the zeroth-order analytic signals of the magnetic sources (Hsu et al., 1998; Hsu, 2002; Doo et al., 2009). When superimposing the distributions of magnetic sources onto the aerial photograph taken before Typhoon Morakot, we can clearly see that all of the magnetic sources correspond well to the buildings located in the southern part of the township (Figure 5B). Next, a sharp boundary of the destruction front of the landslide (thick, dashed lines in Figure 5) can be clearly identified by a comparison between the magnetic anomalies and the aerial photograph. Buildings located to the north of the boundary have been swept off their original sites by a massive impact. The landslide simulation of the Siaolin event suggests the existence of such a destruction front indeed, and its location coincides well with that boundary identified by the magnetic survey. Shown in Figure 5C is the simulated flow-field (arrows) of the landslide at t = 54 seconds in the corresponding region where the magnetic survey was conducted. We find that the impact of the landslide to the south, beyond the identified destruction front, is quite small. The final deposit heights (colors and black contours in Figure 5C) over the southern part of the village are less than one meter, suggesting that many of the buildings to the south may have been spared. The results of the filed magnetic survey thus strongly support the simulated flow pattern of the two-dimensional shallow-water model. The magnetic results also draw the second phase of the disaster that buried the whole Siaolin - the fatal mud flow that followed the breach of the landslide dam.

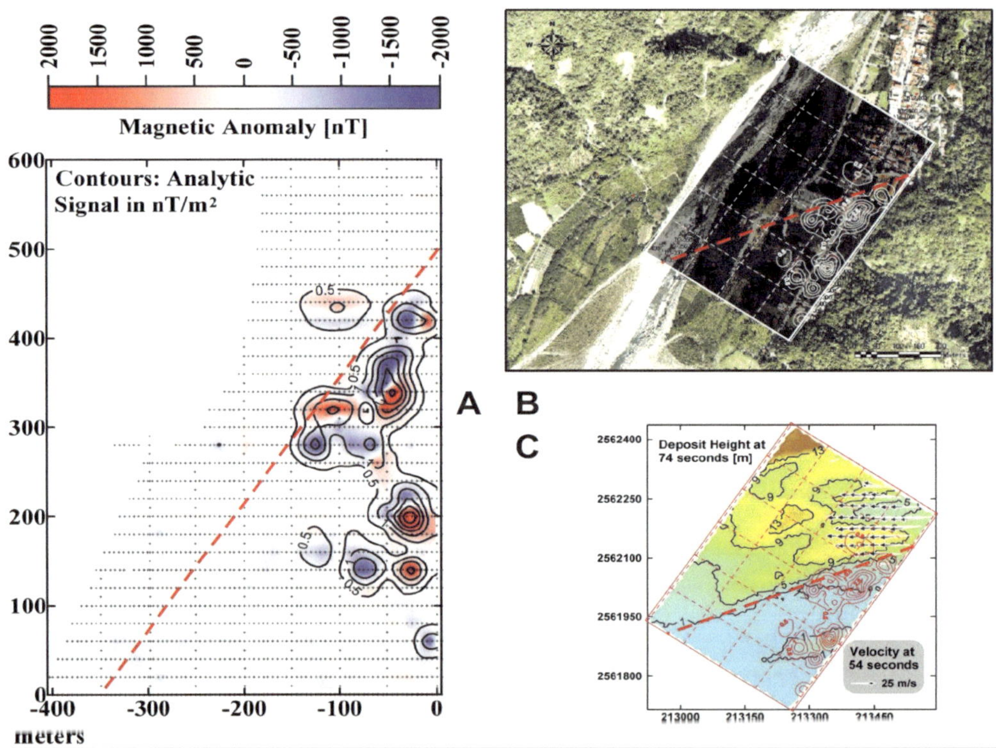

Fig. 5. (A) Magnetic anomaly (colors) and amplitude of the magnetic zeroth-order analytic signal (contours) over the main area of Siaolin Village. Crosses denote the measurement point locations. (B) Aerial photograph taken by AFASI before Typhoon Morakot, superposed by the magnetic analytic signal (white contours). Note that within the area where the magnetic survey was conducted, the photo has been filtered to emphasize the buildings. (C) Simulated velocity (arrows) of debris 54 seconds after the landslide initiation and deposit height (colors and black contours) 74 seconds over the same area of (A), again superposed by the magnetic analytic signal (red contours). The thick red dashed line in each panel depicts the sharp destruction front of the debris suggested from the results of the magnetic survey.

5. Scenario of the breached landslide dam

While the landslide mostly destroyed the northern part of the Siaolin Village, the spared southern village obviously suffered different catastrophic scenario induced by the breach of the landslide dam. For a realistic simulation of the mud flow induced by the dam breach, the US National Weather Service BREACH model (Fread, 1991) was applied. This model generates a landslide dam break hydrograph, which in turn is used as input to drive the Federal Emergency Management Agency-approved FLO-2D model (O'Brien et al., 1993; O'Brien, 2006). The equivalent landslide dam in the model had an along-river width of 300 m and a crest elevation of 420 m. The equivalent cross-valley length of the crest was 213 m after compensating for the differences between the real valley topography and the simplified trapezoidal dam shape required by BREACH. The corresponding debris volume over the main channel of the Cishan River that failed is thus 5.2 million m^3. The model geometry of the landslide dam has been clearly elucidated in the paper of Li et al. (2011), together with the detailed description about the simulation of the dam breach and mud flow.

The initiation time of the dam failure was 6:43 a.m., and it only took about 8 minutes to reach a complete breach. The peak discharge rate of this massive breach was 94,280 m^3/s. The average sediment concentration by volume was 0.38, which was estimated as the ratio of landslide dam volume to the total mud volume of the break hydrograph. Figure 6 depicts the deposited sediment depths over Siaolin as simulated by the FLO-2D model. Muddy floods following the breach covered the entirety of the village around 6:51 a.m., resulting in tragedy for the inhabitants of Siaolin. The deposited mud brought by the breaching dam is around 10 meters thick above the southern village, where was little ruined by the landslide. In terms of spatial distribution and depth, the overall pattern of the deposited sediment obtained from the elevation differences (contours in Figure 6) of the DTMs before and after Typhoon Morakot was well simulated by the model (color patches in Figure 6). Also note that the main river course was slightly diverted to the east, indicating that a significant scour occurred after the dam break deposition. Such a process of mixed scour and deposition is beyond the limitations of current FLO-2D modeling.

6. Concluding remark

Typhoon Morakot, an example of extreme precipitation, dropped more than 2,500 mm of rain over southern Taiwan within the three days following August 7, 2009. The National Disasters Prevention and Protection Commission in Taiwan reported 724 deaths caused by the consequent flooding. Of these victims, 474 were buried alive by the landslide and debris-flow event at Siaolin Village. This disaster had a severe socio-cultural impact on Taiwanese aboriginals of the Pingpu group. Reconstruction of the burying course due to the debris avalanche event will not only bring emotional satisfaction to the surviving villagers but will provide a fundamental understanding of the complex mass movement process (Densmore et al., 1997; Densmore and Hovius, 2000; Stark and Stark, 2001; Bachmann et al., 2004; Bruckl and Parotidis, 2005; Bonnet and Crave, 2006).

Contrary to the fairly precise rupturing simulation of earthquakes, the complex mass movement of landslides/debris-flows is rarely reconstructed by numerical modeling (e.g. Kuo et al., 2009; Crosta et al., 2004; Lin et al., 2005). A quantitative description of the mass movement process of a landslide is thus a challenging task compared to the well-developed techniques associated with earthquake rupturing simulations. The accurate reconstruction of landslide kinematics is generally extremely difficult because of its complexity and of the lack of detailed data on mass movement processes. A landslide/debris-flow simulation is conventionally performed using digital terrain models solely, but one can raise the question whether the results of such simulations are realistic. We here employ interdisciplinary seismological and near-surface magnetic data to validate the results of a simulation of the catastrophic Siaolin, Taiwan landslide triggered by Typhoon Morakot in 2009. We have demonstrated that a realistic reconstruction of the catastrophic Siaolin event can be achieved by a simulation of the frictional continuum debris model validated with the support of geophysical (seismological and magnetic) data. In the first stage of the Siaolin case, a debris avalanche swept away half the village while in the second stage a debris-blocking dam was breached about 30 minutes after it was formed. Then muddy floods shortly covered the entirety of the village. This study shows that the pattern of the Siaolin event can be interpreted as a hybrid of mass movement processes of an avalanche, a landslide dam and a dam breach.

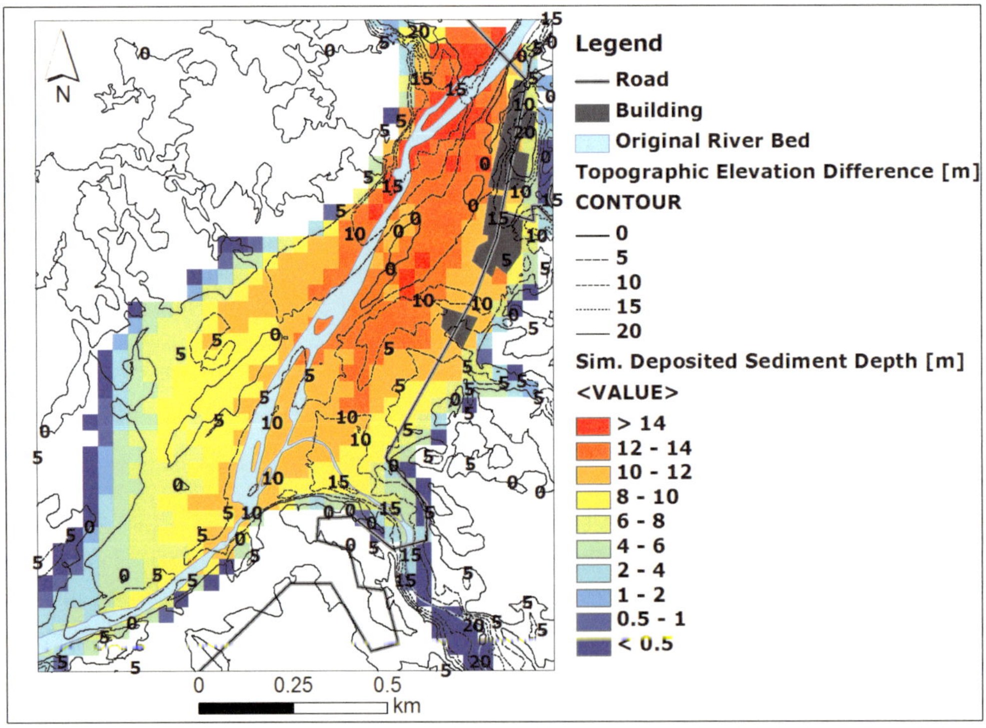

Fig. 6. Simulated deposited sediment depths (color patches) around Siaolin Village due to the muddy flood of the breached landslide dam (Figure 3C). Contour lines show the elevation differences between two versions of DTMs before and after Typhoon Morakot.

7. Acknowledgments

The scientific team of Project Morakot is grateful to the National Science Council (ROC) for its research support from grant NSC 98-2745-M-008-013. Also involved in this project are

Wen-Bin Doo, Hsien-Hsiang Hsieh, Shu-Kun Hsu, Chien-Yin Wang, Horng-Yuan Yen, Wen-Jie Wu, Wei-Lun Yu, Yun-Chen Yu, An-You Siao, Rui-Tang Sung, Chung-Pai Chang, Yue-Gau Chen, Jyr-Ching Hu, Ming-Lang Lin, Kuo-Jen Chang, Ping-Yu Chang, Chun-Yi Yu, Yih-Chin Tai, Chia-Ming Lo, Raehee Han, Toshihiko Shimamoto. C.C.C. thanks K. F. Ma and B. Chao for their encouragement concerning this project. Valuable geophysical surveys of magnetic, electrical and seismic refraction data were immediately collected by the SeeHope Tech. & Eng. Ltd. Co., Taiwan, under dangerous conditions in the aftermath of the disaster.

8. References

Bachmann, D., Bouissou, S., Chemenda, A., 2004. Influence of weathering and pre-existing large scale fractures on gravitational slope failure: insights from 3-D physical modeling, *Nat. Hazards Earth Syst. Sci.* 4, 711-717.

Bonnet, S., Crave, A., 2006. Macroscale dynamics of experimental landscapes, *Geol. Soc. London Special Publications* 253, 327-339.

Bruckl, E., Parotidis, M., 2005. Prediction of slope instabilities due to deep-seated gravitational creep, *Nat. Hazards Earth Syst. Sci.* 5, 155-172.

Crosta, G.B., Chen, H., Lee, C.F., 2004. Replay of the 1987 Val Pola Landslide, Italian Alps, *Geomorphology* 60, 127-146.

Dahlen, F.A., 1993. Single-force representation of shallow landslide sources, *Bull. Seism. Soc. Am.* 83, 130-143.

Densmore, A.L., Anderson, R.S., McAdoo, B.G., Ellis, M.A., 1997. Hillslope Evolution by Bedrock Landslides, *Science* 275, 369-372.

Densmore, A.L., Hovius, N., 2000. Topographic fingerprints of bedrock landslides, *Geology* 28, 371-374.

Doo, W.-B., Hsu, S.-K., Tsai, C.-H., Huang, Y.-S., 2009. Using analytic signal to determine magnetization/density ratios of geological structures, *Geophys. J. Int.* 179, 112-124.

Doo, W.B., Hsu, S.K., Chen, C.C., Hsieh, H.H., Yen, H.Y., Chen, Y.G., Chang, W.Y., 2011. Magnetic signature of Siaolin Village, southern Taiwan, after burial by a catastrophic landslide due to Typhoon Morakot, *Nat. Hazards Earth Syst. Sci.* (accepted).

Ferri, F., Di Toro, G., Hirose, T., Han, R., Noda, H., Shimamoto, T., Pennacchioni, G., 2009. Evolution of the 1963 Vajont landslide (Northern Italy) from low and high velocity friction experiments, *EGU General Assembly Geophysical Research Abstracts* 11, EGU2009-8138.

Fread, D.L., 1991. *BREACH: an erosion model for earthen dam failures* (National Weather Service, Office of Hydrology, Silver Spring, MD).

Gray, J., Tai, Y.C., Noelle, S., 2003. Shock waves, dead-zones and particle-free regions in rapid granular free surface flows, *J. Fluid Mech.* 491, 161-181.

Hsu, S.K., 2002. Imaging magnetic sources using Euler's equation, *Geophysical Prospecting* 50, 15-25.

Hsu, S.K., Coppens, D., Shyu, C.T., 1998. Depth to magnetic source using the generalized analytic signal, *Geophysics* 63, 1947-1957.

Kuo, C.Y., Tai, Y.C., Chen, C.C., Chang, K.J., Siau, A.Y., Dong, J.J., Han, R.H., Shimamoto, T., Lee, C.T., 2011. The landslide stage of the Shiaolin catastrophe: simulation and validation, *J. Geophys. Res. Earth Surface* (submitted)

Kuo, C.Y., Tai, Y.C., Bouchut, F., Mangeney, A., Pelanti, M., Chen, R.F., Chang, K.J., 2009. Smulation of Tsaoling landslide, Taiwan, based on Saint Venant equations over general topography, *Eng. Geol.* 104, 181-189.

Li, M.H., Sung, R.T., Dong, J.J., Lee, C.T., Chen, C.C., 2011. Forming and breaching of a short-lived landslide dam at Siaolin Village, Taiwan - Part II: Simulation of debris flow with landslide dam breach, *EngE Geol.* (submitted)

Lin, M.L., Wang, K.L., Huang, J.J., 2005. Debris flow run off simulation and verification – case study of Chen-You-Lan Watershed, Taiwan, *Nat. Hazards Earth Syst. Sci.* 5, 439-445.

Lee, C.T., Dong, J.J., Lin, M.L., 2009. Geological investigation on the catastrophic landslide in Siaolin Village, southern Taiwan, *Sino-Geotechnics* 122, 87-94 (in Chinese with English abstract).

Miyamoto, Y., Shimamoto, T., Togo, T., Dong, J.J., Lee, C.T., 2009. Dynamic weakening of shale and bedding-parallel fault gouge as a possible mechanism for Tsaoling landslide induced by 1999 Chi-Chi earthquake, *Proceedings of The Next Generation of Research on Earthquake-induced Landslides: An International Conference in Commemoration of 10th Anniversary of the Chi-Chi Earthquake*, 398-401.

Mizoguchi, K., Hirose, T., Shimamoto, T., Fukuyama, E., 2007. Reconstruction of seismic faulting by high-velocity friction experiments: an example of the 1995 Kobe earthquake, *Geophys. Res. Lett.* 34, L01308, doi:10.1029/2006GL027931.

O'Brien, J.S., 2006. *FLO-2D User's Manual* (http://www.flo-2d.com).

O'Brien, J.S., Julien, P.Y., Fullerton, W.T., 1993. Two-dimensional water flood and mudflow simulation, *J Hydraulic Eng* 119, 244-261.

Okura, Y., Kitahara, H., Sammori, T., 2000. Fluidization in dry landslides, *Eng. Geol.* 56, 347-360.

Stark, C.P., Stark, G.J., 2001. A Channelization Model of Landscape Evolution, *Am. J. Sci.* 301, 486-512.

Staron, L., Lajeunesse, E., 2009. Understanding how volume affects the mobility of dry debris flows, *Geo. Res. Lett.* 36, L12402.

Stein, S., Wysession, M., 2003. *An Introduction to Seismology, Earthquake, and Earth Structure.* Blackwell Publishing, UK, 498pp.

Tai, Y.C., Kuo, C.Y., 2008. A new model of granular flows over general topography with erosion and deposition, *Acta Mechanica* 199, 71-96.

Togo, T., Shimamoto, T., Ma, S., Hirose, T., 2009. High-velocity friction of faults: a review and implication for landslide studies, *Proceedings of The Next Generation of Research on Earthquake-induced Landslides: An International Conference in Commemoration of 10th Anniversary of the Chi-Chi Earthquake*, 205-216.

Wang, C.Y., 1981. *Wave theory for seismogram synthesis.* Saint Louis University (Ph. D. dissertation), Saint Louis, USA, 235pp.

Yano, K., Shimamoto, T., Oohashi, K., Dong, J.J., Lee, C.T., 2009. Ultra-low friction of shale and clayey fault gouge at high velocities: implication for Jiufengershan landslide induced by 1999 Chi-Chi earthquake, *Proceedings of The Next Generation of Research on Earthquake-induced Landslides: An International Conference in Commemoration of 10th Anniversary of the Chi-Chi Earthquake*, 402-406.